Edited by
Matthias Dehmer,
Frank Emmert-Streib, and Stefan Pickl

Computational Network Theory

Titles of the Series "Quantitative and Network Biology"

Advisory Board:

Volume 1
Dehmer, M., Emmert-Streib, F., Graber, A., Salvador, A. (eds.)

Applied Statistics for Network Biology

Methods in Systems Biology

2011
ISBN: 978-3-527-32750-8

Volume 2
Dehmer, M., Varmuza, K., Bonchev, D. (eds.)

Statistical Modelling of Molecular Descriptors in QSAR/QSPR

2012
ISBN: 978-3-527-32434-7

Volume 3
Emmert-Streib, F. Dehmer, M. (eds.)

Statistical Diagnostics for Cancer

Analyzing High-Dimensional Data

2013
ISBN: 978-3-527-32434-7

Volume 4
Emmert-Streib, F. Dehmer, M. (eds.)

Advances in Network Complexity

2013
ISBN: 978-3-527-33291-5

Related Titles

Siddique, N., Adeli, H.

Computational Intelligence
Synergies of Fuzzy Logic, Neural Networks and Evolutionary Computing

2013
Hardcover
ISBN: 978-1-118-33784-4

Dehmer, M., Basak, S. C.

Statistical and Machine Learning Approaches for Network Analysis

2012
ISBN: 978-0-470-19515-4

Okabe, A., Sugihara, K.

Spatial Analysis Along Networks
Statistical and Computational Methods

2012
ISBN: 978-0-470-77081-8

Junker, B. H., Schreiber, F.

Analysis of Biological Networks

2008
ISBN: 978-0-470-04144-4

Engelbrecht, A. P.

Computational Intelligence
An Introduction

2007
ISBN: 978-0-470-03561-0

Dunne, R. A.

A Statistical Approach to Neural Networks for Pattern Recognition

2007
ISBN: 978-0-471-74108-4

Quantitative and Network Biology
Series Editors M. Dehmer and F. Emmert-Streib
Volume 5

Computational Network Theory

Theoretical Foundations and Applications

Edited by
Matthias Dehmer, Frank Emmert-Streib, and Stefan Pickl

WILEY-VCH
Verlag GmbH & Co. KGaA

The Editors

Prof. Dr. Matthias Dehmer
UMIT – The Health and Lifesciences University
Eduard Wallnoefer Zentrum 1
A-6060 Hall in Tirol
Austria

and

Universität der Bundeswehr München
Institute for Theoretical Computer Science, Mathematics and Operations Research
Department of Computer Science
Werner-Heisenberg-Weg 39
85577 Neubiberg
Germany

Prof. Dr. Frank Emmert-Streib
Tampere University of Technology
Department of Signal Processing
Korkeakoulunkatu 1
FI-33720 Tampere
Finland

Prof. Dr. Stefan Pickl
Universität der Bundeswehr München
Institute for Theoretical Computer Science, Mathematics and Operations Research
Department of Computer Science
Werner-Heisenberg-Weg 39
85577 Neubiberg
Germany

Library of Congress Card No.: applied for

British Library Cataloguing-in-Publication Data
A catalogue record for this book is available from the British Library.

Bibliographic information published by the Deutsche Nationalbibliothek
The Deutsche Nationalbibliothek lists this publication in the Deutsche Nationalbibliografie; detailed bibliographic data are available on the Internet at <http://dnb.d-nb.de>.

Print ISBN: 978-3-527-33724-8
ePDF ISBN: 978-3-527-69153-1
ePub ISBN: 978-3-527-69154-8
Mobi ISBN: 978-3-527-69152-4
oBook ISBN: 978-3-527-69151-7

Typesetting Laserwords Private Limited, Chennai, India
Printing and Binding Markono Print Media Pte Ltd, Singapore

Printed on acid-free paper

The all-important glue that holds our wonderful complex societies together is cooperation.

Jerry Sabloff, President Santa Fe Institute

Dedicated to the 30th birthday of the Santa Fe Institute and the 85th birthday of Murray Gell-Mann, one of its founders.

What is especially striking and remarkable is that in fundamental physics a beautiful or elegant theory is more likely to be right than a theory that is inelegant.

Murray Gell-Mann.

Contents

Color Plates

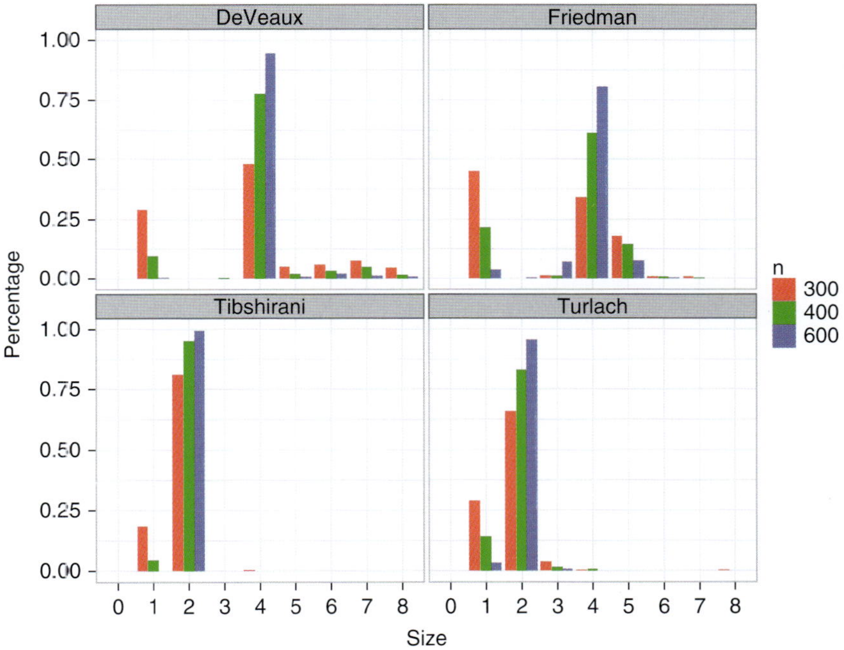

Figure 1.9 Proportion of hidden layer size identification by using the testing procedure for superior predictive ability. (This figure also appears on page 24.)

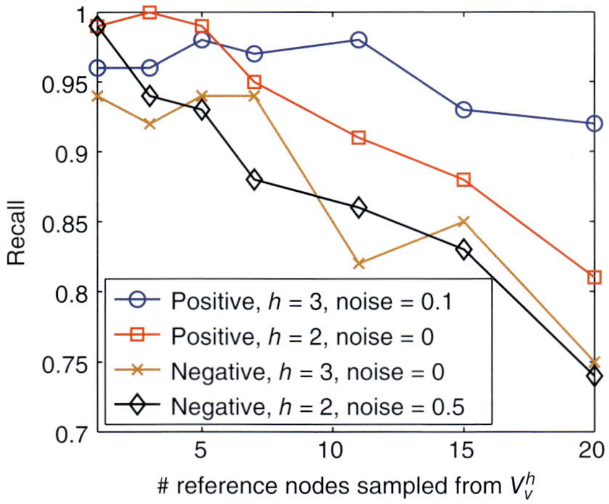

Figure 2.17 Performance of sampling different number of reference nodes from each V_v^h for Importance sampling. (This figure also appears on page 65.)

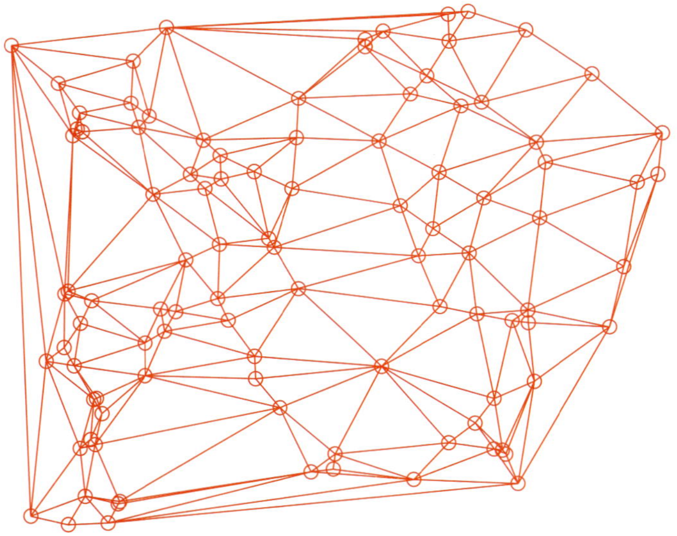

Figure 3.1 The graph G, drawn using its adjacency matrix A and randomly chosen positions for the vertices. (This figure also appears on page 92.)

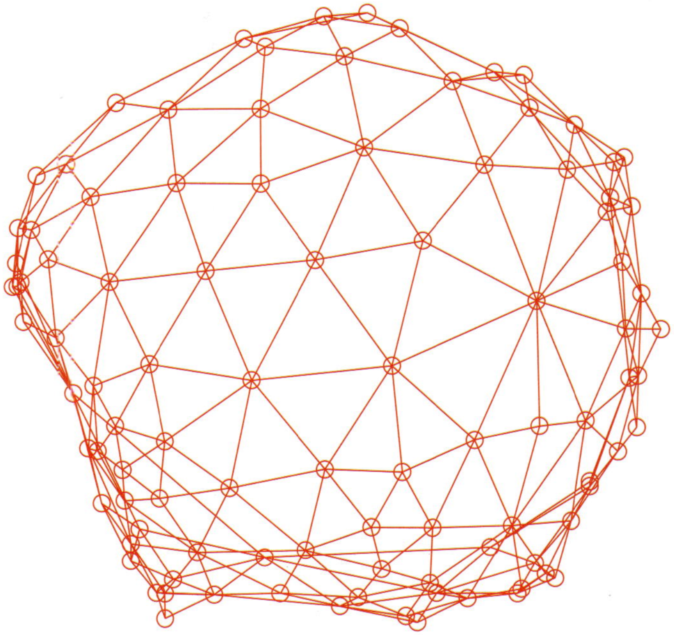

Figure 3.2 The graph G, drawn using its adjacency matrix A and its two Laplace eigenvectors $u_2(L)$ and $u_3(L)$. (This figure also appears on page 92.)

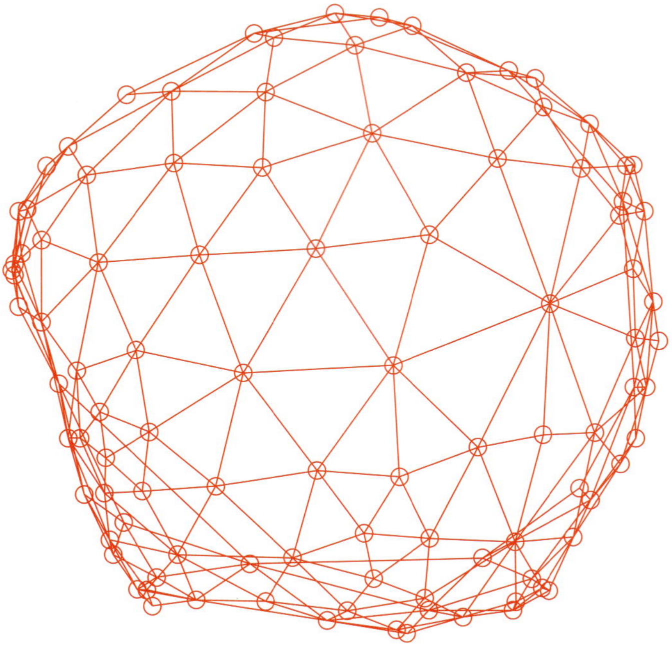

Figure 3.3 The graph G, redrawn using its walk matrix $W = AD^{-1}$ and its two Laplace eigenvectors $u_2(L)$ and $u_3(L)$. (This figure also appears on page 93.)

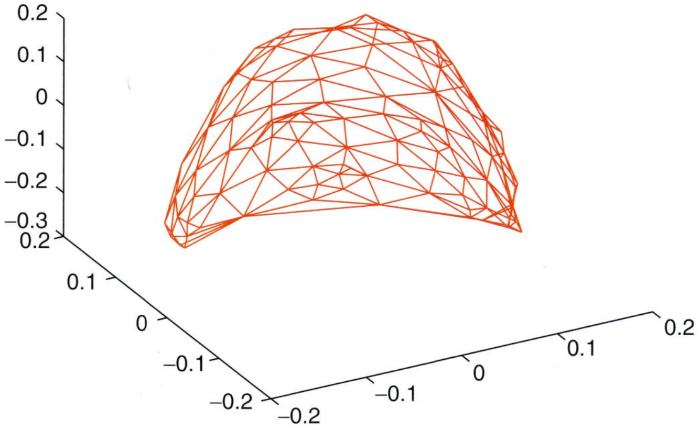

Figure 3.4 The graph G, drawn using its adjacency matrix A and its two Laplace eigenvectors $u_2(L)$, $u_3(L)$, and $u_4(L)$. (This figure also appears on page 93.)

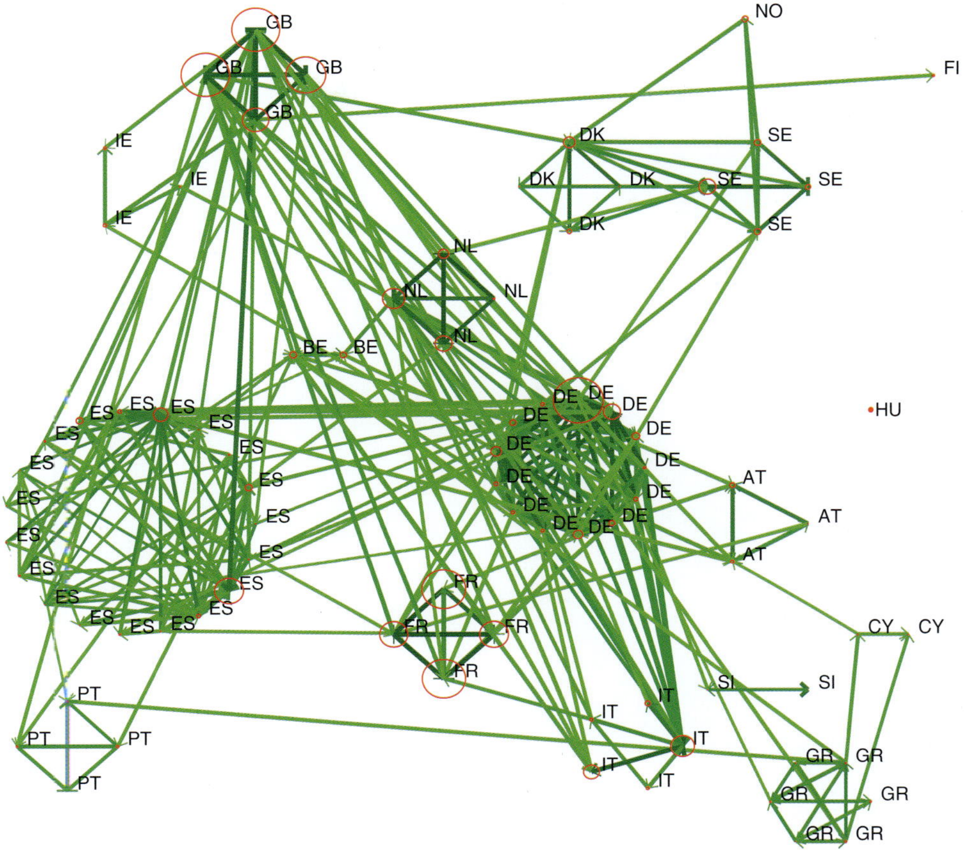

Figure 4.1 A generated interbank network. Note: an arrow between bank A and B indicates an interbank deposit of bank B placed in bank A; the width of an arrow reflects the size of the exposure; the lighter the green color of an arrow, the lower the probability that the arrow joins a given pair of banks. Source: own calculations. (This figure also appears on page 106.)

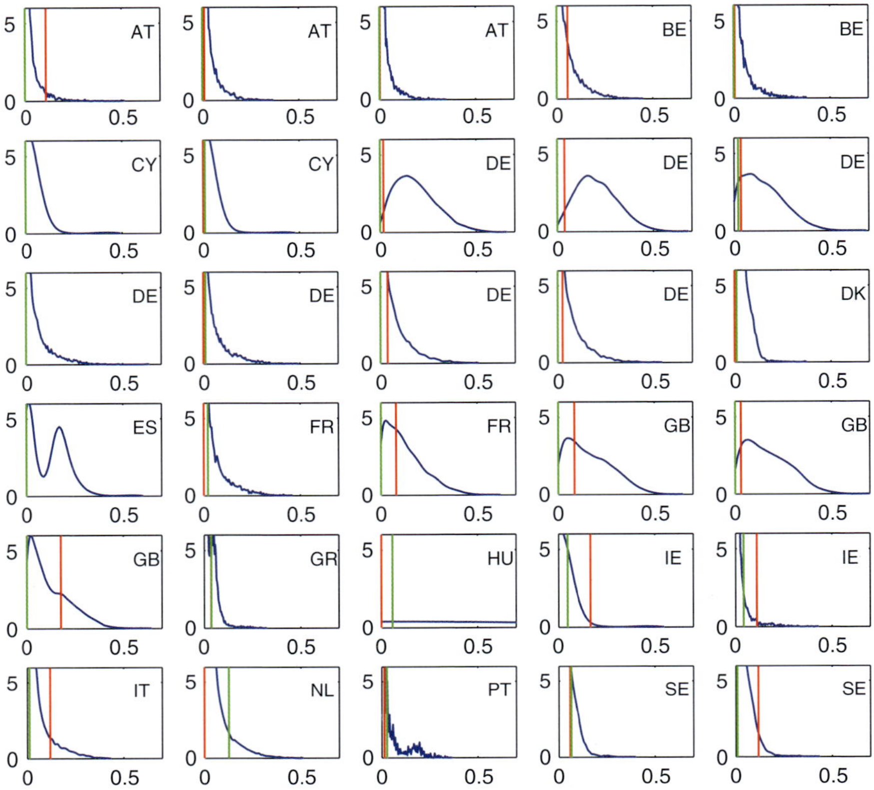

Figure 4.3 Betweenness-centrality measures: distribution on the simulated networks versus the average network. Note: Blue line: distribution on the simulated networks; red (vertical) line: measure for the average simulated network: green (vertical) line: measure for he entropy maximising network. Only internationally active banks are presented. Source: own calculations. (This figure also appears on page 121.)

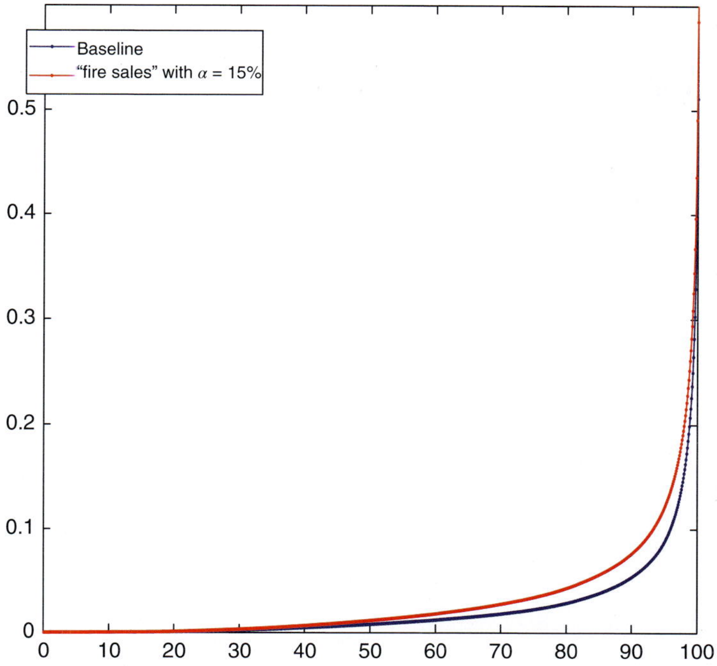

Figure 4.4 Distribution of the average CAR reduction (in p.p.). Source: own calculations. (This figure also appears on page 122.)

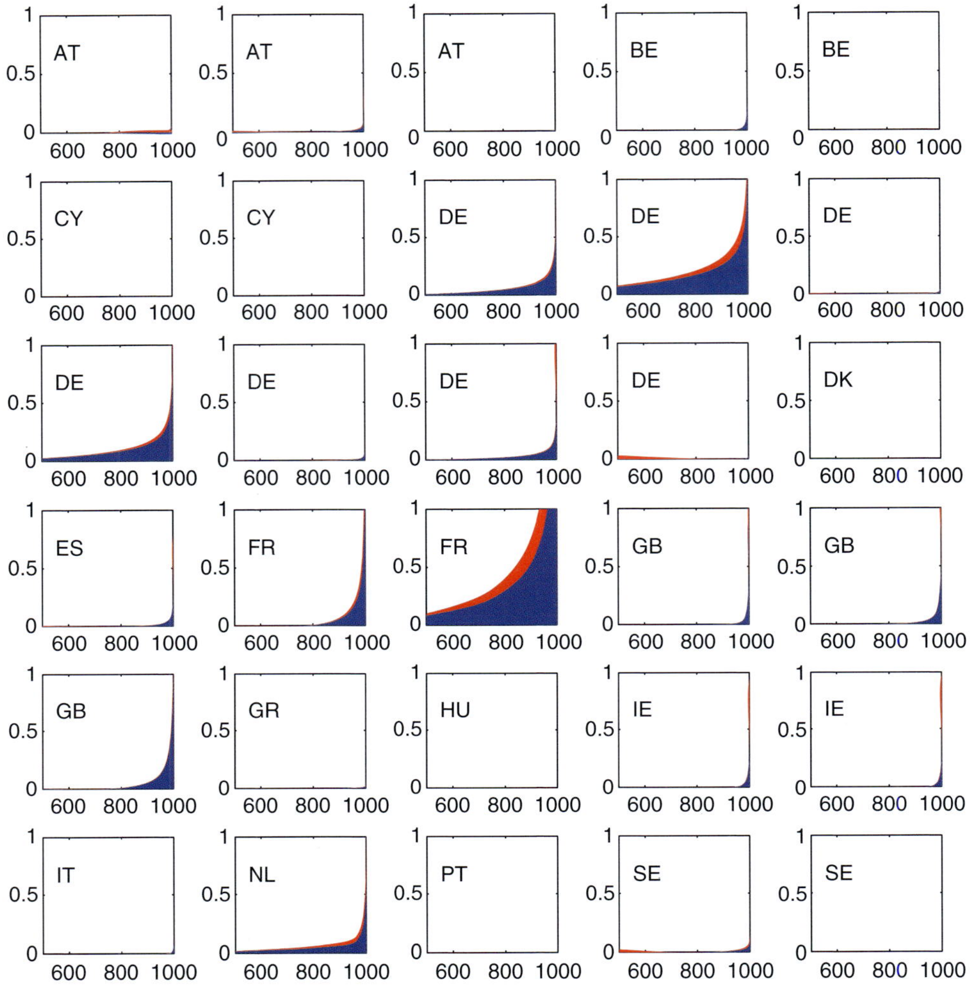

Figure 4.5 Decomposition of the distribution of individual banks' CAR reduction into first-and second-round contagion (in p.p.). Note: blue area – aggregate effect of first-round contagion; red area – second-round contagion. Only internationally active banks are presented. Source: own calculations. (This figure also appears on page 123.)

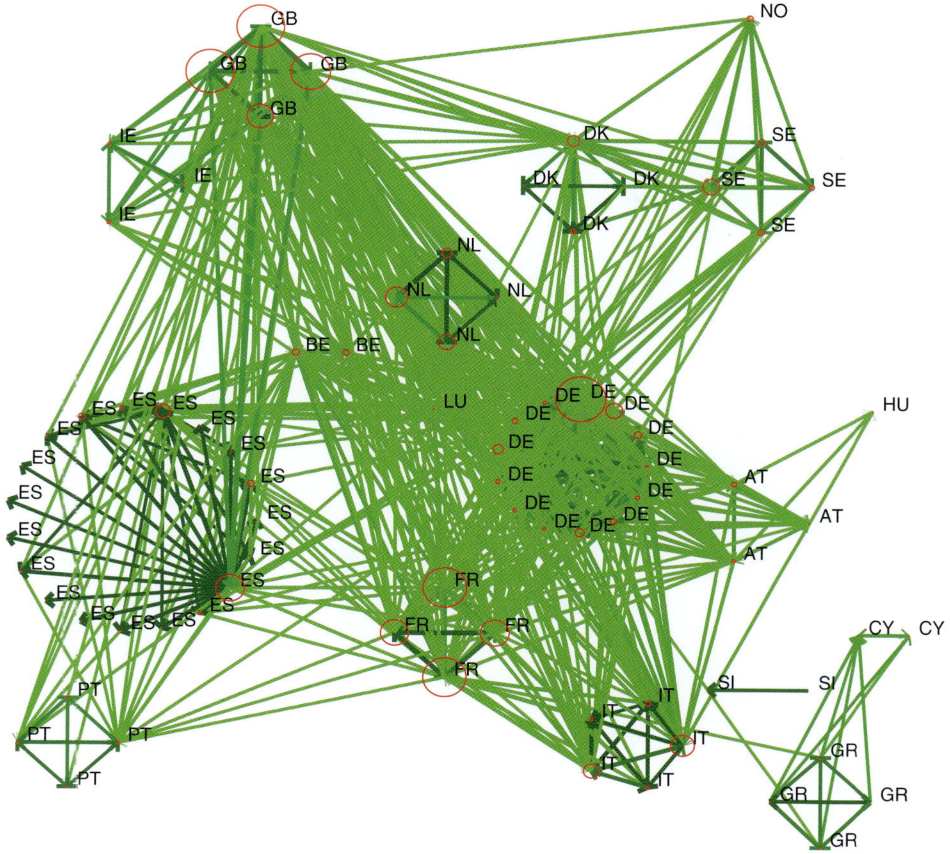

Figure 4.7 Incompleteness of the interbank network structure with the estimated correlation of risks. The darker the line, the higher the probability that the link exists. The circles around the nodes indicate bank sizes (proportional to log of total assets). Source: own calculations. (This figure also appears on page 126.)

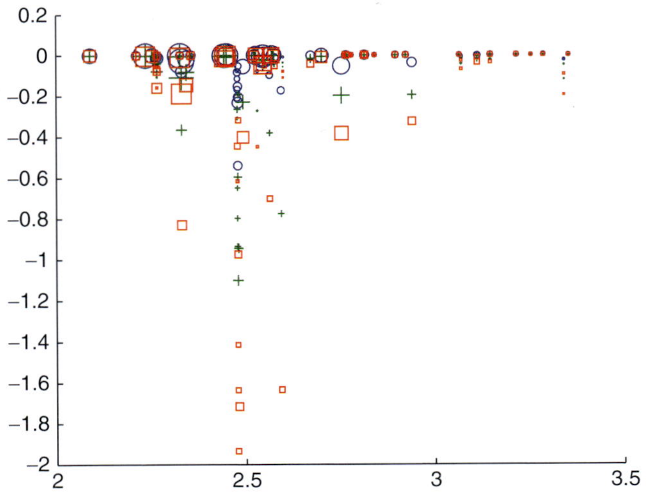

Figure 4.9 Counterparty credit quality and the impact of LE limits on the losses incurred due to contagion. Note: *x*-axis: CDS spread (in bps). *y*-axis: difference of CAR after adverse stress testing shock estimated CVA (◦), 3× estim. CVA (+), 5× estim. CVA (□) versus no CVA regime (in pp, negative number means that introduction of CVA charge decreases contagion losses). No CVA adjustment (i.e., $\gamma \equiv 0$). The size of a circle is proportional to a bank's total assets. Source: own calculations. (This figure also appears on page 129.)

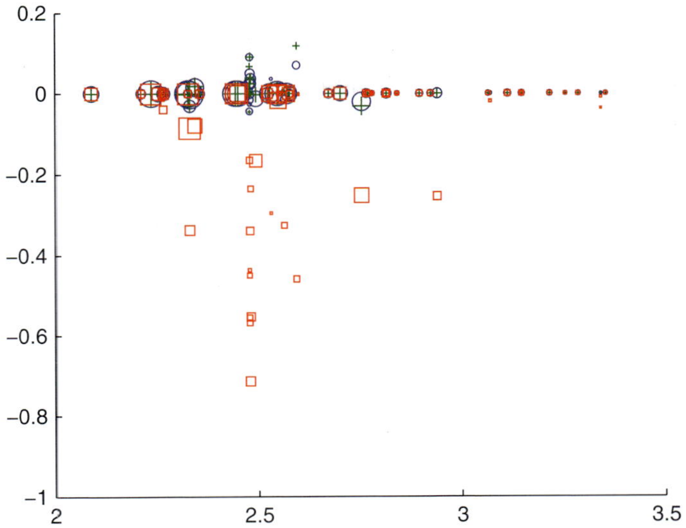

Figure 4.10 Counterparty credit quality and the impact of CVA capital charge on the losses incurred due to contagion. *x*-axis: CDS spread (in bps). *y*-axis: difference of CAR after adverse stress testing shock between no CVA regime calculations and CVA regime with: estimated CVA (∘), 3× estimated CVA (∘) and 5× estimated CVA (∘), in pp, positive number means that introduction of CVA charge increases contagion losses. LE= 25%. The size of a circle is proportional to a bank's total assets. Source: own calculations. (This figure also appears on page 130.)

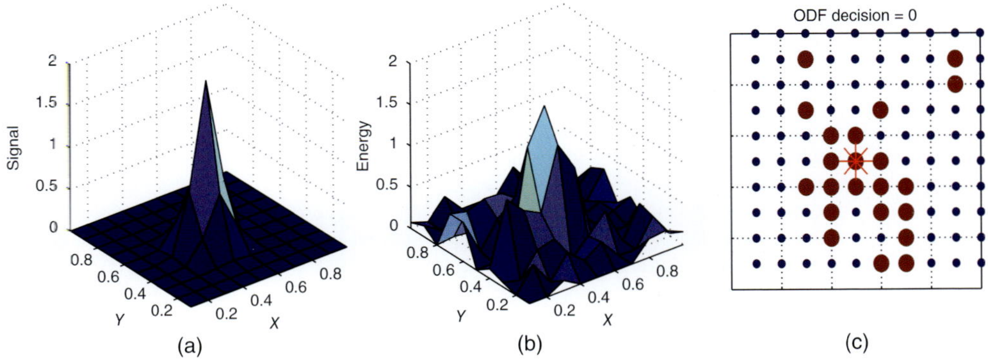

Figure 5.1 (a) Target signal generated by the model $S_i(v) = S_0 \exp(-||s_i - v||^2/\eta^2)$ for a target at location $v = (0.5, 0.5)$ with $S_0 = 2$, $\eta = 0.1$. (b) Target energy contaminated by Gaussian noise of variance $\sigma^2 = 0.16$ (signal-to-noise ratio $S_0/\sigma = 5$). (c) Sensor decisions based on individual false alarm probability (the same for all sensors) $\gamma = 0.2$. (This figure also appears on page 139.)

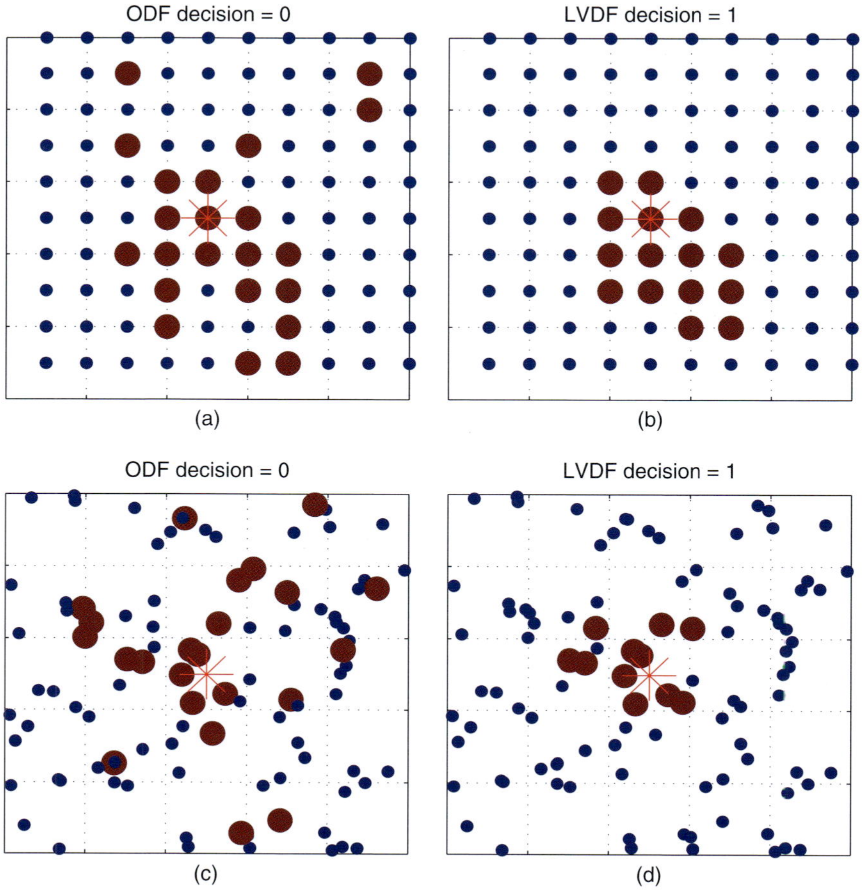

Figure 5.2 Ordinary versus. local vote decision fusion under a square grid design (a,b) and random deployment (c,d). The network is comprised of 100 sensors, with individual sensor false alarm probability $\gamma = 0.2$, system-wide false alarm probability $F = 0.1$ and a target located at the center of the monitored region R. The signal is generated by the model $S_i = S_0 \exp(-||s_i - v||^2/\eta^2)$, with $S_0 = 2$, $\eta = 0.1$, and the measured energy is corrupted by Gaussian noise with $\sigma = 0.4$. (This figure also appears on page 141.)

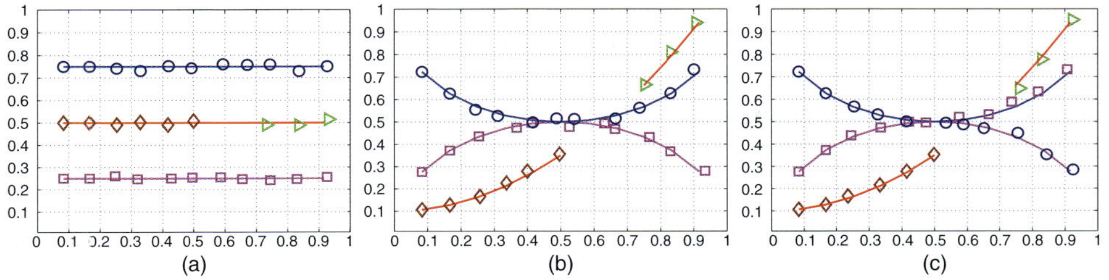

Figure 5.5 True trajectories (solid lines) and positions estimated by ML(Z) at each time point for three targets with SNR = 5. (a) The signal from the second target is briefly lost; (b) Two targets come close together and the third target briefly loses signal; (c) Another noise realization/solution for (b). (This figure also appears on page 165.)

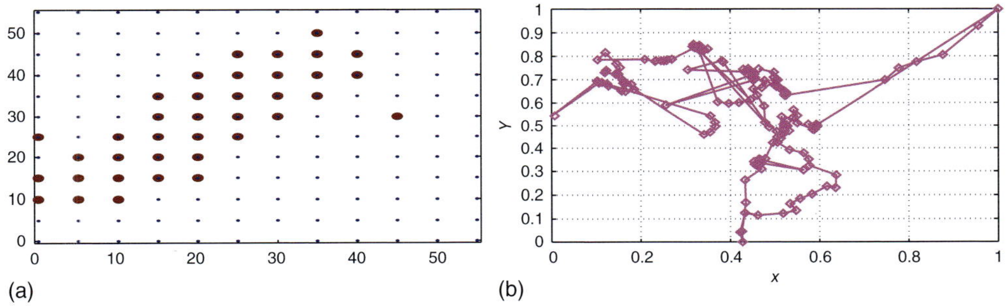

Figure 5.6 (a) The activation pattern of NEST sensors by a person traversing the monitored area. (b) The trajectory of a single zebra in the monitored area. (This figure also appears on page 166.)

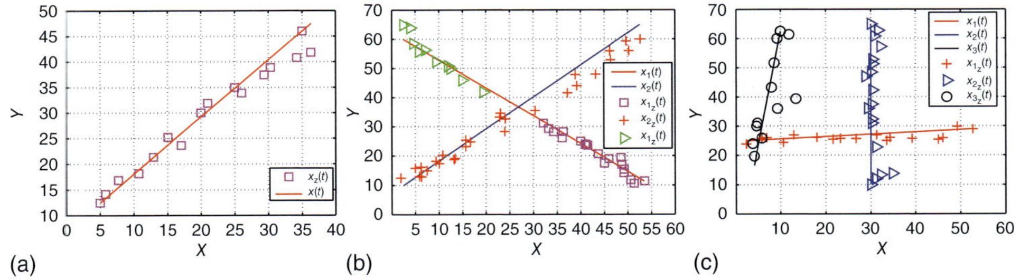

Figure 5.7 Estimated and true trajectories for (a) one, (b) two, and (c) three NEST targets. (This figure also appears on page 167.)

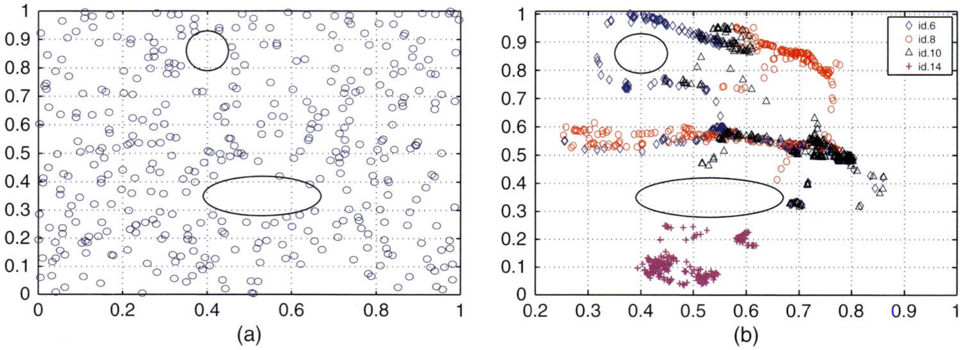

Figure 5.8 (a) Random sensor deployment. (b) The recorded locations of the four zebras scaled and plotted on the unit square. (This figure also appears on page 169.)

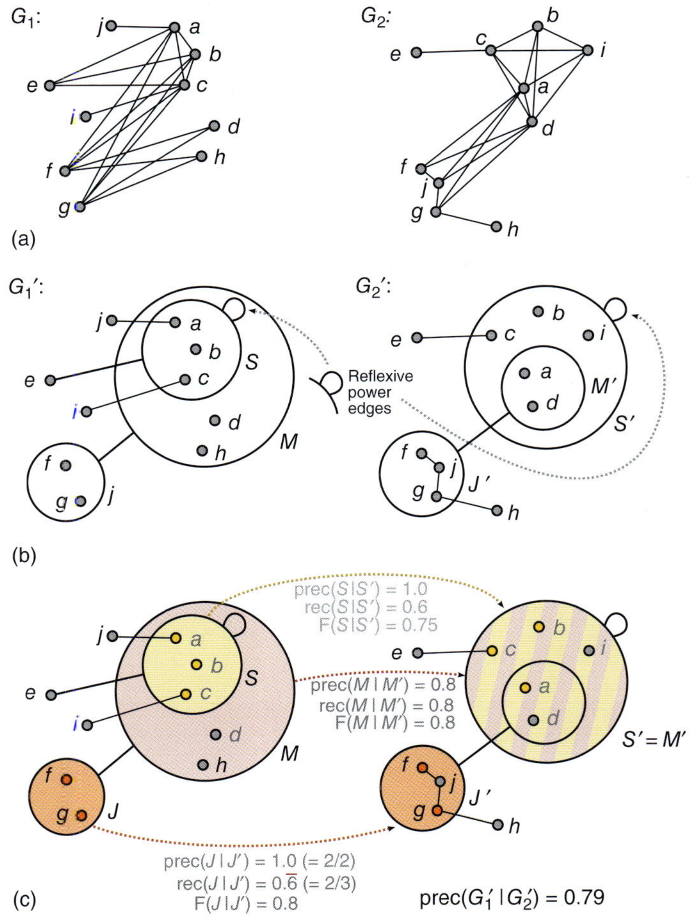

Figure 7.5 Power graph similarity (PG-similarity). (a) Two distinct graphs G_1 and G_2 on the same set of nodes. (b) Power graphs G'_1 and G'_2 for G_1 and G_2, respectively, after applying the power graph algorithm. (C) Power node matching as basis of the similarity measure. Each power node in G'_1 is matched with the power node in G'_2 with the highest F-measure, and vice versa. Precision and recall of those matchings are summarized to precision, recall and F-measure between G'_1 and G'_2. (This figure also appears on page 229.)

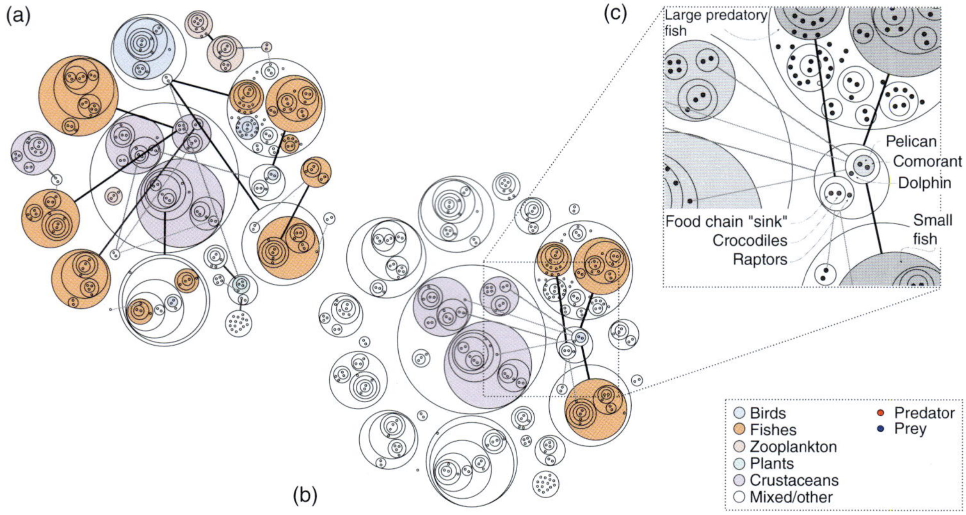

Figure 7.9 Interactive visual analysis of the Florida Food Chain Network. (a) The largest power nodes correspond to relevant groups of animals in the food chain. (b) Selecting power edges around a region of interest – for example here a group of predators – helps to locally explore the network. (C) These predators share many fish species in their diet and are thus in competition in the food chain. Yet it can be seen that crocodiles and raptors prefer larger predatory fish and pelicans, cormorants and dolphins prefer smaller fish. Note: The *food chain sink* corresponds to the output of carbon from the ecological system studied, it is thus not a species but represent exchanges with the outside. (This figure also appears on page 233.)

Preface

Network-based approaches have been used extensively in computational sciences and relevant disciplines such as mathematics, computer science, systems biology, operations research, and decision sciences. In particular, the development of network-based methods has been triggered from the breakthrough of the world wide web and other physics-oriented studies when exploring complex systems. As a consequence, various methods and measures have been developed to analyze structural properties of complex networks. But until now, the present scientific literature lacks a thorough classification of the existing methods for analyzing networks. For instance, computational versus noncomputational, quantitative versus Descriptive, and so forth.

The main goal of the book is to introduce *computational network theory* as a new branch of network sciences. In fact, various computational methods such as machine learning and data analysis techniques have been already applied to structural data sets (network data) extensively. However, data analysis and statistical and machine learning techniques have been introduced on their own, and any of such methods can be applied in an interdisciplinary manner. This calls for a definition and treatment of computational network theory as a branch of network theory to classify the methods developed in this volume correctly. The definition we would like to form views computational network theory as a tool to derive or verify hypotheses by applying computational techniques to network data on a large scale.

The topics addressed in this book cover a wide range of concepts and methods such as:

- Simulation techniques in network theory.
- Exploratory and computational analysis of networks on a large scale.
- Model selection strategies for computational network analysis.
- Computational approaches to graph spectra.
- Computational methods for network visualization.
- Computational analysis of dynamic networks and their stochastic optimization.
- Stochastic optimization on large networks.

Many colleagues, whether consciously or unconsciously, have provided us with input, help and support before and during the preparation of this book. In particular, we would like to thank Ilie Burdujan, Zengqiang Chen, Maria and Gheorghe Duca, Andrey A. Dobrynin, Boris Furtula, Ivan Gutman, Bo Hu, Bernhard Katzy, Xueliang Li, D. D. Lozovanu, Alexei Levitchi, Abbe Mowshowitz, Miriana Moosbrugger, Andrei Perjan, Ricardo de Matos Simoes, Yongtang Shi, Fred Sobik, Shailesh Tripathi, Kurt Varmuza, Chengyi Xia, Dongxiao Zhu, and apologize to all whose names have been inadvertently omitted. Also, we would like to thank our editors Andreas Sendtko and Gregor Cicchetti from Wiley-VCH who have been always available and helpful. Last but not least, Matthias Dehmer thanks the Austrian Science Funds (project P26142) for supporting this work. On top of that, we are grateful for support from the Universität der Bundeswehr München and from the German Federal Ministry of Education and Research (project RiKoV, Grant No. 13N12304).

To date, no book dedicated exclusively to computational network theory has been produced. Existing books dealing with related topics such as complex networks and computational analysis of social networks have limited scope, considering only specialized graph classes and techniques. Therefore, we hope that this book will broaden the scope of scientists who deal with computational network analysis. Finally, we hope this book conveys the enthusiasm and joy we have for this field and inspires fellow researchers in their own practical or theoretical work.

Grace to this enthusiasm, special thanks are given to our friends at the Santa Fe Institute for Complex Systems. The book is influenced by them and tries to cover the spirit of that unique institute and *research network*.

We all, but especially Matthias thanks his sister Marion Dehmer who passed away in 2012. The new approach toward computational network science is dedicated to her.

Neubiberg-München and Tampere
April 2015

Matthias Dehmer
Frank Emmert-Streib
Stefan Pickl

List of Contributors

Ashraf Ahmed
Queen's University Belfast
School of Planning, Architecture and Civil Engineering
University Road
BT7 1NN
Belfast
UK

Ioannis Chatzigiannakis
Computer Technology Institute and Press "Diophantus" (CTI)
Kazantzaki
Patras 265 04
Greece

and

Department of Computer, Control and Management Engineering (DIAG)
Sapienza University of Rome
via Ariosto 25
00185 Rome
Italy

Simone Daminelli
Technische Universität Dresden
Biotechnology Center
Tatzberg 47-51
01307 Dresden
Germany

Ziyu Guan
Northwest University of China
Department of Computer Science and Technology
College of Information and Technology
Xuefu Avenue
Xi'an, 710127
China

Grzegorz Hałaj
European Central Bank
Kaiserstrasse 29
D-60311 Frankfurt am Main
Germany

Natallia Katenka
University of Rhode Island
Department of Computer Science and Statistics
Tyler Hall, 9 Greenhouse Road, Suite 2
Kingston
RI 02881
USA

Christoffer Kok
European Central Bank
Kaiserstrasse 29
D-60311 Frankfurt am Main
Germany

Michele La Rocca
Department of Economics and Statistics
University of Salerno
Via Giovanni Paolo II, 132
84084 Fisciano (Salerno)
Italy

Othon Michail
Computer Technology Institute and Press "Diophantus" (CTI)
Kazantzaki
Patras 265 04
Greece

Salissou Moutari
Queen's University Belfast
School of Mathematics and Physics
University Road
BT7 1NN
Belfast
UK

Cira Perna
Department of Economics and Statistics
University of Salerno
Via Giovanni Paolo II, 132
84084 Fisciano (Salerno)
Italy

Matthias Reimann
Technische Universität Dresden
Biotechnology Center
Tatzberg 47-51
01307 Dresden
Germany

Loïc Royer
Technische Universität Dresden
Biotechnology Center
Tatzberg 47-51
01307 Dresden
Germany

Michael Schroeder
Technische Universität Dresden
Biotechnology Center
Tatzberg 47-51
01307 Dresden
Germany

Paul G. Spirakis
Computer Technology Institute and Press "Diophantus" (CTI)
Kazantzaki
Patras 265 04
Greece

and

University of Liverpool
Department of Computer Science
Ashton Building
Ashton Street
Liverpool L69 3BX
UK

Xifeng Yan
University of California at Santa Barbara
Department of Computer Science
552 University Road
Santa Barbara
CA 93106-5110
USA

1
Model Selection for Neural Network Models: A Statistical Perspective

Michele La Rocca and Cira Perna

1.1
Introduction

It is generally accepted that liner analysis often gives poor performances in approximating real data. Therefore, although it is easy to handle and fast to compute, and many statistical results are available, it cannot be extensively used especially when complex relationships are recognized in the data. In these contexts, it is common the use of non linear analysis which can successfully be employed to reveal these patterns.

However, parametric analysis, both linear and nonlinear, requires an "a priori" specification of the links among the variables of interest, which is not always possible. Therefore, even if the results have the advantage of the interpretability (in the sense that the model parameters are often associated to quantities with a "physical" meaning), misspecification problem can arise and can affect seriously the results of the analysis. In this respect, nonparametric analysis seems to be a more effective statistical tool due to its ability to model non-linear phenomena with few (if any) "a priori" assumptions about the nature of the data generating process. Well-studied and frequently used tools in nonparametric analysis include nearest neighbours regression, kernel smoothers, projection pursuit, alternating conditional expectations, average derivative estimation, and classification and regression trees.

In this context, computational network analysis forms a field of research which has enjoyed rapid expansion and increasing popularity in both the academic and the research communities, providing an approach that can potentially lead to better non-parametric estimators and providing an interesting framework for unifying different non-parametric paradigms, such as nearest neighbours, kernel smoothers, and projection pursuit.

Computational network tools have the advantage, with respect to other non-parametric techniques, to be very flexible tools able to provide, under very general conditions, an arbitrarily accurate approximation to an unknown target the function of interest. Moreover, they are expected to perform better than other non-parametric methods since the approximation form is not so sensitive to the

Computational Network Theory: Theoretical Foundations and Applications, First Edition.
Edited by Matthias Dehmer, Frank Emmert-Streib, and Stefan Pickl.

increasing data space dimension (absence of "curse of dimensionality"), at least within particular classes of functions.

However, a major weakness of neural modeling is the lack of established procedures for performing tests for misspecified models and tests of statistical significance for the various parameters that have been estimated. This is a serious disadvantage in applications where there is a strong interest for testing not only the predictive power of a model or the sensitivity of the dependent variable to changes in the inputs but also the statistical significance of the result at a specified level of confidence. Significant correction for multiple hypothesis testing has been a central concern in many fields of research that deal with large sets of variables and small samples and where, as a consequence, the control of false positives becomes an important problem.

In such context data snooping, which occurs when a given set of data is used more than once for inference or model selection, it can be a serious problem. When such data reuse occurs, there is always the possibility that any satisfactory results obtained may simply be due to chance rather than any merit inherent in the model yielding the result. In other words, looking long enough and hard enough at a given data set will often reveal one or more forecasting models that look good but are in fact useless (see White, 2000; Romano and Wolf, 2005, *inter alia*).

Unfortunately, as far as we know, there are no results addressing the problem just described in a neural network framework. The data snooping can be particularly serious when there is no theory supporting the modeling strategy as it is usual when using computational network analysis, which is basically atheoretical.

The aim of this chapter is to develop model selection strategies useful for computational network analysis based on statistical inference tools. In particular, we propose hypothesis testing procedures both for variable selection and model adequacy. The approach takes into account the problem of data snooping and uses resampling techniques to overcome the analytical and probabilistic difficulties related to the estimation of the sampling distribution of the test statistics involved. The chapter is organized as follows. Section 1.2 describes the structure of the data generating process and the neural network model considered. In Section 1.3, we address the problem of input selection and in Section 1.4 the selection of the hidden layer size. In both cases, application to simulated and real data are considered. Some remarks conclude the papers.

1.2 Feedforward Neural Network Models

Let the observed data be the realization of a sequence $\left\{ \mathbf{Z}_i = \left(Y_i, \mathbf{X}_i^T\right)^T \right\}$ of random vectors of order $(d+1)$, with $i \in \mathbb{N}$ and joint distribution π. Moreover, let μ be the marginal distribution of $\mathbf{X}_i$. The random variables Y_i represent targets (in the neural network jargon) and it is usually of interest the probabilistic relationship with the variables $\mathbf{X}_i$, described by the conditional distribution of the random variable $Y_i | \mathbf{X}_i$. Certain aspects of this probability law play an important role in interpreting what is learned by artificial neural network models. If $\mathbb{E}\left(Y_i\right) < \infty$, then $\mathbb{E}\left(Y_i \left|\mathbf{X}_i\right.\right) = g\left(\mathbf{X}_i\right)$ and we can write

$$Y_i = g\left(\mathbf{X}_i\right) + \varepsilon_i \tag{1.1}$$

where $\varepsilon_i \equiv Y_i - g\left(\mathbf{X}_i\right)$ and $g : \mathbb{R}^d \rightarrow \mathbb{R}$ is a measurable function.

The function g embodies the systematic part of the stochastic relation between Y_i and $\mathbf{X}_i$. On the data-generating process, we assume also that:

1) $\mathbf{Z}_i$ are independent and identically distributed (i.i.d.) random vectors; $\{\varepsilon_i\}$ are independent of $\{\mathbf{X}_i\}$, $\mathbb{E}(\varepsilon_i) = 0$ and $\mathbb{E}(\varepsilon_i^2) = \sigma_\varepsilon^2 < \infty$.
2) The random vectors $\mathbf{X}_i$ have a compact support, say $\chi_1 \in \mathbb{R}^d$.

These conditions guarantee that Y_i has finite variance.

The function g can be approximated by a single hidden layer feed-forward neural network $NN(d, r)$ defined as:

$$f\left(\mathbf{x}, \mathbf{w}\right) = w_{00} + \sum_{j=1}^{r} w_{0j}\psi\left(\tilde{\mathbf{x}}^T \mathbf{w}_{1j}\right) \tag{1.2}$$

where $\mathbf{w} \equiv \left(w_{00}, w_{01}, \dots w_{0r}, \mathbf{w}_{11}^T, \dots, \mathbf{w}_{1r}^T\right)^T$ is a $r(d+2)+1$ vector of network weights, $\mathbf{w} \in \mathbf{W}$ with $\mathbf{W}$ compact subset of $\mathbb{R}^{r(d+2)+1}$, and $\tilde{\mathbf{x}} \equiv \left(1, \mathbf{x}^T\right)^T$ is the input vector augmented by a bias component 1. The network (Eq. (1.2)) has d input neurons, r neurons in the hidden layer and identity function for the output layer. The (fixed) hidden unit activation function ψ is chosen in such a way that $f\left(\mathbf{x}, \cdot\right) : \mathbf{W} \rightarrow \mathbf{R}$ is continuous for each x in the support of μ and $f\left(\cdot, \mathbf{w}\right) : \mathbb{R}^d \rightarrow \mathbb{R}$ is measurable for each $\mathbf{w}$ in $\mathbf{W}$.

On the neural network model, we assume that

1) The activation function, $\psi(\cdot)$, is sigmoidal.
2) The function $\psi(\cdot)$ has all the derivatives.

This latter assumption guarantees (Hornik, Stinchcombe, and Auer, 1994 *inter alia*) that feedforward neural networks with sufficiently many hidden units and properly adjusted parameters can approximate any function arbitrarily well. Moreover, Barron (1993) gives convergence rates for hidden layer feedforward networks with sigmoidal activation functions, approximating a class of functions that satisfy certain smoothness conditions.

Given a training set of N observations, the estimation of the network weights (learning) is obtained by solving the optimization problem

$$\min_{\mathbf{w} \in \mathbf{W}} \frac{1}{N} \sum_{i=1}^{N} q\left(Y_i, f\left(\mathbf{X}_i, \mathbf{w}\right)\right) \tag{1.3}$$

where $q(\cdot)$ is a proper chosen loss function. Under general regularity conditions White (1989), a weight vector $\hat{\mathbf{w}}_n$ solving Eq. (1.3) exists and converges almost surely to $\mathbf{w}_0$, which solves

$$\min_{\mathbf{w} \in \mathbf{W}} \int q\left(y, f\left(\mathbf{x}, \mathbf{w}\right)\right) d\pi\left(\mathbf{z}\right) \tag{1.4}$$

provided that the integral exists and the optimization problem has a unique solution vector interior to $\mathbf{W}$. Observe that this is not necessarily true for neural network models in the absence of appropriate restrictions since the parametrization of the network function is not unique and certain simple symmetry operations applied to the weight vector do not change the value of the output. For a sigmoid activation function, ψ centered around 0, these symmetry operations correspond to an exchange of hidden units and multiplying all weights of connections going into and out of a particular hidden unit by -1. The permutability of hidden units generally results in a non-unique $\mathbf{w}_0$ as there are numerous distinct weight vectors yielding identical network outputs. In any case, this may not be a main concern for different reasons. Firstly, several authors provide sufficient conditions to ensure uniqueness of $\mathbf{w}_0$ in a suitable parameter space $\mathbf{W}$ for specific network configurations. Particularly, for the case of sigmoidal activation functions with $\psi(-a) = -\psi(a)$, it is possible to restrict attention only to weight vectors with $w_{01} \geqslant w_{02} \geqslant \cdots \geqslant w_{0r}$ (see Ossen and Rüugen, 1996). Secondly, the possible presence of multiple minima has no essential effect, at least asymptotically, for solutions to Eq. (1.4) (see White, 1989). Thirdly, several global optimization strategies (simulation annealing, genetic algorithms, etc.) are available to avoid being trapped in local minima and they have been successfully employed in neural network modeling. Finally, when the focus is on prediction, it can be shown that the unidentifiability can be overcome and the problem disappears (Hwang and Ding, 1997).

Asymptotic normality of the weight vector estimator can also be established. In particular, let $l(\mathbf{z}, \mathbf{w}) \equiv q\left(y, f(\mathbf{x}, \mathbf{w})\right)$ and denote by ∇ and ∇^2 the gradient and the Hessian operators, respectively. Assume that $\mathbf{A}^* \equiv \mathbb{E}\left(\nabla^2 l\left(\mathbf{z}, \mathbf{w}_0\right)\right)$ and $\mathbf{B}^* \equiv \mathbb{E}\left(\nabla l\left(\mathbf{z}, \mathbf{w}_0\right) \nabla l\left(\mathbf{z}, \mathbf{w}_0\right)^T\right)$ are nonsingular matrices. If general regularity conditions hold, then

$$\sqrt{n}\left(\hat{\mathbf{w}}_n - \mathbf{w}_0\right) \xrightarrow{d} N\left(\mathbf{0}, \mathbf{C}^*\right)$$

where $\mathbf{C}^* = \mathbf{A}^{*^{-1}}\mathbf{B}^*\mathbf{A}^*$ (White, 1989, theorem 2, p. 457).

These results make it possible to test the hypotheses about the connection strengths, which can be of great help in defining pruning strategies with a strong inferential base. However, focusing on single weights might be misleading due to the black-box nature of the neural network model and better model selection strategies become necessary to select appropriate network architectures for the problem at hand.

1.3
Model Selection

Model selection in neural network models requires selecting both an appropriate number of the hidden units and a suitable set of explicative variables and, as a consequence, the connections thereof. The "atheoretical" nature of the tool, employed for the lack of knowledge about the functional form of the data generating process, and the intrinsic misspecification of the model, makes this problem a hard task.

The problem is not a novel one and a number of different and effective solutions have been proposed. The most popular approaches are pruning, stopped training, and regularization. Although these techniques may lead to satisfactory results, they focus on single weights and this can be misleading due to the black-box nature of the neural network model. Indeed, they do not give any information on the most significant variables, which is useful in any model building strategy and, moreover, different topologies can achieve the same approximation accuracy. Therefore, a proper choice of the network topology cannot be just based on complexity reason and should also take into account model plausibility. All the techniques based on weight selection are much more on the side of computational standpoint than on the side of a statistical perspective. Instead, it would be of some interest to look at the choice of the network topology by including it in the classical statistical model selection approach.

In this perspective, information criteria such as the Akaike information criterion (AIC) and the Schwarz information Criterion (SIC) could be used. These criteria add a complexity penalty to the usual sample log-likelihood, and the model that optimizes this penalized log-likelihood is preferred. Generally, the SIC, imposing a more severe penalty than the AIC, delivers the most conservative models (i.e., least complex) and has been found to perform well in selecting forecasting models in other contexts. Therefore, in the neural network framework, SIC is usually preferred (Franses and Draisma, 1997, *inter alia*). However, many statistical studies agree that these measures should be used with care in choosing the best model in a neural network context. Indeed, Swanson and White (1997) and Qi and Zhang (2001) show that these procedures might lead to over-parameterized models with heavy consequence on overfitting and poor ex-post forecast accuracy. Kuan and Liu (1995) instead propose the predictive stochastic complexity criterion, which is based on forward validation, a better choice for forecasting purposes.

In any case, all these model selection procedures are not entirely satisfactory. Since model selection criteria depends on sample information, their actual values are subject to statistical variations. As a consequence, a model with higher model selection criterion value may not outperform *significantly* its competitors. Moreover, they lack a strong inferential statistical perspective and, usually, they contain a strong judgemental component not giving explicitly any information on the most "significant" variables.

A better model selection strategy should be faced in a statistical framework, relating it to the classical model selection approach, emphasizing the different role in the model of the explanatory variables and of the hidden layer neurons. In a regression framework, input neurons are related to the explanatory variables (useful for identification and interpretation of the model), while the hidden layer size has no clear interpretation, and it should be considered basically as a smoothing parameter taking into account the trade-off between estimation bias and variability.

However, while in principle, the hidden layer size could be chosen according to one of the many results available in the statistical literature, ranging from the information criteria based on the fitting, to the indexes based on prediction accuracy, the input selection should be addressed focusing on procedures for variable selection in regression models.

In this perspective, the model selection strategy discussed in the following, identifies both the input variables and the hidden layer size by using formal test procedures. Particularly, the input variables are selected by using relevance measures, while the hidden layer size is selected by looking at the predictive performance of the neural network model. Both procedures use extensively resampling techniques that are able to deliver consistent results under general assumptions, a very important requirement in a neural network framework.

1.3.1
Feature Selection by Relevance Measures

To select a proper set of input variables, we focus on a selection rule based on relevance measures (White and Racine, 2001; La Rocca and Perna, 2005a,b) following the usual strategy generally employed when selecting a model in the classical regression framework.

There are, of course, a lot of representative criteria that have traditionally been used to quantify the relevance of input variables in neural models. These relevance criteria, often referred to as sensitivity measures, are traditionally obtained by the computation of partial derivatives.

As in White and Racine (2001), the hypotheses that the independent variable X_j has no effect on Y, in model (Eq. 1.1) can be formulated as:

$$\frac{\partial g\left(\mathbf{x}\right)}{\partial x_j} = 0, \forall x. \tag{1.5}$$

Of course, the function g is unknown but we equivalently investigate the hypotheses

$$f_j\left(\mathbf{x}; \mathbf{w}_0\right) = \frac{\partial f\left(\mathbf{x}; \mathbf{w}_0\right)}{\partial x_j} = 0, \forall x. \tag{1.6}$$

since f is known and $\mathbf{w}_0$ can be closely approximated.

The general form for relevance measures is,

$$\theta_j = \mathbb{E}\left[h\lfloor f_j(\mathbf{x}, \mathbf{w}_0\right] \tag{1.7}$$

where $h(.)$ is a proper chosen function and $\mathbb{E}[.]$ is the expected value w.r.t. the probability measure of the vector of the explanatory variables.

As relevance measures, several alternative functions can be used; for example, the average derivative ($h(x) = x$); the absolute average derivative ($h(x) = |x|$); the square average derivative ($h(x) = x^2$); the maximum and the minimum derivative ($h(x) = max(x)$ and ($h(x) = min(x)$)). Each of these measures reflects different aspects of the model and, as a consequence, it can provide different ranks of the variables, according to its magnitudes. The most natural sensitive measure is, of course, the average derivative but, because of cancellations between negative and positive values, the absolute average derivative and the square average derivative are the most used. In many financial applications, such as the construction of the risk neutral portfolios of assets, and in all applicative context where the interest is on inflection points, the maximum and the minimum derivative are also quite common. However, the most general and natural choice in the statistical literature is the square function leading to the following relevance measure for the d input neurons:

$$\theta_j = \mathbb{E}\left[f_j^2\left(\mathbf{x}, \mathbf{w}_0\right)\right], \qquad j = 1, 2, \ldots, d.$$

Therefore, the hypothesis that a given set of variables has no effect on Y can be formulated in a multiple testing framework as

$$H_j : \theta_j = 0 \quad vs \quad H_j' : \theta_j > 0, \quad j = 1, 2, \ldots, d. \tag{1.8}$$

Each null H_j can be tested by using the statistic,

$$\hat{T}_{n,j} = n\hat{\theta}_j = \sum_{i=1}^{n} f_j^2\left(\mathbf{X}_i, \hat{\mathbf{w}}_n\right) \tag{1.9}$$

where $\hat{\theta}_j = \frac{1}{n}\sum_{i=1}^{n} f_j^2\left(\mathbf{X}_i, \hat{\mathbf{w}}_n\right)$ and the vector $\hat{\mathbf{w}}_n$ is a consistent estimator of the unknown parameter vector $\mathbf{w}_0$. Clearly, large values of the test statistics indicate evidence against H_j.

So, the problem here is how to decide which hypotheses to reject, accounting for the multitude of tests. Significant correction for multiple hypothesis testing has been a central concern in many fields of research that deal with large sets of variables and small samples and where, as a consequence, the control of false positives becomes an important problem.

The most relevant methods for significant adjustment are based on standard measures like the familywise error rate (FWE), defined as the probability of finding at least one false positive, that is, rejecting at least one of the true null hypotheses.

The FWE can be controlled by using the well-known Bonferroni method or stepwise procedures proposed by Holm (1979) which are more powerful. Unfortunately, both procedures are conservative since they do not take into account the dependence structure of the individual p-values. A possible solution can be obtained by using the reality check proposed by White (2000) which can be easily extended to our framework.

Here, we use the StepM procedure proposed by Romano and Wolf (2005), suitable for joint comparison of multiple misspecified models.

The step-down procedure begins by constructing a rectangular joint confidence region (with nominal coverage probability $1 - \alpha$), which is used to test the joint null hypothesis that all the nulls $H_j, j = 1, 2, \ldots, d$ are true. If all hypotheses are not rejected, the procedure stops. Otherwise, the rejected hypotheses are removed and a new rectangular joint confidence region with nominal level $1 - \alpha$ is constructed. The process is repeated until no further hypotheses are rejected.

The procedure can be described by Algorithm 1.1.

Algorithm 1.1: Multiple testing algorithm.

Relabel the hypothesis from H_{r_1} to H_{r_d} in redescending order of the value of the test statistics $\widehat{T}_{n,j}$, that is $\widehat{T}_{n,r_1} \geq \widehat{T}_{n,r_2} \geq \ldots \geq \widehat{T}_{n,r_d}$.
Set $L = 1$ and $R_0 = 0$.
for $j = R_{L-1} + 1$ to d **do**
    Fix $c_L\,(1-\alpha)$ such that the joint asymptotic coverage probability is $1-\alpha$
    if $0 \notin \left[\widehat{T}_{n,r_j} - c_L\,(1-\alpha)\,, \infty\right)$ **then**
        reject H_{r_j}
    end if
end for
if no (further) null hypothesis are rejected **then**
    Stop
else
    R_L = number of rejected hypothesis
    $L = L + 1$
    Go to step 3
end if

Of course, in order to apply the StepM procedure, it is necessary to know the distribution of the test statistic $\widehat{T}_{n,j}$. Under general conditions, (Giordano, La Rocca, and Perna 2014) it is straightforward to show that:

$$\widehat{T}_{n,j} \xrightarrow{d} \mathcal{F}\,(0, \mathbf{C}^*, \mathbf{M}^*) \quad \forall j \in \{1, \ldots, d\}$$

where $\mathcal{F}$ denotes the mixture of independent χ^2 random variables and

$$\mathbf{M}^* = \mathbb{E}\left(\left[\nabla f_j(\mathbf{X}_i, \mathbf{w}_0)\nabla f_j(\mathbf{X}_i, \mathbf{w}_0)^T + \nabla^2 f_j(\mathbf{X}_i, \mathbf{w}_0) f_j(\mathbf{X}_i, \mathbf{w}_0)\right]\right).$$

Even if this result is relevant from a theoretical point of view, it does not allow easy the estimation of the quantiles $c_L(1-\alpha)$ in StepM algorithm. Therefore, the sampling distribution can be better approximated by some types of resampling techniques. Here, we propose the use of subsampling. This choice can be justified by its property of being robust against misspecifications, a key property when dealing with artificial neural network models. Moreover, the procedure delivers consistent results under very weak assumptions.

The resampling scheme runs as follows. Fix b such that $b < n$ and let $\mathbf{Y}_1, \ldots, \mathbf{Y}_S$ be equal to $S = \begin{pmatrix} n \\ b \end{pmatrix}$ subsets of $\{\mathbf{Z}_1, \ldots, \mathbf{Z}_n\}$. Let $\widehat{T}^s_{b,j}$ be the test statistic evaluated at $\mathbf{Y}_s, s = 1, \ldots, S$. Then, for $\mathbf{x} \in \mathbb{R}^d$, the true joint cdf of the test statistics evaluated at $\mathbf{x}$ is given by

$$G_n\,(\mathbf{x}) = \Pr\left\{\widehat{T}_{n,1} \leq x_1, \widehat{T}_{n,2} \leq x_2 \ldots, \widehat{T}_{n,d} \leq x_d\right\} \tag{1.10}$$

and it can be estimated by the subsampling approximation

$$\widehat{G}_n(\mathbf{x}) = \binom{n}{b}^{-1} \sum_{s=1}^{S} \mathbb{I}\left\{ \widehat{T}^s_{b,1} \leq x_1, \widehat{T}^s_{b,2} \leq x_2, \ldots, \widehat{T}^s_{b,d} \leq x_d \right\} \tag{1.11}$$

where as usual $\mathbb{I}(\cdot)$ denotes the indicator function.

As a consequence, for $D \subset \{1, \ldots, d\}$, the distribution of the maximum of the test statistics, let's say $H_{n,D}(x)$, can be estimated by the empirical distribution function $\widehat{H}_{n,D}(x)$ of the values $\max\left\{ \widehat{T}^s_{b,j}, j \in D \right\}$, that is

$$\widehat{H}_{n,D}(x) = \binom{n}{b}^{-1} \sum_{s=1}^{S} \mathbb{I}\left\{ \max\left\{ \widehat{T}^s_{b,j}, j \in D \right\} \leq x \right\} \tag{1.12}$$

and the quantile of order $1 - \alpha$ can be estimated as

$$\widehat{c}_L(1-\alpha) = \inf\left\{ x : \widehat{H}_{n,D}(x) \geq 1 - \alpha \right\}. \tag{1.13}$$

The consistency of the subsampling procedure has been proved in (Giordano, La Rocca, and Perna 2014) as a straightforward extension of a result in Romano and Wolf (2005). In particular, under general assumptions, if $b/n \to 0$ when $b \to \infty$ and $n \to \infty$, then

$$\rho\left(\widehat{G}_{n,D}, G_{n,D}\right) \xrightarrow{P} 0$$

for any metric ρ metrizing weak convergence on $\mathbb{R}^{|D|}$ with $|D|$ the cardinality of D. Moreover, the subsampling critical values satisfy

$$\widehat{c}_L(1-\alpha) \xrightarrow{P} c_L(1-\alpha)$$

and

$$\limsup_n FWE \leq \alpha$$

using Algorithm (1.1) with the subsample estimator, $\widehat{c}_L(1-\alpha)$.

The choice of the subsampling as resampling technique can be justified as follows. First, the method does not require any knowledge of the specific structure of the data and so it is robust against misspecifications, a key property when dealing with artificial neural network models. Moreover, the procedure delivers consistent results under very weak assumptions. In our case, by assuming: (i) $b \to \infty$ in such a way that $\frac{b}{n} \to 0$, as $n \to \infty$, (ii) conditions that guarantee asymptotic normality of $\widehat{\mathbf{w}}_n$ are fulfilled (White, 1989), (iii) smoothness conditions on the test statistics $\widehat{T}_{n,j}$ (White and Racine, 2001), the subsampling approximation is a consistent estimate of the unknown (multivariate) sampling distribution of the test statistics (Romano and Wolf, 2005). Observe that, the number of subsets of length b which can be formed out of a sample of size n grows very fast with n. Therefore, usually, just B random selected subsets are considered for computing the subsampling approximation.

Clearly, the main issue when applying the subsampling procedure lies in choosing the length of the block, a problem which is common to all blockwise resamplig techniques. Nevertheless, Politis, Romano, and Wolf (1999) proposed a number of strategies to select b and theorems that ensure that the asymptotic results are still valid for a broad range of choices for the subsample size. More recently, Giacomini, Politis, and White (2013) proposed an approach to reduce the computational effort when conducting Monte Carlo experiments involving resampling techniques. It could be used in the neural network framework to make feasible the block selection calibration algorithm.

1.3.2 Some Numerical Examples

To illustrate the performance of the proposed input selection procedure, we use simulated data sets generated by models with known structure. The aim is to evaluate the ability of the test procedure to select a proper set of explanatory variables for the given data generating process. For the experimental setup, we assume $n = 300$, $b = 100$, $r = 2$, $B = 1000$, $\alpha = 0.05$. The hidden layer size of the neural networks has been determined by using the cross-validation (CV) and all the neural network models have been estimated by using a square loss function in Eq. (1.3), repeating the estimation process with different randomly chosen starting points to avoid being trapped in local minima. The software procedures have been implemented in R.

The simulated data sets have been generated by the following models.

The first model (Model M1) is the same model used in Tibshirani (1996). We assume that Y depends on 10 esplicative variables $\{X_1, X_2, \dots, X_{10}\}$ but just variables $\{X_3, X_4, X_5, X_6\}$ are relevant to the model, that is,

$$Y = 3\psi\left(2X_3 + 4X_4 + 3X_5 + 3X_6\right) + 3\psi\left(2X_3 + 4X_4 - 3X_5 - 3X_6\right) + \varepsilon$$

where ψ is the logistic activation function, $\mathbf{X} = (X_3, X_4, X_5, X_6)^T$ is a vector of multivariate Gaussian random variables with zero mean, unit variance and pair-wise correlation equal to 0.5 and ε Gaussian with zero mean and variance equal to 0.7. This gave a signal-to-noise ratio roughly equal to 1.2. Clearly, a neural network with logistic activation function, four input neurons, and two hidden neurons is a correctly specified model and no misspecification is present.

The results of the multiple-testing procedure for variable selection are reported in Figure 1.1. After the first step, the procedure rejects the hypothesis that variable 4 is not relevant and accepts all others hypotheses. At the second step, variables 5, 3, and 6 are recognized as relevant, as well. At the third step, the remaining variables are recognized as not relevant and the procedure stops. The procedure gives results that are consistent with the data-generating process and the plot reported in Figure 1.1.

The second model (Model M2) is the same model used in De Veaux *et al.* (1998). Again, we assume that Y depends on 10 esplicative variables $\{X_1, X_2, \dots, X_{10}\}$ but

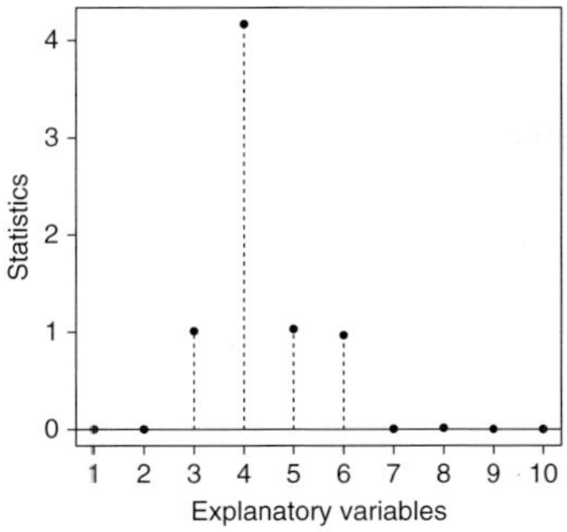

j	$\hat{T}_{n,r_j}$	r_n	$\hat{T}_{n,r_j}-\hat{c}_1$	$\hat{T}_{n,r_j}-\hat{c}_2$	$\hat{T}_{n,r_j}-\hat{c}_3$
1	4.1649	4	**2.8040**	–	–
2	1.0315	5	−0.3295	**0.5303**	–
3	1.0105	3	−0.3505	**0.5092**	–
4	0.9680	6	−0.3930	**0.4667**	–
5	0.0142	8	−1.3468	−0.4871	−0.1836
6	0.0038	7	−1.3571	−0.4975	−0.1940
7	0.0025	9	−1.3585	−0.4988	−0.1952
8	0.0019	10	−1.3590	−0.4993	−0.1958
9	0.0016	2	−1.3594	−0.4997	−0.1962
10	0.0010	1	−1.3599	−0.5002	−0.1967

Figure 1.1 Model M1. Results of the multiple testing procedure ($n = 300$, $b = 100$, $r = 2$, $B = 1000$, $\alpha = 0.05$). Figures in bold refer to the rejection of the corresponding hypotheses H_{r_j}.

just variables $\{X_4, X_5, X_6\}$ are relevant to the model, that is,

$$Y = 1.5\cos\left(\frac{2\pi}{\sqrt{3}}\sqrt{\left(X_4-0.5\right)^2+\left(X_5-0.5\right)^2+\left(X_6-0.5\right)^2}\right)+\varepsilon$$

where $\mathbf{X} = (X_4, X_5, X_6)^T$ is drawn randomly from the unit hypercube. The function is radially symmetric in these three variables. Clearly, the number of the neurons in the hidden layer is unknown and the model we try to identify is, by construction, misspecified. In this latter case, the procedure is able to select the correct set of relevant variables in two steps, as clearly shown in Figure 1.2.

For the third model (Model M3) introduced by Friedman (1991), again, we assume that Y depends on 10 esplicative variables $\{X_1, X_2, \ldots, X_{10}\}$ but just variables $\{X_3, X_4, X_5, X_6, X_7\}$ are relevant, that is

$$Y = \left(10\sin\left(\pi X_3 X_4\right) + 20\left(X_5-0.5\right)^2 + 10X_6 + 5X_7 + \varepsilon\right)/25$$

where $\mathbf{X} = (X_3, X_4, X_5, X_6, X_7)^T$ is drawn randomly from the unit hypercube.

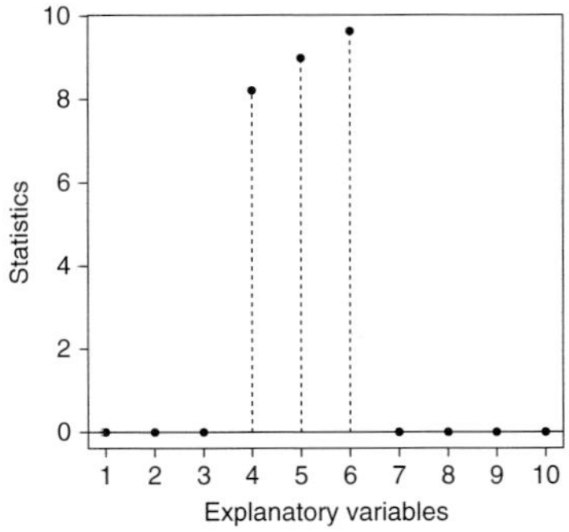

j	$\hat{T}_{n,r_j}$	r_j	$\hat{T}_{n,r_j}-\hat{c}_1$	$\hat{T}_{n,r_j}-\hat{c}_2$
1	10.0015	6	**8.2969**	–
2	9.0820	5	**7.3773**	–
3	8.2477	4	**6.5431**	–
4	0.0020	2	−1.7027	−0.0159
5	0.0010	1	−1.7037	−0.0170
6	0.0007	7	−1.7040	−0.0173
7	0.0006	8	−1.7041	−0.0173
8	0.0003	3	−1.7043	−0.0176
9	0.0003	9	−1.7044	−0.0176
10	0.0002	10	−1.7045	−0.0177

Figure 1.2 Model M2. Results of the multiple-testing procedure ($n = 300$, $b = 100$, $r = 2$, $B = 1000$, $\alpha = 0.05$). Figures in bold refer to the rejection of the corresponding hypotheses H_{r_j}.

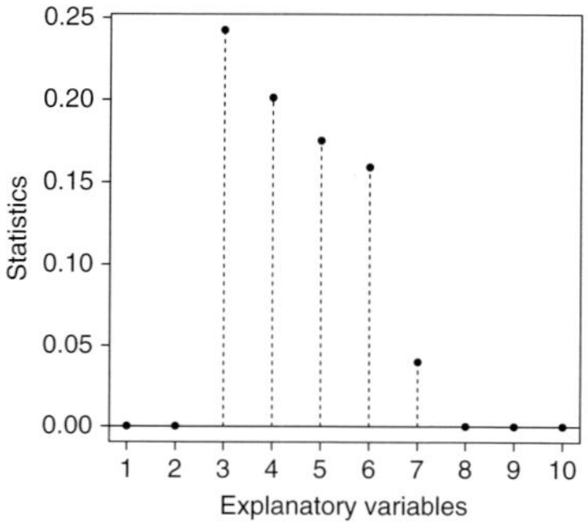

j	$\hat{T}_{n,r_j}$	r_j	$\hat{T}_{n,r_j} - \hat{c}_1$	$\hat{T}_{n,r_j} - \hat{c}_2$	$\hat{T}_{n,r_j} - \hat{c}_3$
1	0.2422	3	**0.1951**	–	–
2	0.2019	4	**0.1548**	–	–
3	0.1750	5	**0.1280**	–	–
4	0.1591	6	**0.1120**	–	–
5	0.0400	7	−0.0070	**0.0354**	–
6	0.0002	1	−0.0470	−0.0045	−0.0020
7	0.0001	2	−0.0470	−0.0045	−0.0020
8	0.0001	8	−0.0470	−0.0045	−0.0020
9	0.00009	10	−0.0470	−0.0045	−0.0020
10	0.00006	9	−0.0470	−0.0045	−0.0020

Figure 1.3 Model M3. Results of the multiple-testing procedure ($n = 300$, $b = 100$, $r = 2$, $B = 1000$, $\alpha = 0.05$). Figures in bold refer to the rejection of the corresponding hypotheses H_{r_j}.

Again, the procedure is able to correctly identify the set of relevant variables in three steps, as clearly shown in Figure 1.3.

The results of a more detailed simulation experiment are reported in Giordano, La Rocca, and Perna (2014) in which we analyse the sensitivity of the CV procedure to omitted or redundant variables and the sensitivity of the StepM testing scheme to hidden layer size error identification. The results show that redundant variables do not appear to be harmful in selecting the hidden layer size; in almost all cases, the true hidden layer size is correctly identified. On the contrary, omitting relevant variables might have negative effects on the hidden layer size. However, this appears to be connected to the number and type of omitted variables. In any case, increasing the sample size may improve the performance. Moreover, the experimental results also suggest the possibility to use the block length as a calibration tool to bring the empirical FWE closer to the nominal FWE. This calibration technique could even be effective if the hidden layer size is not correctly identified but it has been fixed in a neighbourhood of the true value. When the hidden layer size is correctly identified, the procedure correctly identifies the true relevant variables in all cases, for all sample sizes. When the hidden layer size is incorrectly identified (under/over estimation), the results depend on the sample size and on the subsample size. For the case of $n = 400$ and $n = 600$, the proportion is very close to 1. There are some identification problems for the case $n = 200$ for some variables. However, again, the block length of the subsampling can be used to mitigate the problem and to increase the proportion of true relevant variables correctly identified.

1.3.3
Application to Real Data

As an application to real data, we considered a very popular data set, often used to check the performance of non-parametric regression techniques with respect to variable selection. The data are daily measurement of ozone concentration (maximum one hour average) and eight meteorological quantities for 330 observations starting from the beginning of 1976. The data were used by Breiman and Friedman (1985) when introducing the ACE algorithm. The variables considered are: Ozone (Upland ozone concentration, ppm), Temp (Sandburg Air Force Base temperature, °F), Ibh (inversion base height, feet), Dpg (Dagget pressure gradient, mmHg), Vis (visibility, in miles), Vh (Vandenburg 500 millibar height, m), Hum (humidity, percent), Wind (wind speed, mph), Ibt (Inversion base temperature, degrees F), and Day (day of year).

The hidden layer size has been selected by CV, while the input relevant variables have been selected by using the proposed procedure. The StepM procedure has been calibrated with the subsampling, where the subsample size has been fixed by using the minimum volatility method (Politis, Romano, and Wolf , 1999).

The procedure clearly selects the variables Day, Ibt, Vh, Dpg, Hum, and Temp as relevant, while the variables Vis, Wind, and Ibh are classified as not relevant. Note that the variables are selected in six steps (just the variables Day and Ibt are selected in the first step) and so, a multistep procedure appears to be necessary to avoid masking effects (Figure 1.4).

For the sake of comparison with other neural network variable selection schemes, we considered the Bayesian approach proposed in Lee (2004), where it is also reported a comparison with other non-parametric variable selection techniques: The stepwise ACE, the stepwise GAM, adaptive regression splines (TURBO), and adaptive backfitting (BRUTO). All the results are summarized in Table 1.1. The proposed procedure largely agrees with the best network selected by using the Bayesian approach proposed by Lee. Interestingly enough, however, the variable Temp is considered relevant by the multiple testing scheme, while it is never selected by the Bayesian approach when applied to neural networks. Note that this variable is always selected by the other non-parametric techniques, suggesting that the multiple testing scheme is able to uncover possibly masked relationships. Even if the alternative methods disagree about which variable subset is optimal, it does seem clear that some variable selection procedure is necessary.

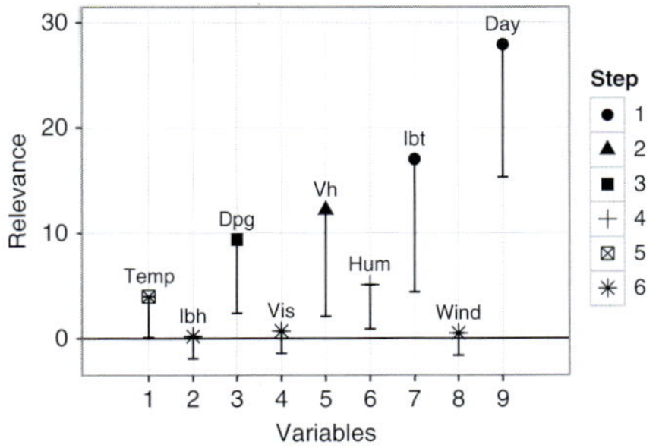

Rank	Variable	$\hat{T}_{n,j}$	C_j	Limit	Step
1	Day	27.9	12.6	15.3	1
2	Ibt	17.0	12.6	4.4	1
3	Vh	12.2	10.1	2.1	2
4	Dpg	9.4	7.0	2.4	3
5	Hum	5.1	4.2	0.9	4
6	Temp	4.0	3.9	0.1	5
7	Vis	0.7	2.1	−1.4	6
8	Wind	0.5	2.1	−1.6	6
9	Ibh	0.2	2.1	−1.9	6

Figure 1.4 IVS for Ozone data via neural networks. The relevance measure is the statistic $\hat{T}_{n,j}$. The hidden layer size has been selected by k-fold CV ($r = 3$). Subsample size selected by using minimum volatility method. The nominal size is $\alpha = 0.05$.

Table 1.1 Comparison of variable selection procedures on the Ozone data.

Method	Vh	Wind	Hum	Temp	Ibh	Dpg	Ibt	Vis	Day
Multiple test (NN(3))	×		×	×		×	×		×
(Ranks)	(3)		(5)	(6)		(4)	(2)		(1)
Lee's Bayes (Best NN(3))	×		×			×	×		×
Lee's Bayes (Second best NN(6))			×			×	×	×	×
Lee's Bayes (Third best NN(3))			×				×		×
ACE (Stepwise)				×	×	×		×	
GAM (Stepwise)	×	×	×	×	×	×		×	×
TURBO	×			×	×	×		×	×
BRUTO				×	×	×		×	×

1.4 The Selection of the Hidden Layer Size

The hidden layer size should be selected by looking at the predictive performance of the neural network model, as in the case of CV. However, in recent years, there is a growing literature addressing the problem of comparing different models and theories via use of predictive performance and predictive accuracy test (Corradi and Swanson, 2006, and references therein). In this Literature, it is quite common to compare multiple models, which are possibly misspecified (they are all approximations of some unknown true model), in terms of their out-of-sample predictive ability for a specified loss function.

Let $(Y_\tau, \mathbf{X}_\tau)$ denote a future observation that satisfies

$$Y_\tau = g(\mathbf{X}_\tau) + \varepsilon_\tau \tag{1.14}$$

Assume then that $k+1$ alternative forecasting neural network models are available, namely $f(\mathbf{x}^j, \mathbf{w}^j), j = 0, 1, \ldots, k$. The models can differ in hidden layer size and/or in number and type of explanatory variables. Model $f(\mathbf{x}^0, \mathbf{w}^0)$ is the benchmark model. In our framework, a sensible choice is the linear model, that is a neural network with skip layer and $r = 0$ neurons in the hidden layer.

Let the generic forecast error be $u_{j,\tau} = Y_\tau - f\left(\mathbf{X}_\tau^j, \mathbf{w}_0^j\right), j = 0, 1, \ldots, k$ where $\mathbf{w}_0$ is defined as in Section 1.2. Let h be a proper chosen loss function (Elliot and Timmermann, 2004) and define

$$\theta_j = \mathbb{E}\left(h(u_{0,\tau}) - h(u_{j,\tau})\right), j = 1, 2, \ldots, k. \tag{1.15}$$

Clearly, if model j beats the benchmark (i.e., shows better expected predictive performances) we have $\theta_j > 0$, otherwise $\theta_j \leqslant 0$ and our goal is to identify as many models for which $\theta_j > 0$. In other words, for a given model j, consider

$$H_j : \theta_j \leqslant 0 \quad vs \quad H_j' : \theta_j > 0 \tag{1.16}$$

and, in a multiple testing framework, make a decision concerning each individual testing problem by either rejecting H_j or not. Also in this case, the data snooping problem can arise and the FWE should be taken under control. In this framework, possible alternative solutions can be obtained by using the reality check and by using test of superior predictive ability, which can be easily extended to our neural network framework.

1.4.1 A Reality Check Approach

To avoid data snooping problems, it is possible to use the reality check as in White (2000) and the modification for nested models as proposed in Clark and McCracken (2012a,b).

For a given loss function, the reality check tests the null hypothesis that a benchmark model (i.e., model 0) performs equal or better than all competitor models (i.e., models 1, … , k). The alternative is that at least one competitor performs better than the benchmark. Formally, we have

$$H_0 : \max_{j=1,\dots,k} \theta_j \leqslant 0 \quad vs \quad H_1 : \max_{j=1,\dots,k} \theta_j > 0. \tag{1.17}$$

Following a common practice often used to select the best predictive model, the sample of size N is split into $N = R + P$ observations where R observations are used for estimation and P observations are used for predictive evaluation. Let $\hat{u}_i = Y_i - f\left(\mathbf{X}_i^j, \hat{\mathbf{w}}_R^j\right)$, $i = R + 1, \dots, N$, where $f\left(\mathbf{X}_i^j, \hat{\mathbf{w}}_R^j\right)$ is the model estimated on the data set $\left\{\left(Y_i, \mathbf{X}_i^j\right), i = 1, \dots, R\right\}$. Following White (2000) define the statistic

$$S_P = \max_{j=1,\dots,k} S_P\left(0, j\right) \tag{1.18}$$

where

$$S_P\left(0, j\right) = \frac{1}{\sqrt{P}} \sum_{i=R+1}^{N} \left\{h\left(\hat{u}_{0,i}\right) - h\left(\hat{u}_{j,i}\right)\right\}, \quad j = 1, \dots, k.$$

It can be shown that, if general regularity conditions hold, under H_0, as $P, R \to \infty$,

$$\max_{j=1,\dots,k} \left\{S_P\left(0, j\right) - \sqrt{P}\theta_j\right\} \xrightarrow{d} \max_{j=1,\dots,k} S\left(0, j\right). \tag{1.19}$$

The $k \times 1$ vector $S = \left(S\left(0, 1\right), S\left(0, 2\right), \dots, S\left(0, k\right)\right)$ has Gaussian distribution with zero mean and covariance matrix defined as

$$V = \lim_{N\to\infty} \text{var}\left(\frac{1}{\sqrt{P}} \sum_{i=R+1}^{N} \mathbf{v}_i\right)$$

where the generic element of vector $\mathbf{v}_i$ is defined as $v_{i,j} = h\left(u_{0,i}\right) - h\left(u_{j,i}\right)$. The matrix V is supposed to be positive semi-definite.

Since it is well known that the maximum of a Gaussian process is not Gaussian in general, standard critical values cannot be used to conduct inference on S_P. Alternatively, resampling techniques such as the subsampling or the bootstrap can be used.

The bootstrap analogue of the statistic S_P can be computed as

$$S_P^* = \max_{j=1,\dots,k} S_P^*(0,j) \tag{1.20}$$

where

$$S_P^*(0,j) = \frac{1}{\sqrt{P}} \sum_{i=R+1}^{N} \left\{ \left(h\left(\hat{u}_{0,i}^*\right) - h\left(\hat{u}_{0,i}\right) \right) - \left(h\left(\hat{u}_{j,i}^*\right) - h\left(\hat{u}_{j,i}\right) \right) \right\} \tag{1.21}$$

with $\hat{u}_{j,i}^* = Y_i^* - f\left(\mathbf{X}_i^{*j}, \hat{\mathbf{w}}_R^j\right)$ and $\left(Y_i^*, \mathbf{X}_i^{*j}\right)$ denote the resampled data. Note that the bootstrap statistics contain only estimators based on the original sample and this is particularly convenient when dealing with neural network models. If an estimation is needed for each bootstrap sample, the procedure will soon become not feasible in our framework.

The bootstrap procedure is consistent in the neural network framework. Under general regularity conditions, it can be shown that, if $q = h$, for $P, R \to \infty$

$$\Pr\left(\sup_{v \in \mathbb{R}} \left| \Pr_*\left(S_P^* \leqslant v\right) - \Pr\left(S_P^\kappa \leqslant v\right) \right| > \varepsilon \right) \to 0 \tag{1.22}$$

where $\Pr_*$ denotes probability induced by the bootstrap resampling scheme and

$$S_P^\kappa = \max_{j=1,\dots,k} \left\{ S_P(0,j) - \sqrt{P}\theta_j \right\}$$

As usual, the bootstrap procedure can be implemented by Monte Carlo. For any bootstrap replication, compute the bootstrap statistics, S_P^*. Perform B bootstrap replications (B large) and compute the quantiles of the empirical distribution of the B bootstrap statistics. Reject the null hypothesis H_0 if S_P is greater than the $(1-\alpha)$th-percentile. Otherwise, do not reject.

The bootstrap procedure can be implemented as described in Algorithm 1.2.

Note that, to estimate a percentile, B should be quite large (usually $B > 1000$) and the indexes are generated just once at the beginning of the procedure. Moreover, we assume that $h = q$.

1.4.2
Numerical Examples by Using the Reality Check

In order to evaluate the ability of the procedure to select a proper model for a given data generating process, we use simulated data sets with known structure.

The first is a linear model (M1) with two regressors defined as:

$$Y = \mathbf{X1} + \varepsilon$$

where $\mathbf{X} = (X_1, X_2)^T$ are drawn from the uniform distribution, ε is a standard Gaussian and $\mathbf{1}$ denotes a column vector of the ones of appropriate length. This

Algorithm 1.2: Bootstrap resampling algorithm.

Fix P, R such that $P + R = N$.
Fix B, the number of bootstrap replicates.
Generate B sets of random observation indexes of length P, namely $\{\theta_b(i), i = R+1, \dots, N; b = 1, \dots, B\}$.
$M_0 \leftarrow -\Delta$, with Δ finite big constant.
$M_0^{(b)} \leftarrow -\Delta$, with Δ finite big constant, $b = 1, \dots, B$.
$\hat{\mathbf{w}}_R^0 \leftarrow \arg\min_{\mathbf{w}\in\mathbf{W}} \frac{1}{R} \sum_{i=1}^{R} q\left(Y_i, f\left(\mathbf{X}_i^0, \mathbf{w}\right)\right)$.
$\hat{u}_{0,i} \leftarrow Y_i - f\left(\mathbf{X}_i^0, \hat{\mathbf{w}}_R^0\right), i = R+1, \dots, N$.
$\hat{h}_{0,i} \leftarrow h\left(\hat{u}_{0,i}\right), i = R+1, \dots, N$.
for $j = 1$ to k **do**
  $\hat{\mathbf{w}}_R^j \leftarrow \arg\min_{\mathbf{w}\in\mathbf{W}} \frac{1}{R} \sum_{i=1}^{R} q\left(Y_i, f\left(\mathbf{X}_i^j, \mathbf{w}\right)\right)$.
  $\hat{u}_{j,i} \leftarrow Y_i - f\left(\mathbf{X}_i^j, \hat{\mathbf{w}}_R^j\right), i = R+1, \dots, N$.
  $\hat{h}_{j,i} \leftarrow h\left(\hat{u}_{j,i}\right), i = R+1, \dots, N$
  $S_P(0,j) \leftarrow \frac{1}{\sqrt{P}} \sum_{i=R+1}^{N} \left\{\hat{h}_{0,i} - \hat{h}_{j,i}\right\}$.
  $M_j \leftarrow \max\left\{S_p(0,j), M_{j-1}\right\}$.
  for $b = 1$ to B **do**
    $\hat{u}_{0,i}^{(b)} = Y_{\theta(b)} - f\left(\mathbf{X}_{\theta(b)}^0, \hat{\mathbf{w}}_R^0\right)$
    $\hat{u}_{i,j}^{(b)} = Y_{\theta(b)} - f\left(\mathbf{X}_{\theta(b)}^j, \hat{\mathbf{w}}_R^j\right)$.
    $S_P^{(b)}(0,j) \leftarrow \frac{1}{\sqrt{P}} \sum_{i=R+1}^{N} \left\{\left(h\left(\hat{u}_{0,i}^{(b)}\right) - \hat{h}_{0,i}\right) - \left(h\left(\hat{u}_{j,i}^{(b)}\right) - \hat{h}_{j,i}\right)\right\}$
    $M_j^{(b)} \leftarrow \max\left\{S_P^{(b)}(0,j), M_{j-1}^{(b)}\right\}$.
  end for
end for
return p-value $\leftarrow \frac{1}{B} \sum_{b=1}^{B} \mathbb{I}\left(M_k^{(b)} > M_k\right)$.

model can be correctly modeled by using a network, with skip layer, two input units, and zero hidden units.

Model M2 is the same model used in Tibshirani (1996) and Model M3 is the same model used in De Veaux *et al.* (1998). Both models have already been used in previous sections.

We have considered $N = 600$, $R = 400$, $P = 200$ and $B = 4999$.

In Table 1.2, we consider values of the test statistics for different input neurons, from X_1 to X_6, and different hidden layer size, from 1 to 6. It is clear that for model M1 and M2, the proposed procedure is able to identify the correct data-generating process. In the first case, the p-values of the tests are all >0.50, and so the benchmark (i.e., the linear model) shows better expected predictive performance with

Table 1.2 Values of the test statistics for different input neuron sets and different hidden layer size.

Model	Inputs/size	1	2	3	4	5	6
M1	1	*−0.116*	*−0.116*	*−0.116*	*−0.116*	*−0.116*	*−0.116*
	2	*−0.290*	*−0.290*	*−0.290*	*−0.290*	*−0.290*	*−0.290*
	3	*−0.721*	*−0.721*	*−0.721*	*−0.721*	*−0.721*	*−0.721*
	4	*−0.986*	*−0.986*	*−0.986*	*−0.986*	*−0.986*	*−0.986*
	5	*−0.844*	*−0.844*	*−0.844*	*-0.844*	*−0.844*	*−0.844*
	6	*−0.873*	*−0.873*	*−0.873*	*−0.873*	*−0.873*	*−0.873*
M2	1	*−0.477*	*−0.477*	*−0.477*	*−0.477*	*−0.477*	*−0.477*
	2	**4.541**	**4.541**	**4.541**	**4.541**	**4.541**	**4.541**
	3	**2.603**	**5.741**	**5.741**	**5.741**	**5.741**	**5.741**
	4	**3.060**	**12.122**	**12.122**	**12.122**	**12.122**	**12.122**
	5	**3.058**	**12.121**	**12.121**	**12.121**	**12.121**	**12.121**
	6	**3.060**	**11.921**	**11.921**	**11.921**	**11.921**	**11.921**
M3	1	**0.748**	**2.159**	**2.159**	**2.159**	**2.159**	**2.159**
	2	**0.752**	**2.143**	**4.857**	**4.857**	**4.857**	**4.857**
	3	**0.807**	**2.722**	**5.391**	**7.215**	**7.222**	**7.249**
	4	**0.824**	**2.737**	**5.402**	**7.226**	**7.232**	**7.246**
	5	**0.886**	**2.811**	**5.531**	**7.264**	**7.269**	**7.277**
	6	**0.816**	**2.826**	**5.520**	**7.262**	**7.267**	**7.295**

The benchmark model is a neural network with skip layer and zero hidden neurons. Values in italics correspond to p-values >0.50. Values in bold correspond to p-values <0.005.

respect to neural networks of all orders and sizes. In the case of model M2, the values of the test statistics do not change significantly starting from a neural network model with 4 inputs and 2 hidden layer neurons. In the case of model M3, clearly test statistics stabilize starting from a model with 3 inputs (as expected) and 4 hidden layer neurons. The small increases in some test statistics possibly are not *significant*.

In order to get a deeper insight, in Figures 1.5–1.7, we report the prediction performance with respect to the benchmark model of neural network models with increasing hidden layer size. The segments refer to bootstrap confidence intervals for the parameters $\max_{j=1,\ldots,r} \theta_j$ with hidden layer size $r = 1, \ldots, 6$ and confidence level equal to 0.95. Each panel refers to different choice of the input variables. Again, for models M1 and M2, the proposed procedure is able to identify the correct data generating process, while for model M3, the identified model is a neural network with 3 input variables and hidden layer size equal to 4. Moreover, an aspect which arises from all the figures is that the predictive performance is improved when relevant variables are included into the model while it remains unchanged when adding irrelevant ones.

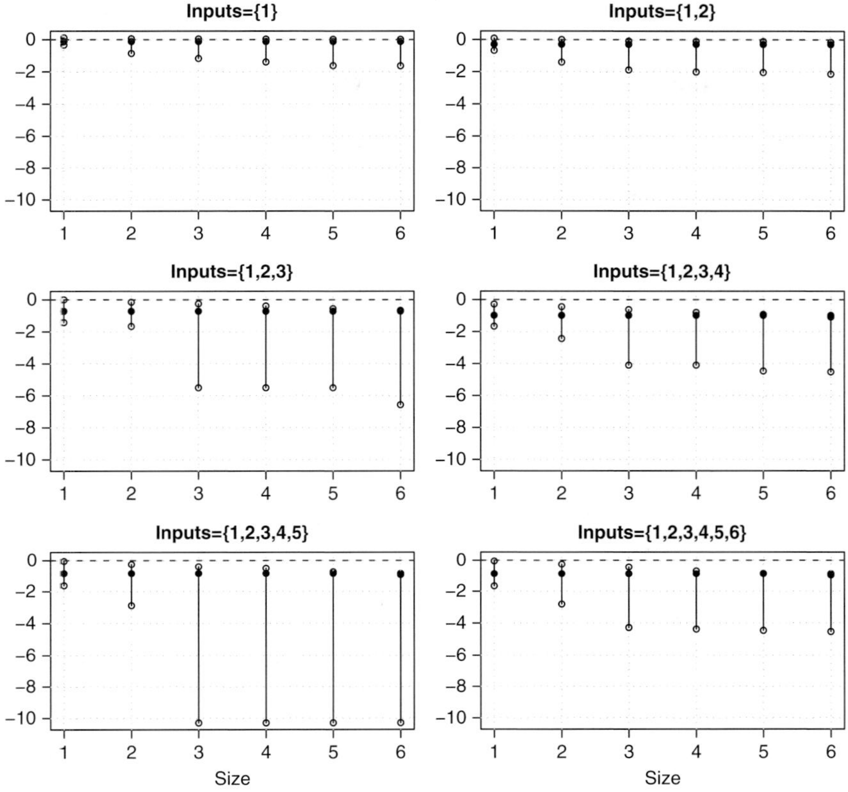

Figure 1.5 Model M1. Bootstrap confidence intervals for the maximum expected predictive performance of the neural networks with respect to the benchmark.

1.4.3 Testing Superior Predictive Ability for Neural Network Modeling

Testing the superior predictive ability approach can be easily extended to our neural network framework (La Rocca and Perna, 2014). Let $\mathcal{S} = \{(Y_i, \mathbf{X}_i), i \in S\}$ and $\mathcal{P} = \{(Y_i, \mathbf{X}_i), i \in P\}$ denote, respectively, the estimation data set and the test data set, where $\mathcal{P}$ is the complement set of $\mathcal{S}$ with respect to $\mathcal{D}$, with $|\mathcal{P}| = N - |\mathcal{S}|$. Let the estimated forecast error be $\hat{u}_{j,\tau} = Y_\tau - f_j(\mathbf{X}_\tau, \hat{\mathbf{w}}), j = 0, 1, \ldots, k$ and let $\text{MPE}_j = \sum_{\tau \in P} h(\hat{u}_{j,\tau})$, where P is the cardinality of the set $\mathcal{P}$.

The test procedure can be based on the F-type statistic defined as

$$Fp_j = P\frac{\text{MPE}_0 - \text{MPE}_j}{\text{MPE}_j}, \qquad j = 1, 2, \ldots, k. \tag{1.23}$$

It has a clear interpretation: large values of Fp_j indicate evidence against the null H_j.

The procedure for testing the system of hypotheses (Eq. 1.16) keeping under control the family wise error rate, runs as follows. Relabel the hypothesis from

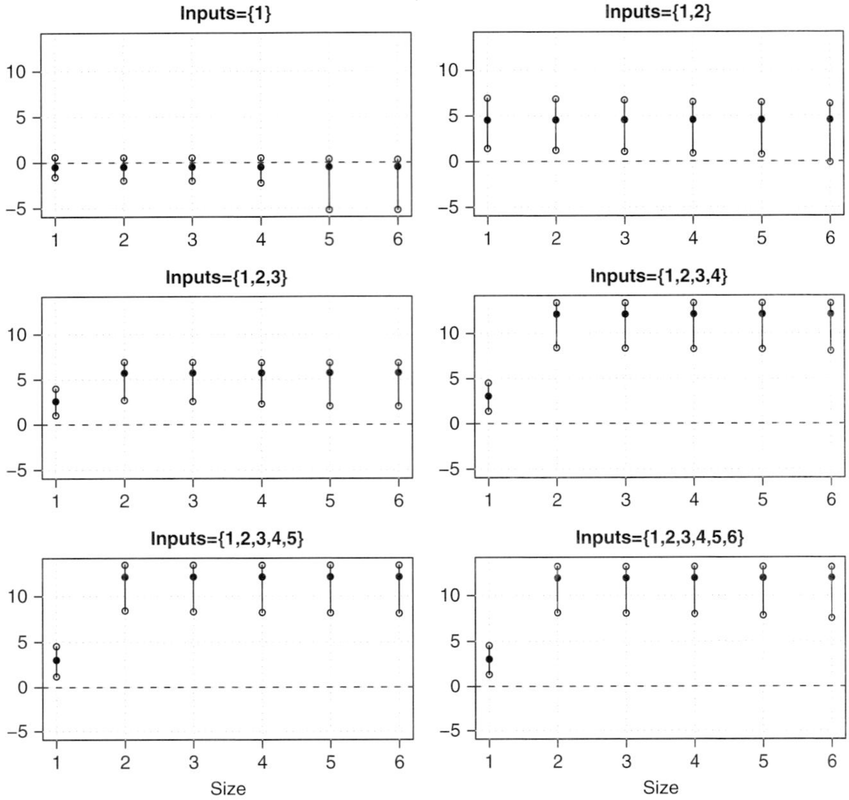

Figure 1.6 Model M2. Bootstrap confidence intervals for the maximum expected predictive performance of the neural networks with respect to the benchmark.

H_{r_1} to H_{r_k} in redescending order with respect to the value of the test statistics Fp_j, that is $Fp_{r_1} \geq Fp_{r_2} \geq \ldots \geq Fp_{r_k}$. The procedure focuses on testing the joint null hypothesis that all hypotheses H_j are true, that is no competing model is able to beat the benchmark model. This hypothesis is rejected if Fp_{r1} is large, otherwise all hypotheses are accepted. In other words, the procedure constructs a rectangular joint confidence region for the vector $\left(Fp_{r_1}, \ldots, Fp_{r_k}\right)^T$, with nominal joint coverage probability $1 - \alpha$. The confidence region is of the form $\left[Fp_{r_1} - c_{1-\alpha}, \infty\right) \times \cdots \times \left[Fp_{r_k} - c_{1-\alpha}, \infty\right)$ where the common value $c_{1-\alpha}$ is chosen to ensure the proper joint (asymptotic) coverage probability. If a particular individual confidence interval $\left[Fp_{r_j} - c_{1-\alpha}, \infty\right)$ does not contain zero, the corresponding null hypothesis H_{r_j} is rejected.

So, the testing procedure will select a set of models which delivers the greatest predictive ability when compared to the benchmark model. All these models are somewhat equivalent and, for a parsimony principle, the one with the smallest hidden layer size should be selected. If all the nulls are not rejected in the first step,

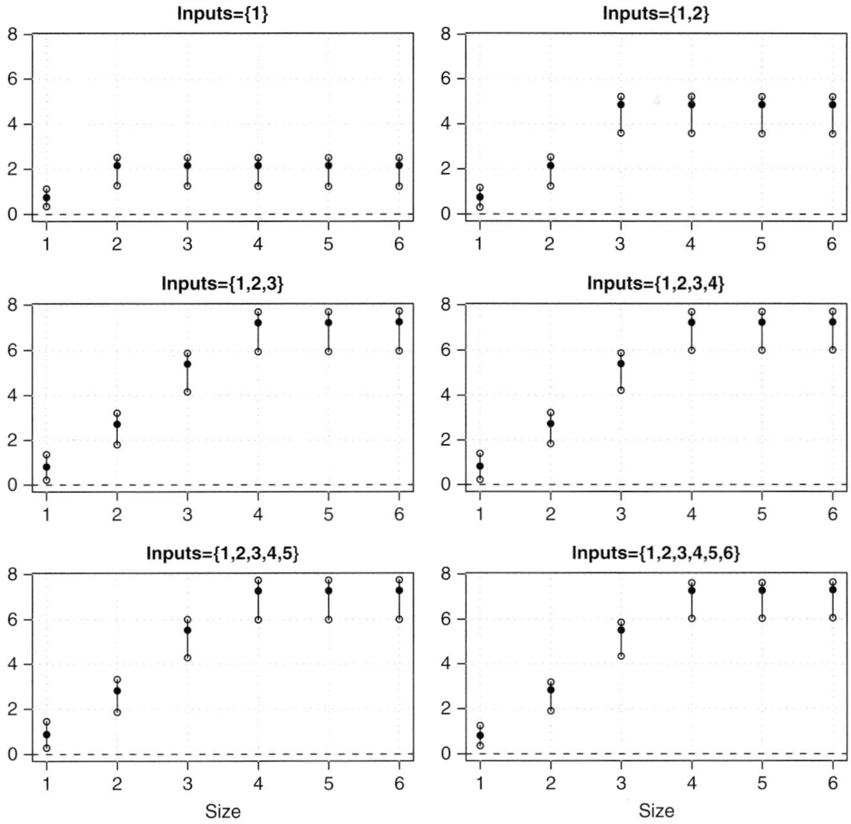

Figure 1.7 Model M3. Bootstrap confidence intervals for the maximum expected predictive performance of the neural networks with respect to the benchmark.

there is no neural network model, which is able to outperform the linear model (assumed as a benchmark) in terms of predictive ability. Again, the quantile of order $c_{1-\alpha}$ can be estimated by using resampling techniques.

The pseudo-code for the complete testing procedure is described in Algorithm 1.3.

1.4.4
Some Numerical Results Using Test of Superior Predictive Ability

To illustrate the performance of the proposed model selection procedure, we use simulated data sets generated by models with known structure. The simulated data sets were generated by using different models, often employed in the neural network literature as data-generating processes.

Again, to generate sintetic data sets, we have used the same models used in De Veaux *et al.* (1998), Friedman (1991), and Tibshirani (1996). We also added as an

Algorithm 1.3: Testing algorithm for superior predictive ability.

Relabel the hypothesis from H_{r_1} to H_{r_k} in redescending order of the value of the test statistics Fp_j, that is $Fp_{r_1} \geq Fp_{r_2} \geq \cdots \geq Fp_{r_k}$.
Generate B bootstrap replicates $\mathbf{Z}^*_{N,1}, \mathbf{Z}^*_{N,2}, \ldots, \mathbf{Z}^*_{N,B}$ as iid samples from $\mathbf{Z}_N$
From each bootstrap data matrix $\mathbf{Z}^*_{N,b}$ with $b = 1, 2, \ldots, B$ compute the bootstrap counterparts of the individual test statistics $F^*p_{j,b}, j = 1, 2, \ldots, k$.
Let $\mathcal{K}$ be the set of indexes of models with better predictive performance
For $b = 1, 2, \ldots, B$ compute $\theta_N^{b,*} = \max_{1 \leq s \leq k} \left(Fp^*_{r_s,b} - Fp_{r_s} \right)$
Compute $\hat{c}_{1-\alpha}$ as the $1 - \alpha$ quantile of the bootstrap values $\theta_N^{b,*}$, $b = 1, 2, \ldots, B$
for $s = 1$ to k **do**
  if $0 \notin \left[Fp_{r_s} - \hat{c}_{1-\alpha}, \infty\right)$ **then**
    reject H_{r_s} and include s in $\mathcal{K}$
  end if
end for
Deliver the set $\mathcal{K}$ (if it is an empty set, no neural network model is able to beat the benchmark model)

additional model the one used by Turlach (2004) and defined as

$$Y = \left(X_1 - 0.5\right)^2 + X_2 + X_3 + X_4 + X_5 + \varepsilon$$

where $\mathbf{X} = (X_1, X_2, X_3, X_4, X_5)^T$ is a vector of multivariate uniform random variables and ε is Gaussian with zero mean and variance equal to 0.05. The model, as in Friedmans's case, includes both linear and nonlinear relationships and it is known to be problematic for other variable selection schemes.

For the numerical examples, we have considered a quadratic loss function h and $N = 600$, $P = 180$, $B = 1000$ and $k = 8$. All neural network models have been estimated by using non linear least squares, including a weight decay in the objective function to control overfitting. Moreover, to avoid to be trapped in local minima, the estimation procedure has been initialized 25 times with random starting values, keeping the estimated network with the lowest residual sum of squares.

The results of the testing procedure for typical realizations are reported in Figure 1.8. In the Tibshirani model case, the hidden layer size is known and equal to 2. The procedure correctly identifies the hidden layer size and indicates that it is not possible to improve accuracy by increasing the hidden layer size. All models with r ranging from 2 to 8 are basically equivalent with respect to the predictive accuracy. Similar remarks apply also to all other models. Note that for the DeVeaux and the Friedman data, simply considering the statistical index would indicate $r = 8$ as the best choice, but this does not give any significant improvement with respect to $r = 4$.

A moderate Monte Carlo experiment has also been performed considering the same data-generating processes as before. We have considered 240 Monte Carlo

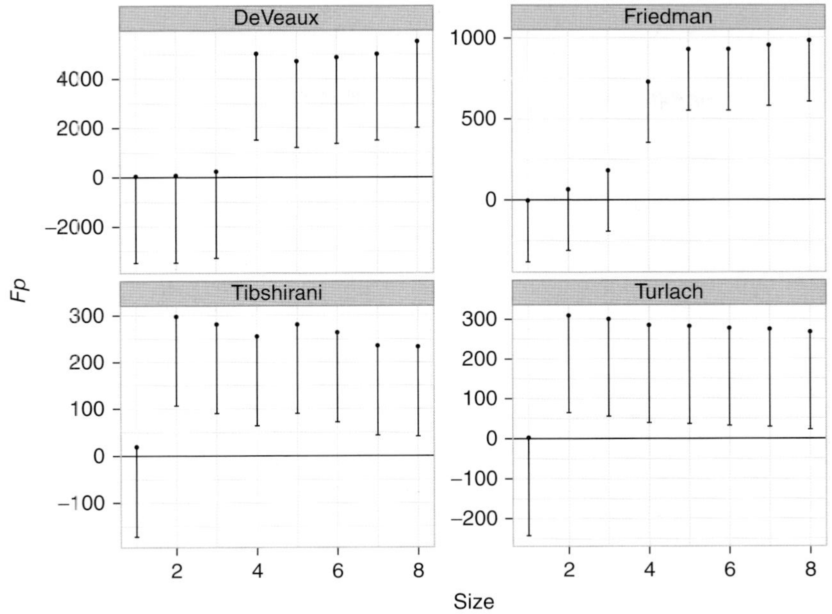

Figure 1.8 Joint confidence regions with nominal coverage probability $1 - \alpha = 0.95$.

runs with three different sample sizes $N = 300, 400, 600$ using the last 30% observations for prediction. The results are reported in Figure 1.9. In the Tibshirani case, the hidden layer size (which is known and equal to 2) the proportion of correct identification is very high for all the sample sizes, reaching 100% for $N = 600$. For the other data sets, the simulations confirm the results shown by the numerical examples and highlight the steep improvement as the sample size increases.

1.4.5
An Application to Real Data

To validate the performance of the proposed procedure, two applications to real data are discussed in La Rocca and Perna (2014).

As a first example, we use the Prostate Cancer data set which comes from a study by Stamey *et al.* (1989). The dependent variable is the level of prostate-specific antigen which depends on 8 clinical measures in men who were about to receive prostatectomy. The data set has 97 observations and it is split in two subsets: 67 observations have been used for the modeling step, while 30 observations have been used for the validation step. By using a linear model and a best subset variable selection rule, just two explanatory variables (out of eight) are identified as relevant: lweight (log prostate weight) and lcavol (log cancer volume). For the sake of comparison, as identification tools for the number of hidden neurons, we also use the k-fold CV selection rule and the Bayesian information criterion, proved

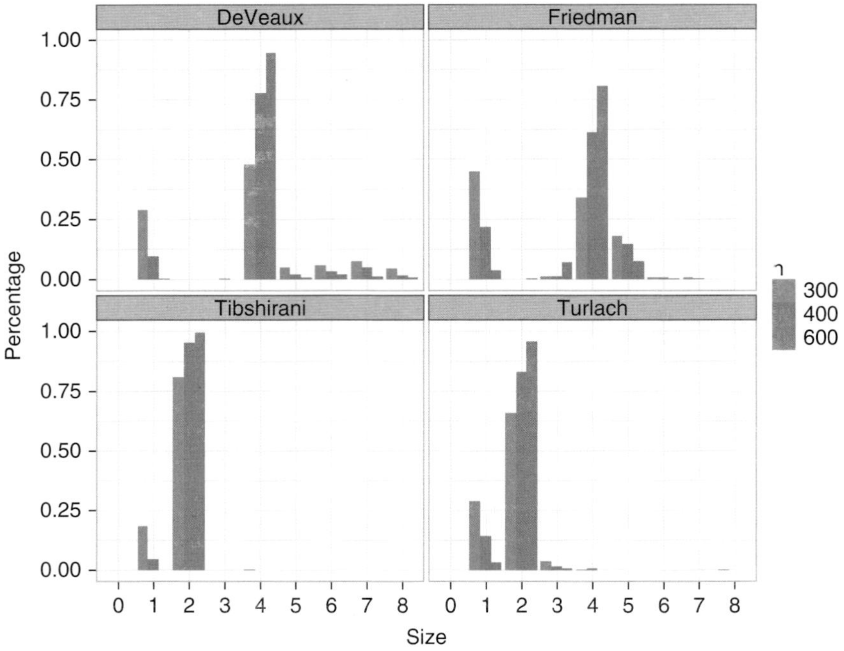

Figure 1.9 Proportion of hidden layer size identification by using the testing procecure for superior predictive ability. (See inset for color representation of this figure.)

to be consistent (almost surely) in the case of multi-layer perceptrons with one hidden layer in White (1990).

Clearly, the BIC identifies a linear model, which is equivalent to a neural network with skip layer and zero hidden neurons, for all the weight decay values considered (Figure 1.10a,b). The latter result is confirmed by looking at the CV and our test of superior predictive ability (Figure 1.11a). To validate these results,

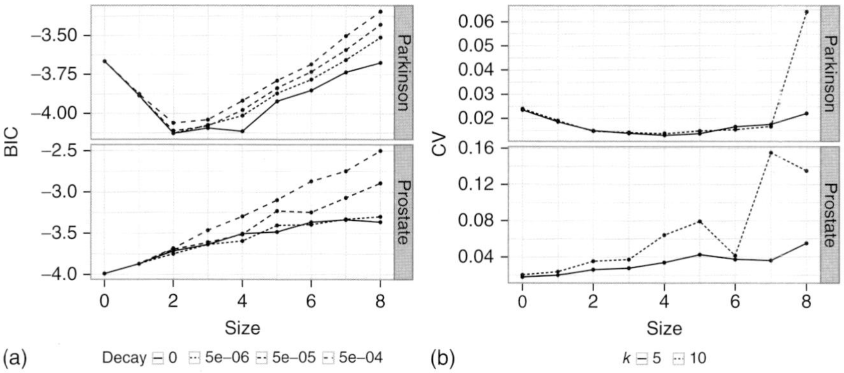

Figure 1.10 Bayesian information criterion values for different hidden layer sizes and different weight decay values (a). k-fold CV values for $k = 5$ and $k = 10$ using a weight decay equal to zero (b).

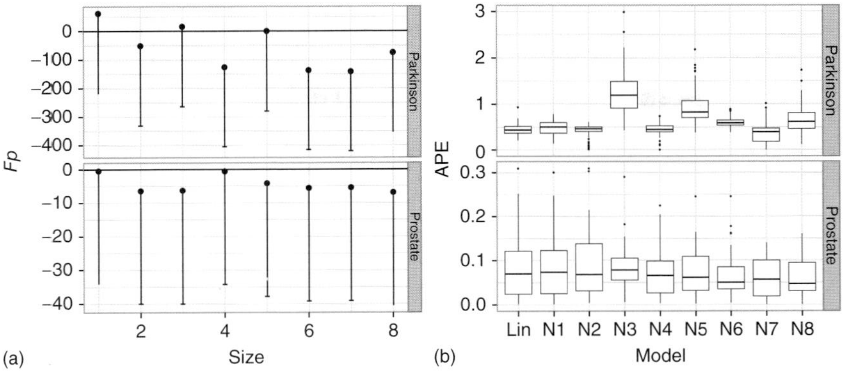

Figure 1.11 Joint confidence regions with nominal coverage probability $1 - \alpha = 0.95$ (a). Absolute prediction error distributions computed on the test set for linear models and neural networks with hidden layer size ranging from 1 to 8 (b).

a linear model and neural networks with hidden neurons ranging from 1 to 8 have been estimated and used to predict the observations in the validation set. The distributions of the absolute prediction errors are reported in Figure 1.11b. The plot shows that the neural networks considered are not able to provide better predictions with respect to the linear model (as predicted by the CV, the BIC and the novel test). These results are confirmed by a formal statistical comparison between the two distributions made by using the Brunner Munzel test and the Wilcoxon rank sum test which give p-values equal to 0.497 and 0.495, respectively.

The second data set is composed of a range of biomedical voice measurements from 42 people with early-stage Parkinson's disease recruited to a 6-month trial of a telemonitoring device for remote symptom progression monitoring (for details see Tsanas *et al.* (2010)). The data set has been downloaded from the UCI Machine Learning Repository and consists of 5875 observations on age, gender, and on 16 biomedical voice measures. The statistical model is used to predict the total UPDRS score. For computational reasons, just the subset of the first 887 observations (corresponding to the first five patients) has been considered. Again, the data set is split into two subsets: 731 observations (the first four individuals) have been used for the modeling step, while 156 observations (corresponding to the fifth patient) are used for validation purposes. In this case, the CV model selection rule identifies a neural network model with four hidden neurons, while the BIC identifies a neural network with two neurons. However, the proposed test shows that there is no superior predictive ability of neural network models with respect to linear models (Figure 1.11a). This is confirmed by the distribution of the absolute predictive errors reported in Figure 1.11b: the linear model and the neural networks with two or four neurons in the hidden layer perform similarly. This latter result is also supported by the Brunner Munzel test and the Wilcoxon rank sum test whose p values are equal to 0.219 and 0.218, respectively. Clearly, neural networks with two or four hidden neurons and 17 explanatory variables appear to be heavily overparametrized, since they have 39 or 77 parameters, respectively. These networks show no clear advantages in terms of predictive ability with respect to the linear model.

1.5 Concluding Remarks

In this chapter, a novel model selection procedure in neural network modeling has been presented and discussed. The basic idea of the proposed approach is that input neurons and hidden neurons play a different role in neural network modeling so that they should be selected by using different criteria. Specifically, the proposed approach identifies the number and the type of input variables by using a formal statistical test focusing the problem on a procedure for variable selection in regression models. On the contrary, the number of the hidden neurons is considered smoothing parameter and it is selected by looking at the predictive performance of the network model.

The proposed strategy addresses the problem of data snooping, which arises when a data set is used more than once for inference and model selection. Moreover, to overcome the analytical and probabilistic difficulties related to the estimation of the sampling distribution of the test statistics involved, the approach uses extensively resampling techniques.

The proposed test procedures have been tested on simulated and real data sets, which confirm their ability to detect correctly the set of input variables and to discriminate among alternative models. Clearly, joint usage of neural network models and resampling techniques are usually quite demanding from a computational point of view. In any case, it is worthwhile to underline that they are suitable to be implemented on parallel and cluster computers almost without any modification of the computing algorithms.

References

Barron, A.R. (1993) Universal approximation bounds for superposition of a sigmoidal function. *IEEE Transactions on Information Theory*, **39**, 930–945.

Breiman, L. and Friedman, J.H. (1985) Estimating optimal transformations for multiple regression and correlation. *Journal of the American Statistical Association*, **80**, 580–619.

Clark, T.E. and McCracken, M.W. (2012a) In-sample tests of predictive ability: a new approach. *Journal of Econometrics*, **170**, 1–14.

Clark, T.E. and McCracken, M.W. (2012b) Reality checks and comparison of nested predictive models. *Journal of Business and Economic Statistics*, **30**, 53–66.

Corradi, V. and Swanson, N.R. (2006) Predictive density evaluation, *Handbook of Economic Forecasting*, Vol. **1**, Elsevier.

De Veaux, R., Schumi, J., Schweinsberg, J., Shellington, D., and Ungar, L.H. (1998) Prediction intervals for neural networks via nonlinear regression. *Technometrics*, **40**, 277–282.

Elliot, G. and Timmermann, A. (2004) Optimal forecast combinations under general loss functions and forecast error distribution. *Journal of Econometrics*, **122**, 47–49.

Franses, P.H. and Draisma, G. (1997) Recognizing changing seasonal patterns using artificial neural networks. *Journal of Econometrics*, **81**, 273–280.

Friedman, J.H. (1991) Multivariate adaptive regression splines. *Annals of Statistics*, **19**, 1–67.

Giacomini, F.R., Politis, D.N., and White, H. (2013) A warp-speed method for conducting Monte Carlo experiments involving bootstrap estimators. *Econometric Theory*, **29**, 567–589.

Giordano, F., La Rocca, M., and Perna, C. (2014) Input variable selection in neural network models. *Communications in Statistics - Theory and Methods*, **43**, 735–750.

Holm, S. (1979) A simple sequentially rejective multiple test procedure. *Scandinavian Journal of Statistics*, **6**, 65–70.

Hornik, K., Stinchcombe, M., and Auer, P. (1994) Degree of approximation results for feedforward networks approximating unknown mappings and their derivatives. *Neural Computation*, **6**, 1261–1275.

Hwang, J.T.G. and Ding, A.A. (1997) Prediction intervals for artificial neural networks. *Journal of the American Statistical Association*, **92**, 748–757.

Kuan, C.-M. and Liu, T. (1995) Forecasting exchange rates using feedforward and recurrent neural networks. *Journal of Applied Econometrics*, **10**, 347–364.

La Rocca, M. and Perna, C. (2005a) Neural network modeling by subsampling, in *Computational Intelligence and Bioinspired Systems* (eds J. Cabestany, A. Prieto, and F. Sandoval), Springer.

La Rocca, M. and Perna, C. (2005b) Variable selection in neural network regression models with dependent data: a subsampling approach. *Computational Statistics and Data Analysis*, **48**, 415–429.

La Rocca, M. and Perna, C. (2014) Designing neural networks for modeling biological data: a statistical perspective. *Mathematical Biosciences and Engineering*, **11**, 331–342.

Lee, H.K.H. (2004) *Bayesian Nonparametrics via Neural Networks*, SIAM.

Ossen, A. and Rügen, S.M. (1996) An analysis of the metric structure of the weight space of feedforward networks and its application to time series modelling and prediction. Proceedings of the 4th European Symposium on Artificial Neural Networks (ESANN96), pp. 315–322.

Politis, D.N., Romano, J.P., and Wolf, M. (1999). *Subsampling*, Springer.

Qi, M. and Zhang, G.P. (2001) An investigation of model selection criteria for neural network time series forecasting. *European Journal of Operational Research*, **132**, 666–680.

Romano, J.P. and Wolf, M. (2005) Stepwise multiple testing as formalized data snooping. *Econometrica*, **73**, 1237–1282.

Stamey, T., Kabalin, J., McNeal, J., Johnstone, I., Freiha, F., Redwine, E., and Yang, N. (1989) Prostate specific antigen in the diagnosis and treatment of adenocarcinoma of the prostate ii radical prostactomy treated patients. *Journal of Urology*, **16**, 1076–1083.

Swanson, N.R. and White, H. (1997) A model selection approach to real time macroeconomic forecasting using linear models and artificial neural networks. *Review of Economics and Statistics*, **79**, 540–550.

Tibshirani, R. (1996) A comparison of some error estimates for neural network models. *Neural Computation*, **8**, 152–163.

Tsanas, A., Little, M.A., McSharry, P.E., and Ramig, L.O. (2010) Accurate telemonitoring of Parkinson's disease progression by non-invasive speech tests. *IEEE Transactions on Biomedical Engineering*, **57**, 884–893.

Turlach, B. (2004) Discussion of least angle regression by Efron, Hastie, Jonstone and Tibshirani. *Annals of Statistics*, **32**, 494–499.

White, H. (1989) Learning in artificial neural networks: a statistical perspective. *Neural Computation*, **1**, 425–464.

White, H. (1990) Connectionist nonparametric regression: multi-layer feedforward networks can learn arbitrary mappings. *Neural Networks*, **3**, 535–549.

White, H. (2000) A reality check for data snooping. *Econometrica*, **68**, 1097–1126.

White, H. and Racine, J. (2001) Statistical inference, the bootstrap, and neural-network modeling with application to foreign exchange rates. *IEEE Transactions on Neural Networks*, **12**, 657–653.

2
Measuring Structural Correlations in Graphs

Ziyu Guan and Xifeng Yan

2.1
Introduction

The rise of the Web, social networks, and bioinformatics has presented the scientists with numerous graphs, each consisting of millions of nodes and edges. Hidden in these large datasets are the answers to important questions in networking, sociology, business, and biology. These graphs not only have topological structures but also contain events/activities that occurred on their nodes. For example, an eBay customer could sell or bid on a product; a Facebook user could play a Zynga game with friends; a computer in the Internet could suffer various attacks from hackers. This complex combination raises new research problems in the graph data analysis (Ester *et al.*, 2006; Moser *et al.*, 2009; Silva, Meira, and Zaki, 2010).

Among different events taking place in a network, some exhibit strong correlations with the network structure, while others do not. Such structural correlations might shed light on viral influence existing in the corresponding network, which is the key to many research problems in product marketing (Chen, Wang, and Wang, 2010), online advertisement (Bao and Chang, 2010), and recommendation (Kautz, Selman, and Shah, 1997). Figure 2.1 shows the distribution of three different events over the same graph. We can easily instantiate Figure 2.1 into different application scenarios. They could be three different products bought by members in a social network, or three different intrusion alerts raised by computers in a computer network. In terms of the products, dark nodes in Figure 2.1a–c represent the members who purchased the products A, B and C, respectively. Intuitively, Figure 2.1 shows that in this network, people who bought product A (or B) are closer to each other. On the contrary, black nodes for C seem to be randomly distributed. We name this kind of correlation as *self structural correlation (SSC)* (Guan *et al.*, 2011). In this scenario, the network would be suitable for promoting A and B, and we can promote A and B to people who have not bought them. While it is hard to derive deterministic relationships between sales and the network structure, it is possible to study how the sales is correlated to the structure. In

Computational Network Theory: Theoretical Foundations and Applications, First Edition.
Edited by Matthias Dehmer, Frank Emmert-Streib, and Stefan Pickl.

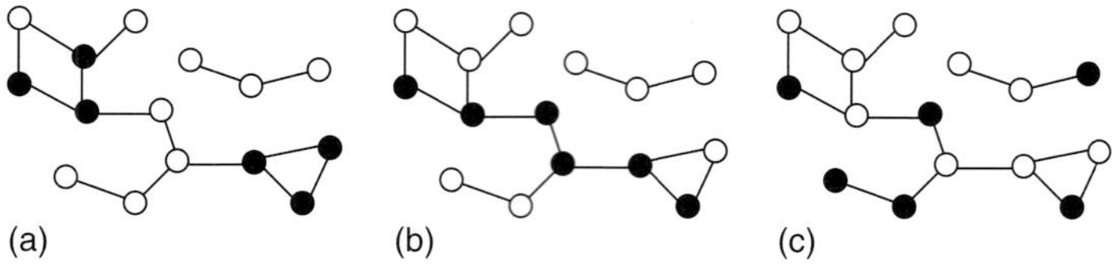

Figure 2.1 Structural correlation.

fact, one can raise several interesting questions related to the structure and events distributed over the structure:

1) In regard to the sales of A and B, which one is more related to the underlying social network?
2) Given two networks G and G' for the same group of users, for example, their email network and Facebook network, are the sales of A more related to G than G'?
3) If we have snapshots of network G during different periods, can we measure how product A was dispersed over the network over time? Was it purchased by a small community at the beginning?

In order to answer the given questions, we need to address the following research problems: (i) *How to define and measure the correlation between the graph structure and events*? (ii) *How to compute the measure efficiently in large graphs, if we want to rank all events according to the measure*? Unfortunately, the classic association-mining concept is not applicable in the graph setting since it only works on homogeneous datasets like transactions and baskets (Agrawal, Imieliński, and Swami, 1993; Srikant and Agrawal, 1995).

Besides, two events occurring on the same graph could be correlated. Two illustrative examples are shown in Figure 2.2. In Figure 2.2a, A and B exhibit a *positive correlation* (attraction). In the context of a social network, they could be two baby-formula brands, Similac and Enfamil. Their distributions could imply that there exists "mother communities" in the social network where different mothers would prefer different baby-formula brands. The two brands attract each other because of the communities. An example of the *negative correlation* (repulsion) could be that people in an Apple fans' community would probably not buy products of ThinkPad

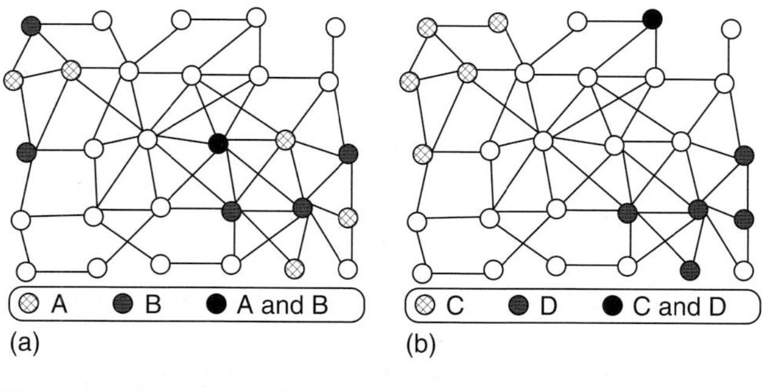

Figure 2.2 Two types of two-event structural correlation: (a) attraction and (b) repulsion.

and vice-versa , as conveyed by Figure 2.2b. We name this kind of structural correlation as the *two-event structural correlation* (TESC) (Guan, Yan, and Kaplan, 2012). TESC is different from correlation in transactions such as market baskets. If we treat nodes of a graph as transactions and assess the transaction correlation (TC) of two events by using measures such as Lift (Han and Kamber, 2006), one can verify that in Figure 2.2a, A and B have a negative TC, although they exhibit a positive TESC. Regarding the baby-formula example, a mother would probably stick to one brand, since switching between different brands could lead to diarrhea. As another example, in terms of computer networks, A and B could be two related intrusion techniques used by hackers to attack target subnets. Since attacks consume bandwidth, there is a tradeoff between the number of hosts attacked and the number of techniques applied to one host. The hackers might choose to maximize coverage by alternating related intrusion techniques for hosts in a subnet, in order to increase the chance of success. Hence, TESC is useful for detecting the structural correlations which might not be detected by TC. It can be used to improve applications such as online advertisement (Bao and Chang, 2010) and recommendation (Kautz, Selman, and Shah, 1997). For instance, most recommendation methods exploit positive TC, while positive TESC provides an alternative recommendation scheme in local neighborhoods. TESC can reveal important relationships between events (the intrusion example) or reflect structural characteristics (communities in the product examples) on a graph.

In one word, efficiently and effectively estimating the structural correlations in graphs is not a trivial problem. Addressing the estimation of SSC and TESC will significantly advance the computational network theory.

2.1.1 Solutions for Measuring Structural Correlations

We develop a novel measure to assess SSC in a graph. The measure aggregates the proximity among nodes on which the same event has occurred, using a proximity function such as hitting time (Lovász, 1993). We propose a modified version of hitting time named *decayed hitting time* (*DHT*) to better and faster calculate SSC. We develop an efficient estimation framework, gScore (Graph structural correlation estimation), to quickly calculate correlation scores in large-scale networks. By estimating the deviation from the expected correlation score of a random situation, our method is able to discover the events of nodes that are highly correlated with the graph structure. We demonstrate gScore's effectiveness by constructing synthetic events on the DBLP coauthor network. We also discover interesting correlated and uncorrelated products discovered from TaoBao.com, the largest online shopping network in China. Scalability of gScore is tested on the Twitter network.

For TESC, we also design a novel measure and then an efficient estimation framework. Specifically, given the occurrences of two events, we choose a sample of "reference nodes" uniformly from the vicinity of all occurrences and compute for each reference node the densities of the two events in its vicinity, respectively. Then, we employ the Kendall's τ rank correlation measure (Kendall and Gibbons, 1990) to compute the average concordance of density changes for the two events, over all pairs of reference nodes. Finally, correlation significance can be efficiently assessed by τ's nice property of being asymptotically normal under the null hypothesis. For efficiently sampling the reference nodes, different sampling techniques are proposed to shorten the statistical testing time. Our framework is scalable to very large graphs. We demonstrate the efficacy of the TESC testing framework by event simulations on the DBLP graph. We further test its scalability in a Twitter graph with 20 million nodes. Case studies of applying the testing framework on real events occurring on real graphs are provided with interesting results.

The rest of the chapter is organized as follows. In Section 2.2, we provide a detailed review of related work. Then, we discuss how to measure SSC and TESC in Sections 2.3 and 2.4, respectively. Finally, the conclusions are in Section 2.5.

2.2 Related Work

Graph mining is a growing research field.. Generally speaking, graph mining works fall into two categories: (i) Mining in a set of graphs and (ii) mining in a single graph. For the former one, various mining problems have been addressed, for example, classification (Saigo *et al.*, 2009), clustering (Dehmer, Emmert-Streib, and Tripathi, 2013). Our work falls into the latter one and is related to a branch of graph mining research which involves both graph structures and node attributes (Ester *et al.*, 2006; Silva, Meira, and Zaki, 2010; Khan, Yan, and Wu, 2010; Moser *et al.*, 2009; Guan *et al.*, 2011; Zhou, Cheng, and Yu, 2009). Zhou, Cheng, and Yu (2009) proposed a graph-clustering algorithm based on both structural and attribute similarities. In Ester *et al.* (2006) also investigated using node-attribute data to improve graph clustering. Moser *et al.* (2009) introduced the problem of mining cohesive graph patterns which are defined as dense and connected subgraphs that have homogeneous node-attribute values. Although these works considered both graph structures and node attributes, they did not explicitly study the relationships between structures and attributes. Recently, Silva, Meira, and Zaki (2010) proposed a structural-correlation pattern mining problem which aims to find pairs (S, V) where S is a frequent attribute set and V induces a dense subgraph. Each node in V contains all the attributes in S. However, this kind of correlation is too restrictive and strong. An attribute which occurs on $|V| - 1$ nodes of V will be discarded, though it also has a positive correlation with attributes in S,while our approaches allow a user to measure the structural correlation within an attribute (SSC) or between any two attributes (TESC) freely. Furthermore, Silva, Meira, and Zaki (2010) and Moser *et al.* (2009) were focused on local patterns but not global structural correlations studied in this work. Khan, Yan, and Wu (2010) studied the problem of mining a set of attributes which frequently co-occurred in local neighborhoods in a graph. They also tried to assess the significance of the discovered patterns. However, compared to the TESC problem, (i) they only consider positive correlations (association) among events, while TESC aims to measure both positive and negative correlations; (ii) they rely on an empirical method for significance testing, while our method is rigorous statistical testing based on the Kendall's τ statistic; (iii) their problem is intrinsically a frequent pattern-mining problem and could miss some rare but positively correlated event pairs, as will be shown in the experiments.

Our work is also related to research in social influence. In social networks, behaviors of two people tend to be related if they are friends. Anagnostopoulos, Kumar, and Mahdian (2008) studied the problem of distinguishing social influence from other sources of correlation using time series of people's behavior. La Fond and Neville (2010) presented a randomization technique for distinguishing social influence and homophily for temporal network data. Our work is different from theirs in two aspects. First, these studies assume the existence of correlation (direct neighborhoods), while we try to determine if there is a correlation. Second, they are concerned with direct friendships, while our structural correlation is defined in a more general graph-proximity notion. The structural correlation problem addressed here can be treated as a step before the above correlation causality (e.g., influence vs homophily) analysis problem, since the latter problem usually assumes the existence of correlations and the goal is to find out whether it is due to influence or not.

Our work is also related to assessing and testing the correlation among spatial point patterns in the field of geographical analysis (Anselin, 1995; Diggle and Cox, 1983; Lotwick and Silverman, 1982; Smith, 2004). However, existing solutions for the analogical continuous problems cannot be applied directly to the graph spaces due to following reasons: (i) Proximity measures for continuous spatial spaces cannot be applied to graph spaces directly. (ii) The fixed and discrete graph structure renders infeasible some popular testing methodologies such as randomly shifting a point pattern around the space (Lotwick and Silverman, 1982). (iii) Focusing on regions where points exist and uniformly sampling reference points are trivial works in continuous spaces (Smith, 2004), while it is not in our TESC problem due to the discrete nature of the graphs. (iv) In our problems, scalability is an important issue, which existing methods for the point-pattern counterparts failed to consider.

There are many graph-proximity measures proposed in the literature. Here, we name a few. *Common neighbors* and *Jaccard's coefficient* are two measures based on node neighborhood (Nowell and Kleinberg, 2003). Common neighbors compute the number of common neighbors of two nodes. Jaccard's coefficient is defined as the number of common neighbors divided by the number of all distinct neighbors of two nodes. Katz (1953) defined a measure which sums over all paths between two nodes, exponentially damped by their length to make short paths more important. *Hitting time* Lovász (1993) and *personalized PageRank* Brin and Page (1998) are random walk-based graph-proximity measures. The two correlation estimation frameworks (for SSC and TESC respectively) are general and can adopt any graph-proximity measure. We instantiate them with proper proximity measures for demonstration.

2.3 Self Structural Correlation

2.3.1 Problem Formulation

An attributed graph $G = (V, E)$ has an event set Q. The event set of a node v is written as $Q(v) \subseteq Q$. We consider undirected and unweighted graphs. Nevertheless, the proposed measure and algorithms could be generalized to weighted and/or directed graphs.

Suppose an event q (e.g., purchasing a specific product) is taking place in G. Each node v can take two values in terms of q: $f_q(v) = 1$ if $q \in Q(v)$; otherwise, $f_q(v) = 0$. Let $m = \sum_v f_q(v)$ denote the number of nodes where q occurred. Let $n = |V|$ be the number of nodes in G. We could formulate the following two research problems for SSC:

Problem 1: Determine whether there is a correlation between q and the graph structure of G (i.e., SSC). If not, q is just randomly distributed in G.

Its associated ranking problem is as follows:

Problem 2: Given a set of different events $Q = \{q_i\}$ on G, rank $\{q_i\}$ according to their SSC strength with respect to the graph structure of G.

To address these problems, we need a measure that captures the distribution of an event in a graph, and then assess the significance of the observed measure score. A simple measure could be to assess the probability that a node's neighbors have q given that the node has q. However, this 1-neighborhood event fraction measure could not well capture the distribution of an event in a graph since it only considers direct neighbors. We will show this drawback in experiments.

2.3.2 The Measure

Intuitively, if the m event nodes of q are close to one another, then the correlation is high. Otherwise, it will be more similar to a random situation where q is randomly distributed in G. Therefore, we propose using the average proximity between one node in those m nodes and the remaining $m - 1$ nodes to assess the SSC of q with respect to G,

$$\rho(V_q) = \frac{\sum_{v \in V_q} s(v, V_q \setminus \{v\})}{|V_q|}, \tag{2.1}$$

where V_q is the set of m nodes on which q occurred and $s(v, V_q \setminus \{v\})$ is the closeness of the remaining nodes to v. $s(\cdot)$ can be any graph-proximity measure that measures the proximity between a node and a set of nodes in a graph topology notion. We could rewrite $s(\cdot)$ as the sum of the contribution of each node in $V_q \setminus \{v\}$,

$$s(v, V_q \setminus \{v\}) = \sum_{u \in V_q \setminus \{v\}} s_q(v \rightsquigarrow u), \tag{2.2}$$

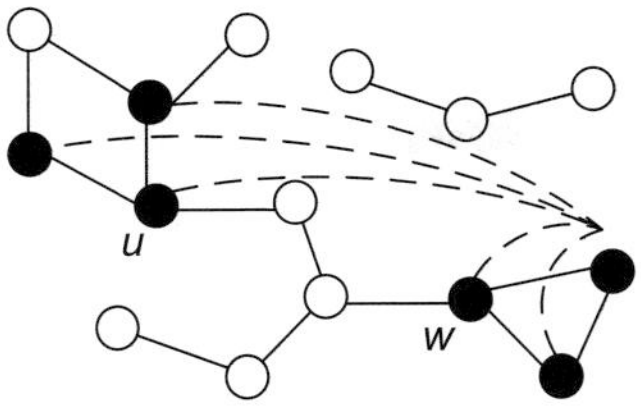

Figure 2.3 Measuring SSC.

where $s_q(v \rightsquigarrow u)$ is the contribution of u to $s(v, V_q \setminus \{v\})$. We can divide proximity measures into two categories: Pairwise and holistic. Pair-wise measures are defined on node pairs, for example, shortest distance and personalized PageRank, while holistic measures explicitly estimate the proximity of a node to a set of nodes. The set of nodes is reached if any node in the set is reached. One example is hitting time: Starting from node v, it calculates the expected number of steps needed to reach one node in $V_q \setminus \{v\}$. For pair-wise measures, $s_q(v \rightsquigarrow u)$ is invariant to q, while for holistic measures, $s_q(v \rightsquigarrow u)$ depends on the distribution of q in G. For example, in Figure 2.3, suppose we want to measure the proximity of v to other black nodes. If we use the shortest distance, u's contribution is 4. If hitting time is adopted, u does not even influence the measure score, since v always hits other black nodes in one step. Hitting time is less influenced by remote black nodes and anomalies, and more focused on graph local areas. This is desirable since we may have the correlation patterns on the graph which are distant to one another, as shown in Figure 2.3. Holistic proximity measures could help detect the correlations better. We will focus on hitting time hereafter. Nevertheless, our framework is applicable to other proximity measures too.

Having the measure ρ, we should also consider how the ρ value is distributed for random cases, that is, there is no correlation. Let $\rho(\widehat{V}_m)$ denote the correlation score for a randomly selected set $\widehat{V}_m$ of m nodes, where $m = |\widehat{V}_m| = |V_q|$. As m increases, there will be an increasing chance for m randomly selected nodes to be close to one another (in the extreme case $m = n$, nodes are obviously very close). Thus, we should estimate the significance (denoted by $\tilde{\rho}(V_q)$) of $\rho(V_q)$ compared to $\rho(\widehat{V}_m)$. Since it is hard to obtain the distribution of $\rho(\widehat{V}_m)$, we propose to estimate the expectation and variance of $\rho(\widehat{V}_m)$ (denoted by $E[\rho(\widehat{V}_m)]$ and $Var[\rho(\widehat{V}_m)]$, respectively), and then estimate $\tilde{\rho}(V_q)$. More details are presented in Section 2.3.4.2. We refer to this framework as gScore .

2.3.2.1 Random Walk and Hitting Time

A *random walk* is a Markov chain. If in the tth step we are at node v_t, we move to a neighbor of v_t with probability $\frac{1}{d(v_t)}$, where $d(v_t)$ is the degree of node v_t. Let $\mathbf{A}$ denote the adjacency matrix of G. A_{ij} equals 1 if there is an edge between node v_i and v_j, and 0 otherwise. We use $\mathbf{P} = [p_{ij}]_{n\times n}$ to denote the transition probability matrix of the random walk. We have $\mathbf{P} = \mathbf{D}^{-1}\mathbf{A}$ where $\mathbf{D}$ is a diagonal matrix with $D_{ii} = d(v_i)$. We use $\Pr(pa_{1\rightsquigarrow l}) = p_{12}p_{23}\cdots p_{(l-1)l}$ to denote the probability that a random walk takes path $v_1, v_2, \ldots, v_l$, starting from v_1 (also called the probability of that path).

Let B be a subset of V. The hitting time (also called access time) (Lovász, 1993) from a node v_i to B is defined as the expected number of steps before a node $v_j \in B$ is visited in a random walk starting from v_i. Let $h(v_i, B)$ denote the hitting time from v_i to B and x_t denote the position of the random walk at time step t. By definition, we have

$$h(v_i, B) = \sum_{t=1}^{\infty} t \Pr(T_B = t | x_0 = v_i), \tag{2.3}$$

where $\Pr(T_B = t|x_0 = v_i)$ is the probability that the random walk starting from v_i first hits a node in B after t steps. Hitting time is a weighted average path length where each considered path is weighted by its probability. One can easily derive (Lovász, 1993)

$$h(v_i, B) = \sum_{v_k \in V} p_{ik} h(v_k, B) + 1. \tag{2.4}$$

Equation (2.4) expresses a one-step look-ahead property of hitting time. The expected time to reach a node in B from v_i is equivalent to one step plus the mean of hitting times of v_i's neighbors to a node in B. By the definition of hitting time (i.e., Eq. (2.3)), we have $h(v_i, B) = 0$ for $v_i \in B$. From this analysis, we obtain a linear system for computing hitting time from all nodes to B:

$$\begin{cases} h(v_i, B) = 0 & v_i \in B \\ h(v_i, B) = \sum_{v_k \in V} p_{ik} h(v_k, B) + 1 & v_i \notin B \end{cases}. \tag{2.5}$$

It can be shown that this linear system has a unique solution (Lovász, 1993). Hitting time can be used to measure a notion of asymmetrical proximity among nodes in a graph with respect to graph topology, with applications such as Web query suggestion (Mei, Zhou, and Church, 2008), recommendation (Brand, 2005), and link prediction (Nowell and Kleinberg, 2003). However, hitting time has one drawback: Its value range is $[1, +\infty)$. That is, if there is no path between two nodes u and v, the hitting time from v to u is $+\infty$. This also results in infinite value of $\rho(V_q)$ (Eq. (2.1)). Hence, we propose DHT, which inverses the range of $[1, +\infty)$ to $[0, 1]$. The meaning of hitting time also changes slightly.

2.3.2.2 Decayed Hitting Time

DHT is defined as follows

$$\tilde{h}(v_i, B) = \sum_{t=1}^{\infty} e^{-(t-1)} \Pr(T_B = t | x_0 = v_i). \tag{2.6}$$

For DHT, when nodes in B are close to v_i, $\tilde{h}(v_i, B)$ will be high. The reason for the substitution of the exponentially decaying term for the number of steps is twofold. First, by doing this, we avoid the infinite-measure score problem mentioned earlier. Second, in our problem setting, we want to emphasize the importance of neighbors and discount the importance of longer range weak relationships. Although hitting time is shown to be empirically effective in practical applications (Brand, 2005; Nowell and Kleinberg, 2003), solving a linear system for a large-scale graph can be computationally expensive (i.e., $O(n^3)$). To address this issue, Sarkar and Moore (2007) proposed *truncated hitting time* (*THT*) and developed sampling techniques to approximate the hitting time (Sarkar, Moore, and Prakash, 2008). However, a drawback of THT is that it sets an arbitrary upper bound T (the truncated path length) for the actual hitting time. For THT, although long paths have small probabilities, they have high weights, that is, their lengths. Therefore, the contribution of these long paths to the hitting time measure may not be negligible. In an extreme case where v_i cannot reach any node in B, THT will return T, while the true hitting time is $+\infty$! In DHT, longer paths not only have lower probabilities but also have lower weights, that is, $e^{-(t-1)}$. Hence, we can properly bound the contribution of long paths. This facilitates approximating DHT.

2.3.3 Computing Decayed Hitting Time

We develop techniques for efficiently estimating DHT. In Section 2.3.3.1, we propose an iterative matrix-vector multiplication algorithm for estimating DHT. Lemma 2.1 and Theorem 2.1 give bounds for real DHT after t iterations. In Section 2.3.3.2, we present a sampling algorithm for estimating DHT and develop its bounds in Theorems 2.2 and 2.3. Finally, Section 2.3.3.3 analyzes the complexity of the two approximation algorithms.

2.3.3.1 Iterative Approximation

From the definition in Eq. (2.6), we know if we can obtain $\Pr(T_B = t|x_0 = v_i), t = 1, 2, \ldots$, then we can compute $\tilde{h}(v_i, B)$. Let $\Pr(T_{\overline{B},v_j} = t|x_0 = v_i)$ be the probability that the random walk starting from v_i hits v_j after t steps without visiting any node in B. We have

$$\Pr(T_B = t|x_0 = v_i) = \sum_{v_j \in B} \Pr(T_{\overline{B},v_j} = t|x_0 = v_i).$$

The problem becomes how to compute $\Pr(T_{\overline{B},v_j} = t|x_0 = v_i)$. In particular, we have

$$\Pr(T_{\overline{B},v_j} = t|x_0 = v_i) = \sum_{v_k \notin B} \Pr(T_{\overline{B},v_k} = t-1|x_0 = v_i)p_{kj},$$

which implies that one can first get to $v_k \notin B$ using $t-1$ steps (without visiting any $v_j \in B$), and then move from v_k to v_j with probability p_{kj}. It takes the sum over all possible v_k's. Therefore, we can derive the following iterative computation method: Let $\mathbf{P}_B$ be a modification of $\mathbf{P}$ where rows corresponding to the nodes in B are set to zeros. Let $\mathbf{u}_t$ be a $n \times 1$ vector containing $\Pr(T_{\overline{B},v_k} = t|x_0 = v_i)$ as its kth element. $\mathbf{u}_0$ is the vector with ith element set to 1 and all other elements to 0. One can easily verify $\mathbf{u}_t = \left(\mathbf{P}_B^T\right)^t \mathbf{u}_0$. In fact, $\mathbf{P}_B$ and $\mathbf{u}_0$ define the corresponding random-walk model for computing DHT from v_i to B. Let $\mathbf{z}_B$ be the vector with elements corresponding to nodes in B set to 1 and all other elements to 0. We can rewrite Eq. (2.6) as

$$\tilde{h}(v_i, B) = e^0 \mathbf{z}_B^T \mathbf{u}_1 + e^{-1} \mathbf{z}_B^T \mathbf{u}_2 + \cdots$$

We can iteratively compute $\mathbf{u}_1, \mathbf{u}_2, \ldots$ and accumulate the elements corresponding to nodes in B from these vectors (multiplied by $e^0, e^{-1}, \ldots$ respectively). If we stop after a number of iterations, it results in an estimate of the actual DHT. In the remaining part of this section, we derive bounds for DHT from such an estimate using Lemma 2.1 and Theorem 2.1. We use $d_B(v_k)$ to denote the number of neighbors of v_k which are in B and $\lambda_{kB} = d_B(v_k)/d(v_k)$ to denote the corresponding fraction.

Lemma 2.1 *Let $pa^l_{i \rightsquigarrow k}$ be a length-l path from v_i to v_k which has not yet hit any node in B. Let $\Pr(pa^l_{i \rightsquigarrow k})$ be the probability that a random walk takes this path. We define subpaths of $pa^l_{i \rightsquigarrow k}$ as length-$(l+1)$ paths sharing $pa^l_{i \rightsquigarrow k}$ as a prefix. The contribution of $pa^l_{i \rightsquigarrow k}$ to $\tilde{h}(v_i, B)$ is upper bounded by $\lambda_{kB} \Pr(pa^l_{i \rightsquigarrow k}) e^{-l} + (1 - \lambda_{kB}) \Pr(pa^l_{i \rightsquigarrow k}) e^{-(l+1)}$ and lower bounded by $\lambda_{kB} \Pr(pa^l_{i \rightsquigarrow k}) e^{-l}$.*

Proof. Since $pa^l_{i \rightsquigarrow k}$ has not hit any node in B, the probability $\Pr(pa^l_{i \rightsquigarrow k})$ will be further distributed to subpaths of $pa^l_{i \rightsquigarrow k}$. By querying the neighbors of v_k, we know the probability $\lambda_{kB} \Pr(pa^l_{i \rightsquigarrow k})$ will be distributed on length-$(l+1)$ hitting subpaths which have certain contribution $\lambda_{kB} \Pr(pa^l_{i \rightsquigarrow k}) e^{-l}$. The remaining probability $(1 - \lambda_{kB}) \Pr(pa^l_{i \rightsquigarrow k})$ will be distributed on length-$(l+1)$ non-hitting subpaths, as illustrated in Figure 2.4. The contribution of this part is lower bounded by 0 and upper bounded by $(1 - \lambda_{kB}) \Pr(pa^l_{i \rightsquigarrow k}) e^{-(l+1)}$ (i.e., all subpaths of those length-$(l+1)$ non-hitting subpaths of $pa^l_{i \rightsquigarrow k}$ hit targets). Combining the two parts, the conclusion follows. □

Theorem 2.1 *Let $\tilde{h}_t(v_i, B) = e^0 \mathbf{z}_B^T \mathbf{u}_1 + \cdots + e^{-(t-1)} \mathbf{z}_B^T \mathbf{u}_t$ be the estimate of $\tilde{h}(v_i, B)$ after $\mathbf{u}_t$ is computed. Then $\tilde{h}(v_i, B)$ can be bounded as follows*

$$\tilde{h}_t(v_i, B) + \sum_{v_k \notin B} \Pr(T_{\overline{B}, v_k} = t | x_0 = v_i) \lambda_{kB} e^{-t} \leq \tilde{h}(v_i, B)$$

$$\leq \tilde{h}_t(v_i, B) + \sum_{v_k \notin B} \Pr(T_{\overline{B}, v_k} = t | x_0 = v_i) \left(\lambda_{kB} e^{-t} + (1 - \lambda_{kB}) e^{-(t+1)} \right).$$

Proof. We can view the random-walk process defined by computing $\tilde{h}(v_i, B)$ as distributing probability 1 onto all paths which start from v_i and end once a node

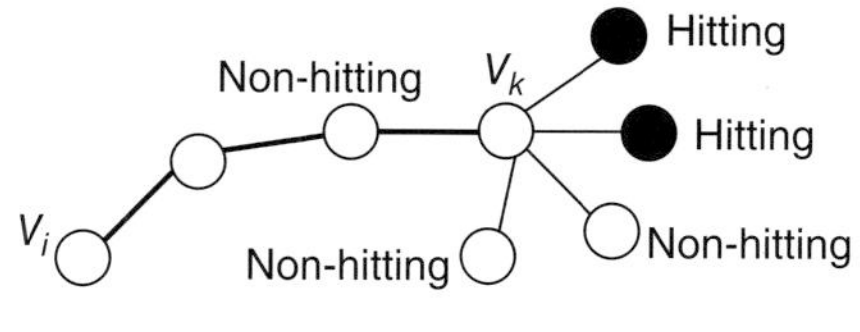

Figure 2.4 Bounds for one path.

$v_j \in B$ is hit. We use paths$_{iB}$ to denote the set of all such paths. Examining the definitions of $\mathbf{u}_t$ and $\mathbf{z}_B$, we can find $\tilde{h}_t(v_i, B)$ is the contribution of all paths in paths$_{iB}$ whose lengths are less than or equal to t. We use paths$_{iB}^{1\to t}$ to denote the set of these paths. The paths in paths$_{iB}^{1\to t}$ have certain contribution to $\tilde{h}(v_i, B)$. In the meantime, let paths$_{i\bar{B}}^{t}$ denote the set of all length-t paths starting from v_i which have not yet hit any node in B. The probabilities accounted for by the paths in paths$_{iB}^{1\to t}$ and paths$_{i\bar{B}}^{t}$ are $\Pr(\text{paths}_{iB}^{1\to t}) = \sum_{v_j \in B} \sum_{l=1}^{t} \Pr(T_{\bar{B},v_j} = l | x_0 = v_i)$ and $\Pr(\text{paths}_{i\bar{B}}^{t}) = \sum_{v_k \notin B} \Pr(T_{\bar{B},v_k} = t | x_0 = v_i)$, respectively. By examining the iterative computation of $\mathbf{u}_t$, one can verify that $\Pr(\text{paths}_{iB}^{1\to t}) + \Pr(\text{paths}_{i\bar{B}}^{t}) = 1$. It means, $\Pr(\text{paths}_{i\bar{B}}^{t})$ is the remaining probability that has not been contributed to $\tilde{h}(v_i, B)$. According to Lemma 2.1, we can derive lower and upper bounds of the contribution for each $pa \in \text{paths}_{i\bar{B}}^{t}$. Aggregating those bounds and adding the certain contribution of paths$_{iB}^{1\to t}$, the conclusion follows. □

2.3.3.2 A Sampling Algorithm for $\tilde{h}(v_i, B)$

We propose a standard Monte Carlo sampling method for estimating $\tilde{h}(v_i, B)$. A straightforward sampling scheme is as follows: We run c independent random-walk simulations from v_i and in each random walk, we stop when a node in B is encountered. Suppose these random walks' path lengths are $l_1, \ldots, l_c$, then, we use the average $\overline{\tilde{h}(v_i, B)} = \sum_{j=1}^{c} e^{-(l_j-1)}/c$ as the estimate of $\tilde{h}(v_i, B)$. However, this scheme is not a wise choice due to the following two reasons: (i) If we cannot reach any node in B from v_i, the random walk will never stop. (ii) For a large-scale graph, if we do not impose a maximum number of steps that a random walk can take, the sampling algorithm will be time consuming. In fact, since we adopt an exponentially damping factor (i.e., $e^{-(t-1)}$), the contribution of long paths are negligible.

With these concerns, we adopt a variant sampling scheme: We run c independent random-walk simulations from v_i and in each random walk, we stop when a node in B is visited or a maximum number of s steps is reached. We provide bounds for $\overline{\tilde{h}(v_i, B)}$ by this sampling scheme in Theorem 2.2.

Theorem 2.2 *Consider a simulation of c independent random walks from node v_i with a maximum number of s steps for each random walk. Suppose out of c runs, c_h random walks hit a node in B and the corresponding path lengths are $l_1, \ldots, l_{c_h}$, let $\bar{c} = c - c_h$ be the number of random walks which reach s steps and do not hit any node in B. Then, the sample mean $\overline{\tilde{h}(v_i, B)}$ in the sampling scheme without the constraint of maximum number of steps can be bounded as follows*

$$\frac{\sum_{j=1}^{c_h} e^{-(l_j-1)}}{c} \leq \overline{\tilde{h}(v_i, B)} \leq \frac{\sum_{j=1}^{c_h} e^{-(l_j-1)} + e^{-s}\bar{c}}{c}.$$

Proof. In the sampling scheme with the constraint of maximum number of steps, only c_h random walks have certain contribution to $\overline{\tilde{h}(v_i, B)}$. Hence, we turn to the bounds for the contribution of the remaining $\bar{c}$ random walks which do not hit any node in B. For each of these $\bar{c}$ random walks, the contribution to $\overline{\tilde{h}(v_i, B)}$ is upper bounded by e^{-s} (i.e., hitting a node in B at $(s+1)$th step). Aggregating those $\bar{c}$ random walks, we have $\overline{\tilde{h}(v_i, B)} \leq (\sum_{j=1}^{c_h} e^{-(l_j-1)} + e^{-s}\bar{c})/c$. A lower bound for the contribution of those $\bar{c}$ random walks is 0. This leads to $\sum_{j=1}^{c_h} e^{-(l_j-1)}/c \leq \overline{\tilde{h}(v_i, B)}$. □

We use $\tilde{h}'_{iB}$ and $\tilde{h}''_{iB}$ to represent the above-mentioned lower and upper bounds for $\overline{\tilde{h}(v_i, B)}$, respectively. The following theorem provides the lower bound for the sample size c in order to obtain an ϵ-correct answer for $\tilde{h}(v_i, B)$ with respect to $[\tilde{h}'_{iB}, \tilde{h}''_{iB}]$ with probability $1-\delta$.

Theorem 2.3 *Suppose we simulate c independent random walks for estimating* $\tilde{h}(v_i, B)$ *and impose a maximum number of s steps for each random walk, then, for any* $\epsilon > 0$ *and* $\delta > 0$*, in order to obtain* $\Pr(\tilde{h}'_{iB} - \epsilon \leq \tilde{h}(v_i, B) \leq \tilde{h}''_{iB} + \epsilon) \geq 1 - \delta$*, c should be at least* $\frac{1}{2\epsilon^2}\ln(\frac{2}{\delta})$.

Proof. From the sampling scheme without the constraint of maximum number of steps, we have $\overline{\tilde{h}(v_i, B)} = \sum_{j=1}^{c} e^{-(l_j-1)}/c$ where l_j is the path length of the jth random walk. It is obvious that $E(\overline{\tilde{h}(v_i, B)}) = \tilde{h}(v_i, B)$. Since random walks are independent and $0 \leq e^{-(l_j-1)} \leq 1$ for $j = 1, \ldots, c$, according to Hoeffding's inequality (Hoeffding, 1963), we have $\Pr(|\overline{\tilde{h}(v_i, B)} - \tilde{h}(v_i, B)| \leq \epsilon) \geq 1 - 2e^{-2c\epsilon^2}$. From Theorem 2.2, we know $\tilde{h}'_{iB} \leq \overline{\tilde{h}(v_i, B)} \leq \tilde{h}''_{iB}$. Therefore, we have

$$-\epsilon \leq \tilde{h}(v_i, B) - \overline{\tilde{h}(v_i, B)} \leq \epsilon$$
$$\Leftrightarrow \tilde{h}'_{iB} - \epsilon \leq \tilde{h}(v_i, B) \leq \tilde{h}''_{iB} + \epsilon$$

We can further obtain

$$\Pr(\tilde{h}'_{iB} - \epsilon \leq \tilde{h}(v_i, B) \leq \tilde{h}''_{iB} + \epsilon)$$
$$= \Pr(|\overline{\tilde{h}(v_i, B)} - \tilde{h}(v_i, B)| \leq \epsilon) \geq 1 - 2e^{-2c\epsilon^2}.$$

Setting $1 - 2e^{-2c\epsilon^2} \geq 1 - \delta$ gives us $c \geq \frac{1}{2\epsilon^2}\ln(\frac{2}{\delta})$. □

2.3.3.3 Complexity

Hereafter, we use *Iterative-alg* and *Sampling-alg* to denote the iterative algorithm and sampling algorithm developed earlier, respectively. Suppose, we use adjacency lists to store graphs and matrices, the space complexity of the two algorithms is $O(|E|)$. The major time-consuming parts of Iterative-alg are the iterative matrix-vector multiplication and the construction of $\mathbf{P}_B$. The corresponding time complexity is $O(t|E|)$, where t is the number of iterations. For Sampling-alg, the major time cost in each random-walk step is the membership judgement of the current node v to B. We can either sort B and use binary search, or build an index array for B. The corresponding time costs are $O(cs \log|B| + |B|\log|B|)$ and $O(cs + |V|)$, respectively.

2.3.4 Assessing SSC

In this section, we first propose a sampling method for estimating $\rho(V_q)$ and develop the lower bound for the sample size in order to get ϵ'-correct answers (Section 2.3.4.1). Then in Section 2.3.4.2, we describe our methodology for assessing the significance of the observed ρ score, that is, $\tilde{\rho}(V_q)$.

2.3.4.1 Estimating $\rho(V_q)$

To compute $\rho(V_q)$, we need to compute $\tilde{h}(v_i, V_q \setminus \{v_i\})$ for all $v_i \in V_q$. However, for large-scale graphs, V_q may also have a large size, posing a challenge for the efficient computation of $\rho(V_q)$. Although these $\tilde{h}$'s are dependent on each other, they form a finite population. We can still use sampling techniques to efficiently estimate $\rho(V_q)$ by applying Hoeffding's inequality for finite populations (Hoeffding, 1963). Specifically, we randomly select c' nodes from V_q, denoted by $v_1, \dots, v_{c'}$, to estimate their DHTs to the remaining nodes and take the average $\overline{\rho(V_q)}$ as an estimate for $\rho(V_q)$. Here, we can use either Iterative-alg or Sampling-alg for estimating each $\tilde{h}(v_i, V_q \setminus \{v_i\})$. If Iterative-alg is used, from Theorem 2.1, we obtain bounds for each $\tilde{h}(v_i, V_q \setminus \{v_i\})$ in the sample set. Aggregating those bounds, we can get bounds for $\overline{\rho(V_q)}$. Following the same manner for the proof of Theorem 2.3 and applying Hoeffding's inequality for finite populations (Hoeffding, 1963), we can obtain the lower bound for c' in order to obtain an ϵ'-correct answer. We omit the details due to the space limitation. When Sampling-alg is used, we provide the lower bound for c' in the following theorem.

Theorem 2.4 *Suppose we randomly select the c' nodes from V_q to estimate their DHTs to the remaining nodes and take the average $\overline{\rho(V_q)}$ as an estimate of $\rho(V_q)$. For the sake of clarity, let $B_i = V_q \setminus \{v_i\}$. Suppose, we have used Sampling-alg to obtain an ϵ-correct answer for each $\tilde{h}(v_i, B_i)$ $(i = 1, \dots, c')$ with respect to $[\tilde{h}'_{iB_i}, \tilde{h}''_{iB_i}]$, then, for any $\epsilon' > 0$ and $\delta' > 0$, in order to obtain*

$$\Pr\left(\frac{\sum_{i=1}^{c'} \tilde{h}'_{iB_i}}{c'} - \epsilon - \epsilon' \le \rho(V_q) \le \frac{\sum_{i=1}^{c'} \tilde{h}''_{iB_i}}{c'} + \epsilon + \epsilon'\right) \ge 1 - \delta',$$

c' should satisfy $(1-\delta)^{c'}(1 - 2e^{-2c'\epsilon'^2}) \ge 1 - \delta'$.

Proof. From the conditions, we have

$$\Pr\left(\tilde{h}'_{iB_i} - \epsilon \le \tilde{h}(v_i, B_i) \le \tilde{h}''_{iB_i} + \epsilon\right) \ge 1 - \delta, \quad i = 1, \dots, c'.$$

Notice $\overline{\rho(V_q)} = \sum_{i=1}^{c'} \tilde{h}(v_i, B_i)/c'$. Since $\tilde{h}$'s are estimated independently, multiplying those probability inequalities together we obtain

$$\Pr\left(\frac{\sum_{i=1}^{c'} \tilde{h}'_{iB_i}}{c'} - \epsilon \le \overline{\rho(V_q)} \le \frac{\sum_{i=1}^{c'} \tilde{h}''_{iB_i}}{c'} + \epsilon\right) \ge (1-\delta)^{c'}.$$

Since $0 \le \tilde{h}(v_i, B_i) \le 1$ for $i = 1, \ldots, c'$, according to Hoeffding's inequality for finite populations (Hoeffding, 1963), we know $\Pr(|\overline{\rho(V_q)} - \rho(V_q)| \le \epsilon') \ge 1 - 2e^{-2c'\epsilon'^2}$. Since the underlying estimation of $\overline{\rho(V_q)}$ is independent from Hoeffding bounds, we have

$$\Pr\left(\frac{\sum_{i=1}^{c'} \tilde{h}'_{iB_i}}{c'} - \epsilon - \epsilon' \le \rho(V_q) \le \frac{\sum_{i=1}^{c'} \tilde{h}''_{iB_i}}{c'} + \epsilon + \epsilon'\right)$$
$$\ge (1 - \delta)^{c'}(1 - 2e^{-2c'\epsilon'^2}).$$

Setting $(1 - \delta)^{c'}(1 - 2e^{-2c'\epsilon'^2}) \ge 1 - \delta'$, we get the inequality c' should satisfy. Note δ should be large enough so that $(1 - \delta)^{c'}(1 - 2e^{-2c'\epsilon'^2})$ can go beyond $1 - \delta'$ as c' increases. □

2.3.4.2 Estimating the Significance of $\rho(V_q)$

After obtaining the estimate of $\rho(V_q)$, we need to measure the deviation of $\rho(V_q)$ from the expected ρ value of $\widehat{V}_m$ (i.e., a set of randomly selected m nodes from the graph), in order to distinguish SSC from random results. In particular, we have

$$E[\rho(\widehat{V}_m)] = \frac{\sum_{V_m \subseteq V} \rho(V_m)}{C_n^m}, \tag{2.7}$$

where V_m is any set of m nodes. The ideal solution is to obtain the distribution of $\rho(\widehat{V}_m)$ and use the ratio between the number of node sets with size m whose ρ values are greater than or equal to $\rho(V_q)$ and C_n^m as the significance score for q. However, for a large-scale graph, it is very hard to get the distribution since C_n^m is very large. Here, we propose an approximation method. Notice $\rho(\widehat{V}_m)$ is defined as the average of $\tilde{h}(v_i, \widehat{V}_m \setminus \{v_i\})$ where $v_i \in \widehat{V}_m$. If we assume these $\tilde{h}$'s are independent, according to Central Limit Theorem, $\rho(\widehat{V}_m)$ can be approximated by a normal distribution, where $Var[\rho(\widehat{V}_m)] = Var[\tilde{h}(v_i, \widehat{V}_m \setminus \{v_i\})]/m$. If we obtain $E[\rho(\widehat{V}_m)]$ and $Var[\rho(\widehat{V}_m)]$, we can calculate the **adjusted SSC** $\tilde{\rho}$ for q as follows:

$$\tilde{\rho}(V_q) = \frac{\rho(V_q) - E[\rho(\widehat{V}_m)]}{\sqrt{Var[\rho(\widehat{V}_m)]}}. \tag{2.8}$$

This idea is similar to using z scores to assess the significance of data-mining results (Gionis *et al.*, 2006). Equation (2.8) can be used to derive the significance of q for a hypothesis that q is not randomly distributed over G. The independence assumption should work well as long as the dependence between those $\tilde{h}$'s is weak. This could be true because DHT focuses on local areas of the graph. Each black node only has high dependence on the nearby black nodes. In the remaining part of this section, we provide efficient methods for estimating $E[\rho(\widehat{V}_m)]$ and $Var[\rho(\widehat{V}_m)]$.

We propose two methods to efficiently estimate $E[\rho(\widehat{V}_m)]$. The first one is a sampling method. Equation (2.7) suggests sampling ρ values for different sets of m nodes. However, computing ρ is costly since m could be large. Using the sampling method proposed in Section 2.3.4.1 to estimate ρ, introduces another layer of sampling and would introduce more estimation error. Therefore, we propose to sample $\tilde{h}$ directly. Specifically, we have

$$E[\rho(\widehat{V}_m)] = \frac{\sum_{V_m} \frac{\sum_{v_i \in V_m} \tilde{h}(v_i, V_m \setminus \{v_i\})}{m}}{C_n^m} = \frac{\sum_{v_i, V_{m-1}} \tilde{h}(v_i, V_{m-1})}{C_n^1 C_{n-1}^{m-1}}$$

where $v_i \notin V_{m-1}$. It means $E[\rho(\widehat{V}_m)]$ is equal to the expected DHT from a random v_i to a random V_{m-1} which does not contain v_i. Thus, we can directly sample (v_i, V_{m-1}) pairs and take the average DHT among those pairs as an estimate of $E[\rho(\widehat{V}_m)]$. Given that we have already obtained ϵ-correct $\tilde{h}$'s by Theorem 2.3, we can derive a very similar sample size lower bound in the same manner for Theorem 2.4's proof, by applying Hoeffding's inequality (Hoeffding, 1963). We omit the details due to the space limitation. For a fixed graph, we can pre-compute $E[\rho(\widehat{V}_m)]$ for a number of different m values and employ interpolation to estimate $E[\rho(\widehat{V}_m)]$ for arbitrary m.

Alternatively, we can derive an approximation method for $E[\rho(\widehat{V}_m)]$ by a *geometric distribution*. This approximation method is empirical. A geometric distribution is a probability distribution of the number t of Bernoulli trials needed for one to get success. When we randomly generate $\widehat{V}_m$, each node of G has probability $\frac{m}{n}$ to be chosen. In the following discussion, we assume each node of G is chosen independently with probability $\frac{m}{n}$. With this relaxation, $|\widehat{V}_m|$ becomes a binomial random variable with m as its expected value. Consider we start from a node $v_i \in \widehat{V}_m$ to hit the remaining nodes in $\widehat{V}_m$. Let $p = \frac{m-1}{n-1}$ be the probability of each node other than v_i being in $\widehat{V}_m$. The probability that we first hit (i.e., stop) a target node after one step is $\sum_j p_{ij} p = p$. The probability that we stop after two steps is $\sum_{j,k} p_{ij} p_{jk} (1-p)p = (1-p)p$. We do not consider cases where the surfer comes back to v_i in this approximation. This forms a geometric distribution where the probability that we "succeed" after t steps is $(1-p)^{t-1}p$. By the definition of DHT (i.e., Eq. (2.6)), $\tilde{h}(v_i, \widehat{V}_m \setminus \{v_i\})$ is actually the expectation of $e^{-(t-1)}$ under the geometric distribution described earlier:

$$\tilde{h}(v_i, \widehat{V}_m \setminus \{v_i\}) = \sum_{t=1}^{\infty} e^{-(t-1)} (1-p)^{t-1} p = \frac{p}{1 - e^{-1}(1-p)}. \tag{2.9}$$

Since v_i is an arbitrary node in $\widehat{V}_m$, we have

$$\rho(\widehat{V}_m) = \frac{p}{1 - e^{-1}(1-p)}. \tag{2.10}$$

Since we assume each node of G is chosen independently, the obtained $\rho(\widehat{V}_m)$ is an approximation of $E[\rho(\widehat{V}_m)]$. In case the graph contains 0-degree nodes, we just need to multiply Eq. (2.10) by the probability that a randomly selected node is not a 0-degree node. We empirically compare this heuristic approximation method

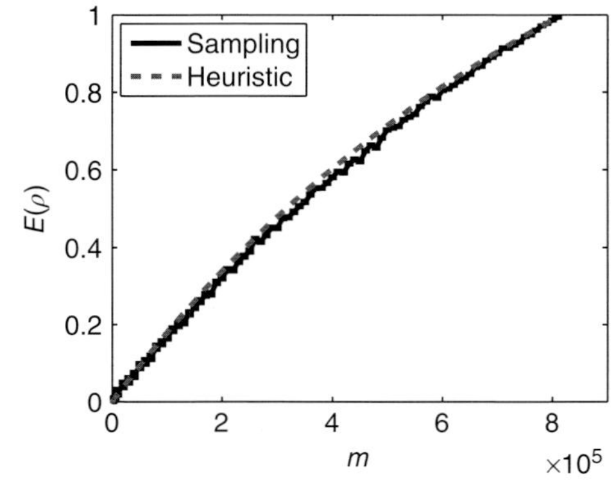

Figure 2.5 Comparison of sampling and geometric distribution heuristic for estimating $E[\rho(\widehat{V}_m)]$.

with the sampling method on the DBLP coauthor network. The results are shown in Figure 2.5. Regarding the sampling method, we sample 1500 (v_i, V_{m-1}) pairs for each m and use Sampling-alg to estimate DHT. The error bars on the curve of the sampling method represent lower and upper bounds for the estimates of $E[\rho(\widehat{V}_m)]$. We can see that results obtained by sampling roughly fit the curve of the heuristic method. Therefore, we can either use sampling method and interpolation or the heuristic method to estimate $E[\rho(\widehat{V}_m)]$. In our experiments, we employ the heuristic method.

Regarding $Var[\rho(\widehat{V}_m)]$, we also propose a sampling method. Directly estimating $Var[\rho(\widehat{V}_m)]$ by sample variance again requires computing ρ for each sampled V_m and is time consuming since m could be large. Recall that we assume $\tilde{h}(v_i, \widehat{V}_m \setminus \{v_i\})$'s in the numerator of the definition of $\rho(\widehat{V}_m)$ are independent. We approximate $Var[\rho(\widehat{V}_m)]$ by $Var[\tilde{h}(v_i, \widehat{V}_m \setminus \{v_i\})]/m$. For a given m, we just sample (v_i, V_{m-1}) pairs and take the sample variance of the corresponding DHTs divided by m as an estimate of $Var[\rho(\widehat{V}_m)]$. Again, pre-computation and interpolation can be used here to estimate $Var[\rho(\widehat{V}_m)]$ for arbitrary m.

We summarize the gScore framework in Algorithm 2.1. Steps 2 and 4 can use either Iterative-alg or Sampling-alg to estimate DHT.

Algorithm 2.1: The gScore framework

Offline Phase

1 Choose a set of m values.
2 For each m value, sample (v_i, V_{m-1}) pairs to estimate $Var[\tilde{h}(v_i, V_{m-1})]/m$ as the variance estimate.

Online Phase

3 Randomly choose a sample of c nodes from V_q.
4 For each of c nodes, estimate its DHT to the remaining nodes in V_q.
5 Compute sample mean $\overline{\rho(V_q)} = \sum_{i=1}^{c} \tilde{h}(v_i, V_q \setminus \{v_i\})/c$.
6 Estimate $E[\rho(\widehat{V}_m)]$ by Eq. (2.10).
7 Estimate $Var[\rho(\widehat{V}_m)]$ by interpolating the pre-computed variances.
8 Compute approximate z-score by Eq. (2.8).

2.3.5 Empirical Studies

This section presents experimental results on three real-world datasets: DBLP, TaoBao, and Twitter, as well as synthetic events. We first investigate the performance of two DHT approximation algorithms proposed in Section 2.3.3. We then verify gScore's effectiveness on synthetic events. In Section 2.3.5.4, we report interesting correlated and uncorrelated products discovered from the TaoBao network. Finally, we analyze the scalability of gScore with the Twitter network. All experiments are run on a PC with Intel Core i7 CPU and 12GB memory. The source code of gScore can be downloaded at *http://www.cs.ucsb.edu/~xyan/software/gScore.html.*

2.3.5.1 Datasets

DBLP The DBLP snapshot was downloaded on Oct. 5th, 2010 (*http://www.informatik.uni-trier.de/~ley/db*). Its paper records were parsed to obtain the coauthor social graph. Keywords in the paper titles are treated as events associated with nodes (authors) on the graph. The first time an author used a keyword was also recorded. It contains 815 940 nodes, 2 857 960 edges, and 171 614 events.

TaoBao The TaoBao dataset was obtained from China's most famous customer-to-customer shopping Website named TaoBao (*http://www.taobao.com*). By the end of 2009, TaoBao has about 170 million users and 1 billion products. We extracted users from three cities (Beijing, Shanghai and Hangzhou) with their product purchase history and constructed the friend-social graph among them. It consists of 794 001 nodes, 1 370 284 edges. We selected 100 typical products from TaoBao to show the effectiveness of our measure.

Twitter The Twitter dataset has about 40 million nodes and 1.4 billion edges (*http://twitter.com*). We do not have events for this dataset. It is mainly used to test the scalability of gScore.

2.3.5.2 Performance of DHT Approximation

We investigate the convergence and running time of the two DHT approximation algorithms: Iterative-alg and Sampling-alg. Iterative-alg has one parameter (number of iterations t) and Sampling-alg has two parameters (maximum number of steps s and number of random walks c). For Iterative-alg, we investigate its converging speed with respect to t. For Sampling-alg, we find when $c > 600$, increasing c hardly improves the obtained bounds. Thus, we set $c = 600$ and investigate the converging speed of Sampling-alg with respect to s. The results are shown in Figure 2.6 with various m values (the number of nodes that have the same event). For each m value, we randomly select a node v and a set B of $m-1$ nodes and apply the two algorithms to estimate $\tilde{h}(v, B)$. This process is repeated 50 times and the averaged results are reported. As shown in Figure 2.6, both algorithms converge quickly after about five iterations. Note that Iterative-alg gives

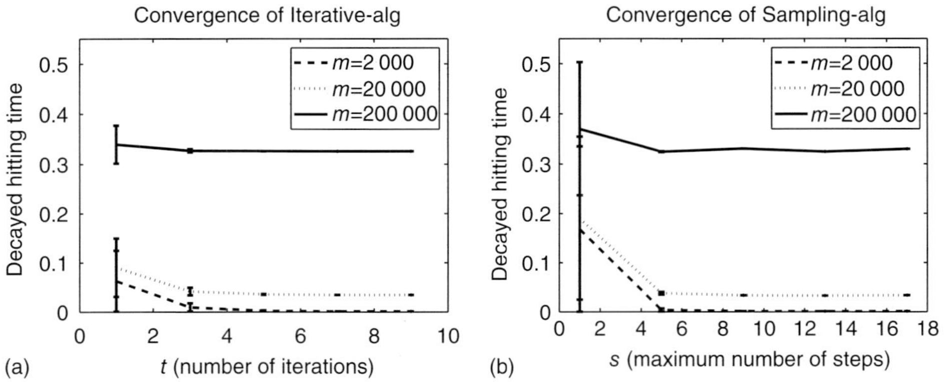

Figure 2.6 Exploring the convergence of (a) Iterative-alg, and (b) Sampling-alg.

lower and upper bounds for $\tilde{h}$, while Sampling-alg gives bounds for an estimate of $\tilde{h}$, that is, $\bar{\tilde{h}}$. Comparing Figure 2.6a and b, one can find that the two algorithms converge to roughly the same values. It means empirically Sampling-alg provides a good estimation of $\tilde{h}$.

The running time of Iterative-alg and Sampling-alg for estimating one DHT under different m values is shown in Figure 2.7. For Iterative-alg, we report the running time for $t = 1$ and $t = 9$ and for Sampling-alg, $s = 1$ and $s = 17$. It shows that Sampling-alg is much faster than Iterative-alg. Note that regarding Iterative-alg, the time cost of "$s = 9$" is not 9 times as that of "$s = 1$" This is because not only matrix-vector multiplication but also the construction of $\mathbf{P}_B$ accounts for time cost. In fact, Iterative-alg runs even faster when m increases: Less rows of $\mathbf{P}$ are needed to construct the desired matrix. Since Sampling-alg is much faster

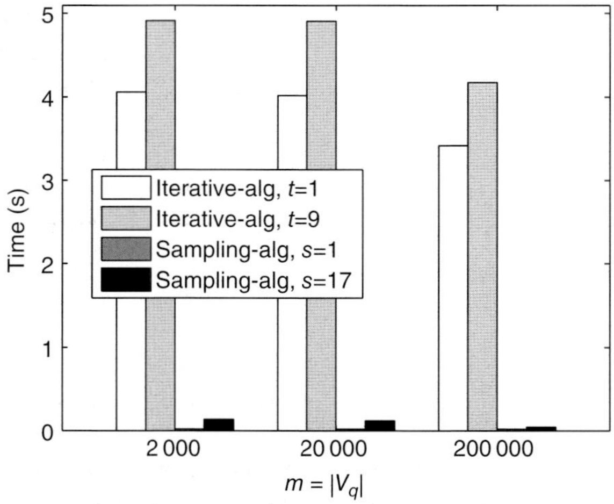

Figure 2.7 Comparison of Iterative-alg and Sampling-alg with respect to the time used to estimate one DHT.

than Iterative-alg and also provides reasonable estimates for DHTs, for the following experiments, we employ Sampling-alg to estimate DHT. gScore also refers to Sampling-alg. Hereafter, we set $s = 12$ and $c = 600$.

2.3.5.3 Effectiveness on Synthetic Events

To evaluate the effectiveness of our measure, we generate synthetic events on the DBLP graph using the cascade model for influence spread (Kempe, Kleinberg, and Tardos, 2003): At first, a random set of 100 nodes is chosen as the initial V_q; then in each iteration nodes joining V_q in the last iteration can activate each currently inactive neighbor with probability p_{ac}; we stop when $|V_q| > 10\,000$. p_{ac} can be regarded as representing the level of participation in an event. Intuitively, higher p_{ac} would lead to higher correlation. *For all the following experiments, we report the significance estimates as the measure of SSC, that is, $\tilde{p}$ in Eq. (2.8). $\tilde{p}$ can be regarded as approximate z scores. Higher scores mean higher (more significant) correlations, while a score close to 0 indicates that there is no correlation.*

The results are shown in Figure 2.8. "Random" means we expand the initial 100 random nodes with randomly selected nodes from the remaining nodes in order to match the corresponding event sizes of cascade model. We can see as p_{ac} increases, the curve of cascade model goes up, while that of "Random" remains around 0.

We further test the performance of gScore by adding noises to the earlier cascade model. p_{ac} is set to 0.2. Specifically, we break the correlation structure by relocating each black node to a random node in the remaining graph with probability p_n (noise level). $p_n = 1$ means all black nodes are randomly redistributed. We report results for different event sizes (m), that is, spread levels.

gScore is applicable to other graph proximity measures. Here, we also instantiate gScore with pair-wise shortest distance for comparison. In this case, Eq. (2.1) becomes the average shortest distance among all pairs of black nodes. For large-scale graphs, computing shortest distances for all pairs of black nodes is usually

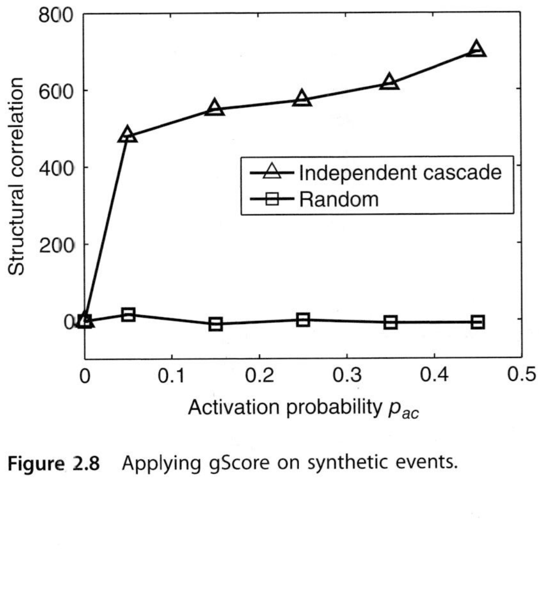

Figure 2.8 Applying gScore on synthetic events.

very costly. Pre-computing and storing pair-wise shortest distances for the whole graph is not practical either. Hence, we sample black node pairs to estimate the correlation measure. By applying the Hoeffding's inequality for finite populations (Hoeffding, 1963), we can easily derive a lower bound for the sample size in order to get ϵ-correct answers. The significance estimation methodology in Section 2.3.4 is also applicable. The expectation of the correlation measure for $\hat{V}_m$) is the average shortest path length of the graph. Its variance can be approximated by the variance of shortest path length divided by the event size m. We use sampling to estimate mean and variance. We use the reciprocal of shortest distances to avoid infinite distances when no path exists between two nodes.

We show results in Figure 2.9. For a good proximity measure, the correlation significance should decrease smoothly, as the noise level increases. As we see, the curves of DHT gradually decrease with increasing noises and get around 0 when $p_n = 1$, indicating DHT can well capture structural correlations, while the curves of shortest distance are not stable and fluctuate a bit with increasing noises. The reason should be that pair-wise shortest distance is affected by long distances among black nodes, as mentioned in Section 2.3.2. The relocation of one node will affect all remaining $m - 1$ nodes equally, and the independent assumption in normal approximation and variance estimation may not work very well. In Figure 2.9b, we find the correlation scores for $m = 500$ is much lower than that for $m = 50\,000$. This is also due to long distances. Recall that the cascade model chooses initial black nodes randomly, which means different threads of influence spread could be distant from one another. When m is small, long distances could show high impact.

We also compare DHT with the 1-neighborhood event-fraction measure described in Section 2.3.1. We find the 1-neighborhood measure performs as well as DHT with the cascade model. This is because the cascade model always generates correlation patterns in 1-neighborhood. However, more general correlation patterns can occur in a graph, for example, products can be attractive

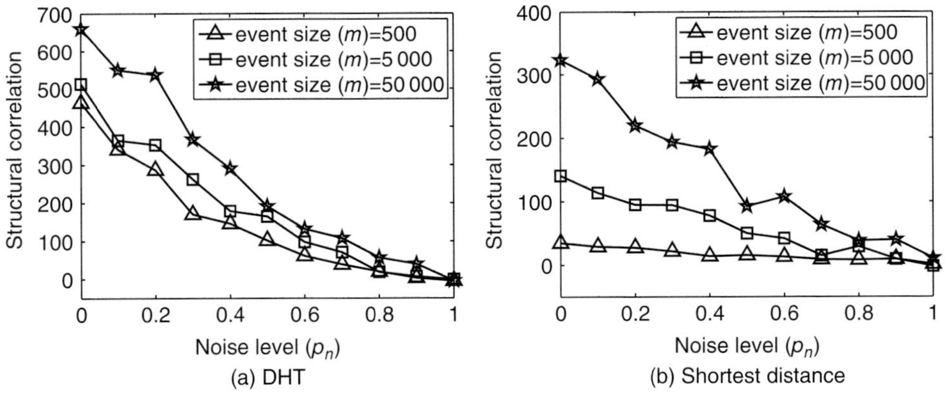

Figure 2.9 Comparison of DHT and pair-wise shortest distance as the proximity measure by adding noises into the cascade model.

to specific communities in a graph, but purchases may not always occur among direct neighbors. We use a new model to generate more general correlation patterns: We first randomly select 500 nodes as seed black nodes; then $a\%$ nodes (minimum 1) in each seed node's k-hop neighborhood are painted black. Their distances to the seed node are distributed as Gaussian with mean μ and variance 0.25. Distances out of range are reset to the nearest value within range. We explore $k = 2, 3$ and set $a = 0.1$. μ controls the average distance to seed nodes. The results are shown in Figure 2.10. As μ increases, the curves of the 1-neighborhood measure drop to around 0 (no correlation), while those of DHT stay around 22 (2-hop) and 9 (3-hop). This means DHT can detect more general correlations, while the 1-neighborhood measure cannot. If the user only considers the correlations in 1-hop neighborhoods, 1-neighborhood event fraction is preferred since it is more efficient.

2.3.5.4 SSC of Real Event

We apply gScore on real events occurring on graphs and report Interesting, highly correlated events and uncorrelated events. Using Eq. (2.8), we obtain an estimate (lower and upper bounds) of $\tilde{p}$ for each event. A ranked list of events can be generated according to these bounds. If the bounds of two events overlap, we increase sample numbers and the maximum steps to break a tie. For this experiment, we omit the results for DBLP keywords due to the space limitation. The readers can refer to (Guan *et al.*, 2011) for details. We group the products from TaoBao into two categories: *Laptops and tablets* and *Other* and show top-five products for each case. Before presenting the results, we would like to emphasize that our correlation findings are just for the specific social networks involved in this study.

Table 2.1 shows the ranked lists for top-five products in "Laptops and tablets". We also show ρ values in Table 2.1. ThinkPad and Apple products usually have high correlation with the underlying network, indicating there are fan communities for these brands. An interesting exception is ThinkPad T410, which is a new

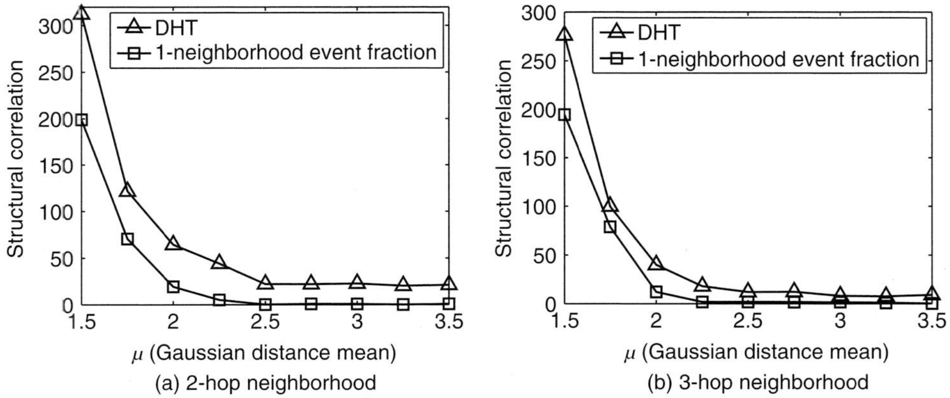

Figure 2.10 Comparison of DHT and 1-neighborhood event fraction as the proximity measure by generating more general SSC in local neighborhoods.

Table 2.1 SSC for top five correlated products in category "Laptops and tablets" in TaoBao.

#	Product	Bounds for $\tilde{\rho}$	ρ ($\times 10^{-2}$)	$\|V_q\|$
1	ThinkPad T400	[554.43, 554.47]	[6.2396, 6.2400]	47
2	Apple iPad	[227.56, 227.57]	[6.7979, 6.7984]	698
3	ThinkPad X200	[91.39, 91.42]	[1.0799, 1.0802]	60
4	Toshiba L600	[20.36, 20.41]	[0.2009, 0.2014]	31
5	ThinkPad T410	[−1.13, −1.09]	[0.0004, 0.0009]	72

Table 2.2 SSC for top-five correlated products in category "Other" in TaoBao.

#	Product	Bounds for $\tilde{\rho}$	$\|V_q\|$
1	Mamy Poko baby diapers	[238.50, 238.51]	4892
2	Beingmate Infant milk powder	[227.71, 227.72]	163
3	EVE game cards	[198.56, 198.58]	374
4	Mabinogi game cards	[189.56, 189.58]	446
5	Gerber cookies	[149.51, 149.52]	1491

Table 2.3 SSC for the five most uncorrelated products in category "Other" in TaoBao.

#	Product	Bounds for $\tilde{\rho}$	$\|V_q\|$
1	Tiffany rings	[2.71,2.72]	1092
2	Jack&Jones suits	[−0.48, −0.46]	311
3	Ray-Ban sunglasses	[−0.78, −0.77]	4958
4	Swarovski anklets	[−0.88, −0.84]	72
5	Jack&Jones shirts	[−3.28, −3.27]	1606

version of Thinkpad T400. In comparison with T400, its correlation score is very close to that of random cases. The reason may be that people in the fan community already bought T400 and they would not further buy a new version for T400 since they are quite similar and not cheap.

The ranked list for top-five products from category "Other" is shown in Table 2.2. Here "EVE" and "Mabinogi" are two online games and players in China must buy *game cards* to obtain gaming time. We find products for infants, like diapers and powder tend to be correlated with the network. This indicates people tend to follow friends' recommendations when choosing this kind of products. Game card is also highly correlated with the network. Intuitively, playing with friends is an important attractive feature of the online games.

Finally, we show the $\tilde{\rho}$ scores for the five most uncorrelated products from category "Other" in Table 2.3. These products' scores are very close to those of random cases (some scores deviate a little from random cases due to estimation errors in variance). This indicates that for clothing and accessories, people usually follow their own preferences.

2.3.5.5 Scalability of Sampling-alg

Finally, we investigate the scalability of Sampling-alg when the graph size n increases. The Twitter graph is used to perform this experiment. We extract subgraphs with different sizes (i.e., n) and for each n, different values of m are tested. The results are averaged over 50 sampled DHTs. Figure 2.11 shows that Sampling-alg is scalable and only needs 0.17 s to estimate one DHT on a graph with 10 million nodes. Although the time cost of Sampling-alg is linear in n, it only involves creating an index array of size n in memory. Regarding ρ, the estimation time is only 8.5 s on a graph with 10 million nodes if we set the number of samples $c' = 50$. Note that this can also be regarded as the time used for computing one adjusted correlation $\tilde{\rho}$ since $E(\rho)$ and $Var(\rho)$ can be obtained from pre-computed results. Intuitively, when n is fixed and m increases, the running time should decrease since it is easier to hit a target node (most random walks do not need to reach the maximum steps, s). This is the reason that the curve of $m = 0.05n$ is below that of $m = 0.005n$. Since we only store the adjacency list, the memory cost is linear in the number of edges in the graph. We do not show the curve here due to space limitation.

2.3.6 Discussions

We proposed a novel concept, SSC, to measure how an event is distributed in a graph and address a key research problem in analyzing the relation between the structures and the contents. While many studies have demonstrated that social links could significantly influence the behavior of human beings (Brown and Reingen, 1987; Chen, Wang, and Wang, 2010; Kempe, Kleinberg, and Tardos, 2003), we suspect that such influence should be further scrutinized for more fine-grained knowledge: *In which kind of social links (e.g., phone networks, email networks,*

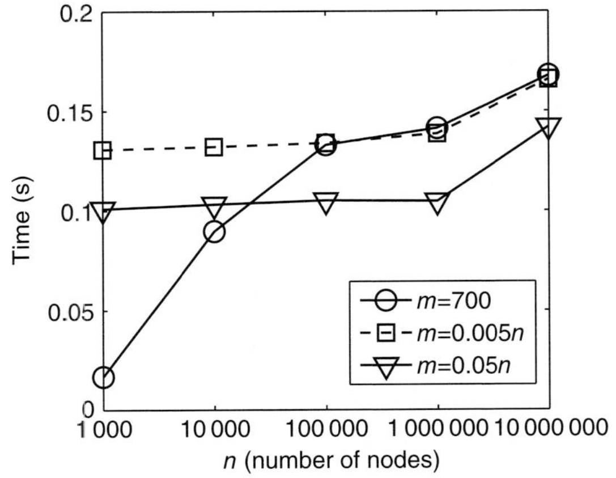

Figure 2.11 Running times of Sampling-alg for estimating one DHT when varying the graph size.

employee networks) and for which kind of behaviors (e.g., shopping, hobby, interest, and opinion) social influence is observed, and how strong the influence is. We quantify the correlation between link structures and human behaviors, and make different behaviors' correlations comparable using statistical significance.

2.4 Two-Event Structural Correlation

2.4.1 Preliminaries and Problem Formulation

As in Section 2.3, we have an attributed graph $G = (V, E)$ with an event set Q containing all events that occur on V. Each node v possesses a set of events $Q_v \subseteq Q$ which have occurred on it. For an event $a \in Q$, we denote the set of nodes having a as V_a. In this section, we use a and b to denote the two events for which we want to assess the structural correlation. For the sake of simplicity, we assume G is undirected and unweighted. Nevertheless, the proposed approach could be extended for graphs with directed and/or weighted edges.

Problem Statement *Given two events a and b and their corresponding occurrences V_a and V_b, to determine whether a and b are correlated (if correlated, positive or negative) in the graph space with respect to a vicinity level h.*

We formally define the notion of vicinity on a graph as follows.

Definition 2.1 (Node Level-h Vicinity) *Given graph $G = (V, E)$ and a node $u \in V$, the level-h vicinity (or h-vicinity) of u is defined as the subgraph induced by the set of nodes whose distances from u are less than or equal to h. We use V_u^h and E_u^h to denote the sets of nodes and edges in u's h-vicinity, respectively.*

Definition 2.2 (Node Set h-Vicinity) *Given a graph $G = (V, E)$ and a node set $V' \subseteq V$, the h-vicinity of V' is defined as the subgraph induced by the set of nodes which are within distance h from at least one node $u \in V'$. For event a, we use V_a^h and E_a^h to denote the sets of nodes and edges in V_a's h-vicinity, respectively.*

Let $V_{a\cup b} = V_a \cup V_b$ denote the set of nodes having at least one of the events a and b, that is, all the event nodes. The sets of nodes and edges in the h-vicinity of $V_{a\cup b}$ is denoted by $V_{a\cup b}^h$ and $E_{a\cup b}^h$, respectively. To assess the structural correlation between a and b, we employ a set of reference nodes.

Definition 2.3 (Reference Nodes) *Given two events a and b on G, a node $r \in V$ is a reference node for assessing level-h TESC between a and b iff $r \in V_{a\cup b}^h$.*

Definition 2.3 indicates that we treat $V_{a\cup b}^h$ as the set of all reference nodes for assessing level-h TESC between a and b. The reason will be explained in Section 2.4.2.2. We define the notion of concordance for a pair of reference nodes as follows.

Definition 2.4 (Concordance) *Two reference nodes r_i and r_j for assessing level-h TESC between a and b are said to be concordant if both a's density and b's density increase (or decrease) when we move from r_i's h-vicinity to r_j's h-vicinity.*

Mathematically, the concordance function $c(r_i, r_j)$ is defined as

$$c(r_i, r_j) = \begin{cases} 1 & (s_a^h(r_i) - s_a^h(r_j))(s_b^h(r_i) - s_b^h(r_j)) > 0 \\ -1 & (s_a^h(r_i) - s_a^h(r_j))(s_b^h(r_i) - s_b^h(r_j)) < 0 \\ 0 & \text{otherwise} \end{cases}, \tag{2.11}$$

where $s_a^h(r_i)$ is the density of event a in r_i's h-vicinity:

$$s_a^h(r_i) = \frac{|V_a \cap V_{r_i}^h|}{|V_{r_i}^h|}. \tag{2.12}$$

$c(r_i, r_j)$ encodes the concordance as 1 and the discordance as −1. 0 means r_i and r_j are in a tie, that is, $s_a^h(r_i) = s_a^h(r_j)$ or $s_b^h(r_i) = s_b^h(r_j)$, which means the pair indicates neither concordance nor discordance. Regarding $s_a^h(r_i)$, the reason that we use $|V_{r_i}^h|$ to normalize the occurrence number is that different nodes could have quite different sizes of h-vicinities. $|V_{r_i}^h|$ can be regarded as an analogue to the area in spatial spaces. The normalization makes all reference nodes' h-vicinities have the same "area". The computation of $s_a^h(r_i)$ is simple: We do a breadth-first search (BFS) up to h hops (hereafter to be called h-hop BFS) from r_i to count the number of occurrences of the event. More sophisticated graph-proximity measures could be used here, such as hitting time (Lovász, 1993) and personalized PageRank (Brin and Page, 1998). However, the major issue with these sophisticated measures is the high computational cost. As will be demonstrated in experiments, our density measure is not only much more efficient but also effective.

2.4.2 Measuring TESC

This section presents our TESC testing framework. First, we show the intuition behind using reference nodes to assess TESC. If events a and b are positively correlated on G, a region where a appears tends to also contain occurrences of b, and vice-versa. Furthermore, more occurrences of one event will tend to imply more occurrences of the other one. On the contrary, when a and b are negatively correlated, the presence of one event is likely to imply the absence of the other one. Even if they appear together, an increase of occurrences of one event is likely to imply a decrease of the other. Figure 2.12 shows the four typical scenarios described earlier. r_1 and r_2 are two reference nodes. Here, let us assume h-vicinities (denoted by dotted circles) of r_1 and r_2 have the same number of nodes so that we can treat the number of occurrences as density. We can see in Figure 2.12a and b, when a and b attract each other, r_1 and r_2 are concordant, implying an evidence of positive correlation. In the repulsion cases (Figure 2.12c, d), r_1 and r_2 are discordant, showing an evidence of negative correlation. Therefore, the idea is to aggregate all these evidences from all pairs of reference nodes to assess TESC.

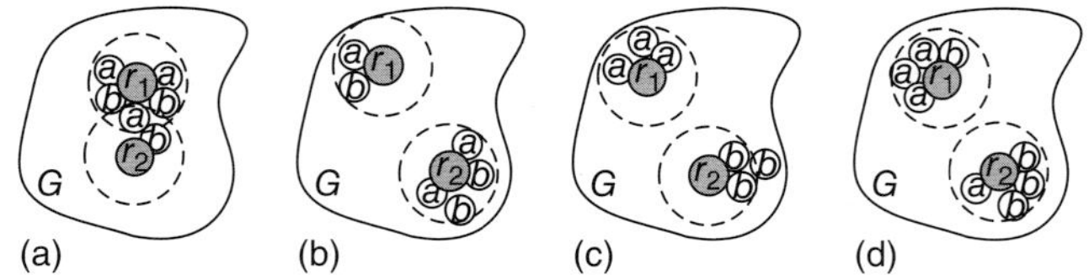

Figure 2.12 Four illustrative examples showing that density changes of the two events between two reference nodes show an evidence of correlation.

The natural choice for computing the overall concordance among the reference nodes with regard to density changes of the two events is the Kendall's τ rank correlation (Kendall and Gibbons, 1990), which was also successfully applied to the spatial point pattern-correlation problem (Diggle and Cox, 1983; Smith, 2004). For clarity, let $N = |V^h_{a\cup b}|$. We have N reference nodes: $r_1, r_2, \ldots, r_N$. The Kendall's τ measure is defined as an aggregation of $c(r_i, r_j)$'s

$$\tau(a,b) = \frac{\sum_{i=1}^{N-1} \sum_{j=i+1}^{N} c(r_i, r_j)}{\frac{1}{2}N(N-1)}. \tag{2.13}$$

$\tau(a,b)$ lies in $[-1, 1]$. A higher positive value of $\tau(a,b)$ means a stronger positive correlation, while a lower negative value means a stronger negative correlation. $\tau(a,b) = 0$ means there is no correlation between a and b, that is, the number of evidences for positive correlation is equal to that of negative correlation.

2.4.2.1 The Test

If N is not large, we can directly compute $\tau(a,b)$ and judge whether there is a correlation (and how strong) by $\tau(a,b)$. However, real-life graphs usually have very large sizes and so does N. It is often impractical to compute $\tau(a,b)$ directly. We propose to sample reference nodes and perform hypothesis testing (Wasserman, 2004) to efficiently estimate TESC. In a hypothesis test, a *null hypothesis* H_0 is tested against an *alternative hypothesis* H_1. The general process is that we compute from the sample data a statistic measure $\mathcal{X}$, which has an associated *rejection region* C such that, if the measure score falls in C, we reject H_0, otherwise H_0 is not rejected. The *significance level* of a test, α, is the probability that $\mathcal{X}$ falls in C when H_0 is true. The *p-value* of a test is the probability of obtaining a value of $\mathcal{X}$ at least as extreme as the one actually observed, assuming H_0 is true. In our case, $\mathcal{X}$ is τ and H_0 is "events a and b are independent with respect to G's structure". The test methodology is as follows: First, we uniformly choose a random sample of n reference nodes from $V^h_{a\cup b}$; then we compute the τ score over sampled reference nodes (denoted by $t(a,b)$):

$$t(a,b) = \frac{\sum_{i=1}^{n-1} \sum_{j=i+1}^{n} c(r_{k_i}, r_{k_j})}{\frac{1}{2}n(n-1)}, \tag{2.14}$$

where $r_{k_1}, \ldots, r_{k_n}$ are the n sampled reference nodes; finally, we estimate the significance of $t(a,b)$ and reject H_0 if the p value is less than a predefined significance level. We use $\mathbf{s}^h_a$ to represent the vector-containing densities of a measured in all n

sample reference nodes' h-vicinities where the ith element is $s_a^h(r_{k_i})$. Under H_0, $\tau(a,b)$ is 0. Consequently, for a uniformly sampled set of reference nodes, any ranking order of $\mathbf{s}_b^h$ is equally likely for a given order of $\mathbf{s}_a^h$. It is proved that the distribution of $t(a,b)$ under the null hypothesis tends to the normal distribution with mean 0 and variance

$$\sigma^2 = \frac{2(2n+5)}{9n(n-1)}. \tag{2.15}$$

The idea of the proof is to show the moments of t's distribution under H_0 converge to those of the normal distribution, and then apply the second limit theorem (Fréchet and Shohat, 1931). The readers could refer to Chapter 5 of (Kendall and Gibbons, 1990) for details. A good normality approximation can be obtained when $n > 30$ (Kendall and Gibbons, 1990). When $s_a^h(r_{k_i}) = s_a^h(r_{k_j})$ or $s_b^h(r_{k_i}) = s_b^h(r_{k_j})$, $c(r_{k_i}, r_{k_j})$ can be 0. This means there could be ties of reference nodes where pairs in a tie show evidences of neither concordance nor discordance. When ties are present in $\mathbf{s}_a^h$ and/or $\mathbf{s}_b^h$ (often, the case is that a set of reference nodes only have occurrences of a or b in their h-vicinities), σ^2 should be modified accordingly. Let l/m be the number of ties in $\mathbf{s}_a/\mathbf{s}_b$. The variance of the numerator of Eq. (2.14) becomes (Kendall and Gibbons, 1990):

$$\begin{aligned}\sigma_c^2 = &\frac{1}{18}\left[n(n-1)(2n+5) - \sum_{i=1}^{l} u_i(u_i-1)(2u_i+5)\right.\\ &\left.-\sum_{i=1}^{m} v_i(v_i-1)(2v_i+5)\right] + \frac{1}{9n(n-1)(n-2)}\\ &\times\left[\sum_{i=1}^{l} u_i(u_i-1)(u_i-2)\right]\left[\sum_{i=1}^{m} v_i(v_i-1)(v_i-2)\right]\\ &+\frac{1}{2n(n-1)}\left[\sum_{i=1}^{l} u_i(u_i-1)\right]\left[\sum_{i=1}^{m} v_i(v_i-1)\right],\end{aligned} \tag{2.16}$$

where u_i and v_i are the sizes of the ith ties of $\mathbf{s}_a^h$ and $\mathbf{s}_b^h$, respectively. When these sizes all equal to 1, Eq. (2.16) reduces to Eq. (2.15) multiplied by $[\frac{1}{2}n(n-1)]^2$, that is, the variance of the numerator of Eq. (2.14) when no ties exist. By grouping terms involving u_i/v_i together, one can verify that more (larger) ties always lead to smaller σ_c^2. σ^2 is then modified as σ_c^2 divided by $[\frac{1}{2}n(n-1)]^2$. Once the variance is obtained, we compute the significance (z-score) of the observed $t(a,b)$ by

$$z(a,b) = \frac{t(a,b) - E(t(a,b))}{\sqrt{Var(t(a,b))}} = \frac{t(a,b)}{\sigma}. \tag{2.17}$$

For τ, we do not substitute the alternative normalization term (see Chapter 3 of Kendall and Gibbons (1990)) for $[\frac{1}{2}N(N-1)]$ when ties are present, since it makes no difference on the significance result, that is, simultaneously dividing $\sum_{i=1}^{n-1}\sum_{j=i+1}^{n} c(r_{k_i}, r_{k_j})$ and σ_c by the same normalization term. t is an unbiased and consistent estimator for τ. In practice, we do not need to sample too many reference nodes since the variance of t is upper bounded by $\frac{2}{n}(1 - \tau^2)$ (Kendall and Gibbons, 1990), regardless of N.

2.4.2.2 **Reference Nodes**

Given the occurrences of two events a and b on graph G, not all nodes in G are eligible to be reference nodes for the correlation estimation between a and b. We do not consider areas on G where we cannot "see" any occurrences of a or b. That is, we do not consider nodes whose h-vicinities do not contain any occurrence of a or b. We refer to this kind of nodes as out-of-sight nodes. The reasons are: (i) We measure the correlation of presence, but not the correlation of absence. The fact that an area does not contain a and b currently does not mean it will never have a and/or b in the future. (ii) If we incorporate out-of-sight nodes into our reference set, we could get unexpected high z scores, since in that case we take the correlation of absence into account. Out-of-sight nodes introduce two 0 ties containing the same set of nodes into $\mathbf{s}_a^h$ and $\mathbf{s}_b^h$, respectively. As shown in the toy example of Figure 2.13, the two 0 ties contain r_6 through r_9. Adding r_6 through r_9 to the reference set can only increase the number of concordant pairs, thus increasing $\sum_{i=1}^{n-1}\sum_{j=i+1}^{n} c(r_{k_i}, r_{k_j})$. Moreover, the variance of $\sum_{i=1}^{n-1}\sum_{j=i+1}^{n} c(r_{k_i}, r_{k_j})$ under the null hypothesis is relatively reduced (Eq. (2.16)). These two factors tend to lead to an overestimated z-score. Therefore, given two events a and b, we treat $V_{a\cup b}^h$ as the set of all reference nodes for assessing level-h TESC between a and b. It means, we should sample reference nodes within $V_{a\cup b}^h$, otherwise we would get out-of-sight nodes. This is different from the spatial point pattern correlation problem where point patterns are assumed to be isotropic, and we can easily identify and focus on regions containing points. In the next section, we study how we can do reference node sampling efficiently.

2.4.3 **Reference Node Sampling**

In this section, we present efficient algorithms for sampling reference nodes from $V_{a\cup b}^h$. We need to know which nodes are within $V_{a\cup b}^h$, but only have $V_{a\cup b}$ in hand. For continuous spaces, we can perform range search efficiently by building R-tree indexes Arge *et al.* (2004) or k-d tree indexes Bentley (1990). However, for graphs it is difficult to build efficient index structures for answering range queries, for example, querying for all nodes in one node's h-vicinity. Pre-computing and storing pair-wise shortest distances is not practical either, since it requires $O(|V|^2)$

	r_1	r_2	r_3	r_4	r_5	r_6	r_7	r_8	r_9
$\mathbf{s}_a$	[0.0,	0.3,	0.1,	0.0,	0.4,	0.0,	0.0,	0.0,	0.0]
$\mathbf{s}_b$	[0.4,	0.6,	0.0,	0.7,	0.8,	0.0,	0.0,	0.0,	0.0]

Figure 2.13 $\mathbf{s}_a$ and $\mathbf{s}_b$ when we incorporate nodes whose h-vicinities do not contain any occurrence of a or b.

storage. In the following section, we first propose an approach which employs BFS to retrieve all nodes in $V^h_{a\cup b}$, and then randomly chooses n nodes from $V^h_{a\cup b}$. Then, we present efficient sampling algorithms which avoid enumerating all nodes in $V^h_{a\cup b}$. Finally, we analyze time complexity of these algorithms.

2.4.3.1 Batch_BFS

The most straightforward method for obtaining a uniform sample of reference nodes is to first obtain $V^h_{a\cup b}$, and then simply sample from it. $V^h_{a\cup b}$ can be obtained by performing a h-hop BFS search from each node $v \in V_{a\cup b}$ and doing set unions. However, this strategy would perform poorly since the worst case-time complexity is $O(|V_{a\cup b}|(|V| + |E|))$. The problem is that the h-vicinities of nodes in $V_{a\cup b}$ could have many overlaps. Therefore, we adopt a variant of h-hop BFS search which starts with all nodes in $V_{a\cup b}$ as source nodes. For clarity, we show the algorithm Batch_BFS in Algorithm 2.2. It is similar to the h-hop BFS algorithm for one source node, except that the queue *Queue* is initialized with a set of nodes. The correctness of Batch_BFS can be easily verified by imagining that we do a $(h + 1)$-hop BFS from a virtual node which is connected to all nodes in $V_{a\cup b}$. By means of Batch_BFS, the worst case-time complexity is reduced from $O(|V_{a\cup b}|(|V| + |E|))$ to $O(|V| + |E|)$, which means for each node in the graph, we do adjacency list examination utmost once. As we will show in experiments, though simple, Batch_BFS is a competitive method for reference node selection.

Algorithm 2.2: Batch_BFS

```
Input: Adj_lists: Adjacency lists for all nodes in G, V_{a∪b}: the set of all event nodes,
       h: # of hops
Output: V_out: all nodes in h-vicinity of V_{a∪b}
begin
    Initialize V_out = ∅.
    Initialize queue Queue with all v ∈ V_{a∪b} and set v.depth = 0.
    while Queue is not empty do
        v = Dequeue(Queue)
        foreach u in Adj_lists(v) do
            if u ∉ V_out and u ∉ Queue then
                u.depth = v.depth + 1
                if u.depth ⩾ h then
                    | V_out = V_out ∪ {u}
                else
                    | Enqueue(Queue,u)
                end
            end
        end
        V_out = V_out ∪ {v}
    end
end
```

Procedure RejectSamp($V_{a\cup b}$)

1 Select a node $v \in V_{a\cup b}$ with probability $|V_v^h|/N_{sum}$.
2 Sample a node u from V_v^h uniformly.
3 Get the number of event nodes in u's h-vicinity: $c = |V_u^h \cap V_{a\cup b}|$.
4 Flip a coin with success probability $\frac{1}{c}$. Accept u if we succeed, otherwise a failure occurs.

2.4.3.2 Importance Sampling

Though Batch_BFS algorithm is efficient in that its worst case-time cost is *linear* in the number of nodes plus the number of edges in the graph, it still enumerates all N reference nodes. In practice, the sample size n is usually much smaller than N and can be treated as a constant since we can fix n for testing different pairs of events. Hence, the question is, can we develop reference node selection algorithms which have time costs depending on n, rather than N?

The idea is that we directly sample reference nodes without first enumerating the whole set of reference nodes. It is challenging since we want to sample from the uniform probability distribution over $V_{a\cup b}^h$, but only have $V_{a\cup b}$ in hand. The basic operation is randomly picking an event node in $V_{a\cup b}$ and peeking at its h-vicinity. It is not easy to achieve uniform sampling. On one hand, the h-vicinities of event nodes could have many overlapped regions, as illustrated by Figure 2.14. The circles represent h-vicinities of the corresponding nodes and the shadowed regions are overlaps. Nodes in the overlapped regions are easier to be selected if we sample nodes uniformly from a random event node's h-vicinity. On the other hand, different nodes have h-vicinities with different node set sizes, that is, $|V_v^h|$, conveyed by the circle sizes in Figure 2.14. If we pick event nodes uniformly at random, nodes in small circles tend to have higher probabilities to be chosen.

We can use rejection sampling (Gilks and Wild, 1992) to achieve uniform sampling in $V_{a\cup b}^h$, if we know $|V_v^h|$ for each $v \in V_{a\cup b}$. Let $N_{sum} = \sum_{v\in V_{a\cup b}} |V_v^h|$ be the sum of node set sizes of all event nodes' h-vicinities. It is easy to verify $N_{sum} \geq N$ due to the overlaps. The sampling procedure is shown in Procedure RejectSamp. Proposition 2.1 shows that `RejectSamp` generates samples from the uniform probability distribution over $V_{a\cup b}^h$. $|V_v^h|$'s ($h = 1, \ldots, h_m$) can be pre-computed offline by doing a h_m-hop BFS from each node in the graph. The space cost is only $O(|V|)$ for each vicinity level and once we obtain the index, it can be efficiently

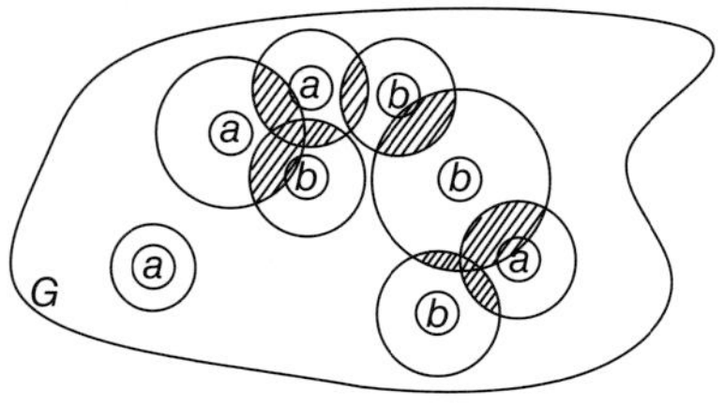

Figure 2.14 h-vicinities of event nodes.

updated as the graph changes. The time cost depends on $|V|$ and the average size of node h_m-vicinities, that is, average $|V_v^{h_m}| + |E_v^{h_m}|$. Fortunately, we do not need to consider too-high values of h since (i) correlations of too-broad scales usually do not convey useful information and (ii) in real networks like social networks, increasing h would quickly let a node's h-vicinity cover a large fraction of the network due to the "small world" phenomenon of real-life networks (Albert and Barabási, 2002). Therefore, we focus on relatively small h values, such as $h = 1, 2, 3$.

Proposition 2.1 `RejectSamp` *generates each node in* $V^h_{a\cup b}$ *with equal probability.*

Proof. Consider an arbitrary node $u \in V^h_{a\cup b}$. In step 2 of `RejectSamp`, u has a chance to be sampled if a node $v \in V_u^h \cap V_{a\cup b}$ is selected in step 1. Thus, the probability that u is generated after step 2 is $\sum_{v\in V_u^h \cap V_{a\cup b}} \frac{|V_v^h|}{N_{sum}} \times \frac{1}{|V_v^h|} = \frac{|V_u^h \cap V_{a\cup b}|}{N_{sum}}$. This is a non-uniform probability distribution over $V^h_{a\cup b}$. Then, by the discount in step 4, u is finally generated with probability $\frac{1}{N_{sum}}$, which is independent of u. □

Each run of `RejectSamp` incurs a cost of two h-hop BFS searches (steps 2 and 3). Simply repeating `RejectSamp` until n reference nodes are obtained will generate a uniform sample of reference nodes. However, each run of `RejectSamp` could fail. The success probability of a run of `RejectSamp` is $p_{\text{succ}} = N/N_{sum}$, which can be easily derived by aggregating success probabilities of all nodes in $V^h_{a\cup b}$. When there is no overlap among event nodes' h-vicinities, $p_{\text{succ}} = 1$ since $N_{sum} = N$. The expected time cost in terms of h-hop BFS is $2n/p_{\text{succ}}$. It means, the heavier the overlap among the different event nodes' h-vicinities is, the higher the cost is. Considering the "small world" property of real-life networks (Albert and Barabási, 2002), it would be easy to get a heavy overlap as $V_{a\cup b}$ and h grow. Preliminary experiments confirm `RejectSamp` is inefficient.

We propose a weighting technique to address the earlier problem. The idea is similar to the importance sampling (Hesterberg, 1995). In particular, we use the same sampling scheme with `RejectSamp` except that we do not reject any sampled nodes. This leads to samples generated from the non-uniform distribution $\mathbb{P} = \{p(v)\}_{v\in V^h_{a\cup b}}$, where $p(v) = |V_v^h \cap V_{a\cup b}|/N_{sum}$. Notice that $t(a, b)$ is intrinsically an estimator of the real correlation score $\tau(a, b)$. The idea is, if we can derive a proper estimator for $\tau(a, b)$ based on samples from $\mathbb{P}$, we could use it as a surrogate to $t(a, b)$. Let $S = \{(r_1, w_1), \dots, (r_n, w_n)\}$ be a set consisting of n distinct reference nodes sampled from $\mathbb{P}$, where w_i is the number of times r_i is generated in the sampling process. We denote the sample size of S as $n' = \sum_{i=1}^n w_i$. We define a new estimator for $\tau(a, b)$ based on S

$$\tilde{t}(a, b) = \frac{\sum_{i=1}^{n-1} \sum_{j=i+1}^{n} c(r_i, r_j) \frac{w_i w_j}{p(r_i)p(r_j)}}{\sum_{i=1}^{n-1} \sum_{j=i+1}^{n} \frac{w_i w_j}{p(r_i)p(r_j)}}. \tag{2.18}$$

This estimator is a consistent estimator of $\tau(a, b)$, which is proved in Theorem 2.5.

Theorem 2.5 *$\tilde{t}(a,b)$ is a consistent estimator of $\tau(a,b)$.*

Proof. To prove $\tilde{t}(a,b)$ is a consistent estimator for $\tau(a,b)$, we need to show that $\tilde{t}(a,b) \xrightarrow{P} \tau(a,b)$, that is, $\tilde{t}(a,b)$ converges to $\tau(a,b)$ in probability as the sample size $n' \to \infty$. For each r_i, we define a Bernoulli random variable X_{r_i} which is 1 if a run of sampling from $\mathbb{P}$ outputs node r_i, and 0 otherwise. $\frac{w_i}{n'}$ is the sample mean for X_{r_i}. By the Law of Large Numbers, $n' \to \infty$, $\frac{w_i}{n'}$ converges in probability to the expectation $E(X_{r_i}) = p(r_i)$. Moreover, all nodes in $V^h_{a\cup b}$ will be added into S when $n' \to \infty$, which means $n = N$. Therefore, as $n' \to \infty$, we can obtain:

$$\tilde{t}(a,b) = \frac{\frac{1}{n'^2}\sum_{i=1}^{N-1}\sum_{j=i+1}^{N} c(r_i,r_j)\frac{w_i w_j}{p(r_i)p(r_j)}}{\frac{1}{n'^2}\sum_{i=1}^{N-1}\sum_{j=i+1}^{N} \frac{w_i w_j}{p(r_i)p(r_j)}}$$

$$= \frac{\sum_{i=1}^{N-1}\sum_{j=i+1}^{N} c(r_i,r_j)\frac{p(r_i)p(r_j)}{p(r_i)p(r_j)}}{\sum_{i=1}^{N-1}\sum_{j=i+1}^{N} \frac{p(r_i)p(r_j)}{p(r_i)p(r_j)}} = \tau(a,b),$$

which completes the proof. □

It is easy to verify that $\tilde{t}(a,b)$ is a biased estimator by considering a toy problem and enumerating all possible outputs of a sample of size n' (together with their probabilities) to compute $E(\tilde{t}(a,b))$. However, unbiasedness used to receive much attention but nowadays is considered less important (Wasserman, 2004). We will empirically demonstrate that $\tilde{t}(a,b)$ can achieve acceptable performance in the experiments. For clarity, we show the Importance sampling algorithm in Algorithm 2.3. In each iteration of the sampling loop, the major cost is one h-hop BFS

Algorithm 2.3: Importance sampling

Input: $V_{a\cup b}$: the set of all event nodes, $|V^h_v|$: h-vicinity node set sizes for all $v \in V_{a\cup b}$, h: # of hops
Output: S: a set of n sampled reference nodes, W: the set of weights (frequencies) for each $r \in S$

```
begin
    Initialize S = ∅.
    while |S| < n do
        Randomly select a node v ∈ V_{a∪b} with probability |V^h_v|/N_sum.
        Do a h-hop BFS search from v to get V^h_v and sample a node r from V^h_v
        uniformly.
        if r ∈ S then
            W(r) = W(r) + 1
        else
            S = S ∪ {r}
            W(r) = 1
        end
    end
end
```

search (line 5). The number of iterations n', though $\geq n$, is typically $\approx n$ in practice. This is because when N is large, the probability of selecting the same node in different iterations is very low. Thus, the major cost of Importance sampling could be regarded as depending on n. Once S and W are obtained, we can then compute $\tilde{t}(a, b)$ as a surrogate for $t(a, b)$ and assess the significance accordingly.

Improving Importance Sampling Although the time cost of Importance sampling depends on n rather than N, in practice n h-hop BFS searches could still be slower than one Batch_BFS search as h increases. This is because the overlap among different event nodes' h-vicinities tends to become heavier as h increases. We can alleviate this issue by sampling reference nodes in a *batch* fashion. That is, when V_v^h is obtained for a sampled $v \in V_{a\cup b}$ (line 5 of Algorithm 2.3), we sample more than one reference nodes from V_v^h. In this way, the ratios between different reference nodes' probabilities of being chosen do not change. However, this also introduces dependence into S. Sampling too many nodes from one V_v^h would degrade the performance since the number of event nodes peeked at decreases and consequently, we are more likely to be trapped in local correlations. This is a tradeoff between efficiency and accuracy. We will test this approximation idea in experiments.

2.4.3.3 Global Sampling in Whole Graph

When $|V_{a\cup b}|$ and h increase, the chance that a random node selected from the whole graph is in $V_{a\cup b}^h$ also increases. In this situation, we can simply sample nodes uniformly in the whole graph and the obtained nodes which are within $V_{a\cup b}^h$ can be regarded as a uniform sample from $V_{a\cup b}^h$. We use an iterative process to harvest reference nodes: (i) First a node is chosen uniformly from the whole graph; (ii) test whether the selected node is within $V_{a\cup b}^h$; (iii) if it is in $V_{a\cup b}^h$, keep it. (iv) Another node is selected uniformly from the remaining nodes and we go to step 2. This process continues until n reference nodes are collected. For completeness, the Whole graph-sampling algorithm is shown in Algorithm 2.4. The major cost is incurred by one h-hop BFS search in each iteration (line 5), where the purpose is to examine whether v is an eligible reference node.

2.4.3.4 Complexity Analysis

The major space cost is $O(|E|)$, for storing the graph as adjacency lists. Regarding time complexity, we have mainly three phases: Reference node sampling, event density computation (Eq. (2.12)), and measure computation (z-score, Eq. (2.17)). Let c_B be the average cost of one h-hop BFS search on graph G, which is linear in the average size of node h-vicinities, that is, average $|V_v^h| + |E_v^h|$. Let n be the number of sample reference nodes. The event density computation for a reference node has time complexity $O(c_B)$. The cost of z-score computation is $O(n^2)$. Fortunately, we do not need to select too many reference nodes, as discussed in Section 2.4.2.1. We will demonstrate the efficiency of the earlier two phases in the experiments.

Algorithm 2.4: Whole graph sampling

Input: $V_{a\cup b}$: the set of all event nodes, V: all nodes in graph G
Output: $\mathcal{S}$: a set of n sampled reference nodes
begin
  $\mathcal{S} = \emptyset$
  while $|\mathcal{S}| < n$ **do**
    Randomly pick a node $v \in V$
    Do a h-hop BFS search from v to get V_v^h
    if $V_v^h \cap V_{a\cup b} \neq \emptyset$ **then**
      $\mathcal{S} = \mathcal{S} \cup \{v\}$
    end
    $V = V - \{v\}$
  end
end

For reference node sampling, we have three methods. The time complexity of Batch_BFS is $O(|V_{a\cup b}^h| + |E_{a\cup b}^h|)$ where $|V_{a\cup b}^h| = N$. The cost of Importance sampling is $O(nc_B)$. For Whole-graph sampling, the time cost is $O(n_f c_B)$, where n_f is the number of nodes examined, which are not in $V_{a\cup b}^h$. The cost incurred by examined nodes which are in $V_{a\cup b}^h$ is counted in the event density computation phase. n_f is a random variable. Treating Whole graph sampling as sampling with replacement, the probability of selecting a node in $V_{a\cup b}^h$ in each iteration is $N/|V|$. The expected total number of iterations is $n|V|/N$ and therefore $E(n_f) = n|V|/N - n$. When N is small, Batch_BFS can be used. For large N, Importance sampling and Whole-graph sampling are better candidates. We will empirically analyze their efficiency in the experiments.

2.4.4 Experiments

This section presents the experimental results of applying our proposed TESC testing framework on several real-world graph datasets. First, we verify the efficacy of the proposed TESC testing framework by event simulation on the DBLP graph. Then, we examine the efficiency and scalability of the framework with a Twitter network. The third part of the experiments, concentrates on analyzing highly correlated real-event pairs discovered by our measure in real-graph datasets. All experiments are run on a PC with Intel Core i7 CPU and 12GB memory. The source code of our TESC estimation framework can be downloaded at *http://grafia.cs.ucsb.edu/strucorr/*.

2.4.4.1 Graph Datasets

We use three datasets to evaluate our TESC testing framework: DBLP, Intrusion, and Twitter.

DBLP The DBLP dataset was downloaded on Oct. 16th, 2010 (*http://www.informatik.uni-trier.de/~ley/db*). Its paper records were parsed to obtain the coauthor social network. Keywords in the paper titles are treated as events associated with nodes (authors) on the graph. The DBLP graph contains 964 677 nodes and 3 547 014 edges. Totally, it has around 0.19 million keywords.

Intrusion The Intrusion dataset was derived from the log data of intrusion alerts in a computer network. It has 200 858 nodes and 703 020 edges. There are 545 different types of alerts which are treated as events in this network.

Twitter The Twitter dataset has 20 million nodes and 0.16 billion edges, which is a bidirectional subgraph of the whole twitter network (*http://twitter.com*). We do not have events for this dataset. It is used to test the scalability of the proposed TESC testing framework.

2.4.4.2 Event Simulation Methodology

A suitable method for evaluating the efficacy of our approach is to simulate correlated events on graphs and see if we can correctly detect correlations. Specifically, we adopt similar methodologies as those used in the analogous point pattern problem (Diggle and Cox, 1983) to generate pairs of events with positive and negative correlations on graphs. The DBLP network is used as the test bed. We investigate correlations with respect to different vicinity levels $h = 1, 2, 3$. Positively correlated event pairs are generated in a linked pair fashion: We randomly select 5000 nodes from the graph as event a, and each node $v \in V_a$ has an associated event b node whose distance to v is described by a Gaussian distribution with mean zero and variance equal to h (distances go beyond h are set to h). When the distance is decided, we randomly pick a node at that distance from v as the associated event b node. This represents strong positive correlations since wherever we observe an event a, there is always a nearby event b. For negative correlation, again we first generate 5000 event a nodes randomly, after which we employ Batch_BFS to retrieve the nodes in the h-vicinity of V_a, that is, V_a^h. Then, we randomly color 5000 nodes in $V \setminus V_a^h$ as having event b. In this way, every node of b is kept at least $h + 1$ hops away from all nodes of a and the two events exhibit a strong negative correlation. For each vicinity level, we generate 100 positive event pairs and 100 negative event pairs from the simulation processes, respectively. We use *recall* as the evaluation metric which is defined as the number of correctly detected event pairs divided by the total number of event pairs (100). We report results obtained from one-tailed tests with significance level $\alpha = 0.05$. In our experiments, we empirically set the sample size of reference nodes $n = 900$.

2.4.4.3 Performance Comparison

We investigate the performance of three reference node sampling algorithms, namely, Batch_BFS, Importance sampling, and Whole-graph sampling, under different vicinity levels and different noise levels. Noises are introduced as follows. Regarding positive correlation, we introduce a sequence of independent

Bernoulli trails, one for each linked pair of event nodes, in which with probability p the pair is broken and the node of b is relocated outside V_a^h. For negative correlation, given an event pair each node in V_b has probability p to be relocated and attached with one node in V_a. The probability p controls to what extent noises are introduced and can be regarded as noise level.

We show the experimental results in Figures 2.15 and 2.16, for positive correlation and negative correlation, respectively. As can be seen, overall the performance curves start from 100% and fall off as the noise level increases. This indicates that the proposed statistical testing approach is efficacious for measuring TESC. Among the three reference node sampling algorithms, Batch_BFS achieves relatively better performance. Importance sampling, though not as good as Batch_BFS, can also achieve acceptable recall, especially for $h = 1, 2$. We shall show in Section 2.4.4.6 that Importance sampling is more efficient than Batch_BFS in many cases. Whole-graph sampling also shows good recall in most cases, as expected. However, its running time can vary drastically and therefore it can only be applied in limited scenarios. An interesting phenomenon is that positive correlations for higher vicinity levels (e.g., 3) are harder to break than those for lower levels, while for negative correlations it is the reverse: Lower level ones are harder to break. Note that the noise level ranges in figures of Figures 2.15 and 2.16 are not the same.

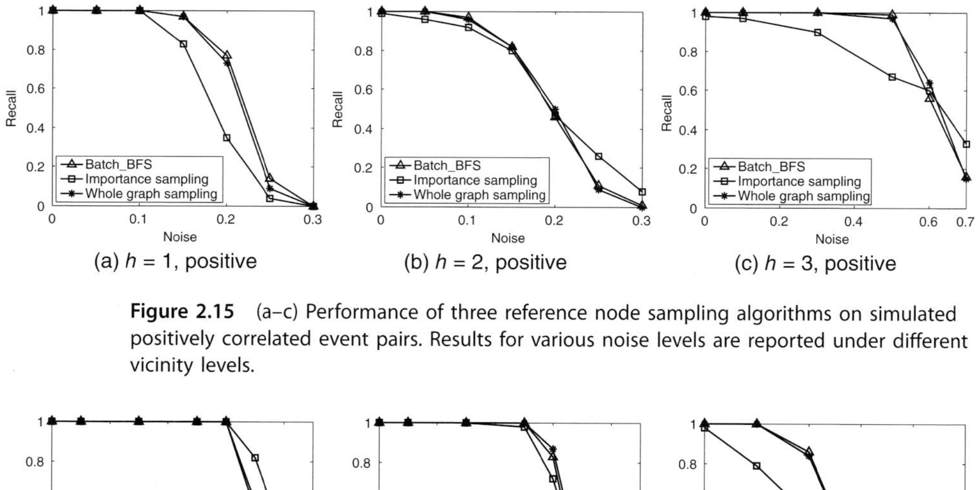

(a) $h = 1$, positive (b) $h = 2$, positive (c) $h = 3$, positive

Figure 2.15 (a–c) Performance of three reference node sampling algorithms on simulated positively correlated event pairs. Results for various noise levels are reported under different vicinity levels.

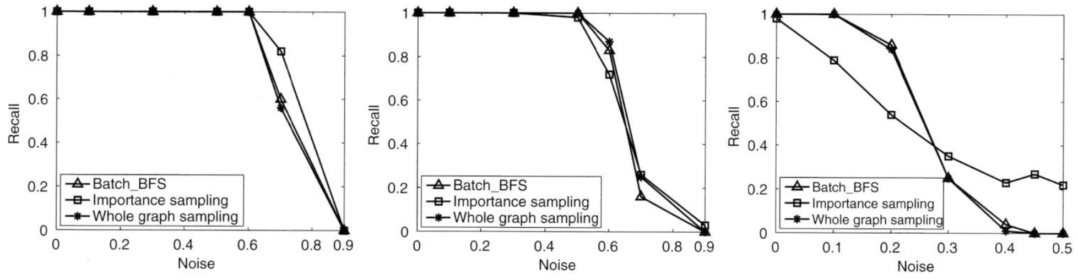

(a) $h = 1$, negative (b) $h = 2$, negative (c) $h = 3$, negative

Figure 2.16 (a–c) Performance of three reference node sampling algorithms on simulated negatively correlated event pairs. Results for various noise levels are reported under different vicinity levels.

This is intuitive. Consider the size of V_a^h. When h increases, $|V_a^h|$ usually increases exponentially. For example, among our synthetic events in DBLP graph, the typical size of V_a^1 is 60k, while that of V_a^3 is 700k (7/10 of the whole graph), for $|V_a| = 5000$. Hence, it is much harder for event b to "escape" event a for the higher vicinity levels. On the contrary, for $h = 1$ it is easier to find a node whose 1-vicinity does not even overlap with V_a^1. Hence, low vicinity level positive correlations and high vicinity level negative correlations are hard to maintain and consequently more interesting than those in other cases. In the following experiment on real events, we will focus on these interesting cases.

2.4.4.4 Batch Importance Sampling

In Importance Sampling, when V_v^h is obtained for a sampled $v \in V_{a \cup b}$ (line 5 of Algorithm 2.3), we could sample more than one node from V_v^h as reference nodes, in order to reduce the cost. However, sampling too many nodes from one V_v^h would degrade the performance since the number of event nodes peeked at decreases and consequently, we are more likely to be trapped in local correlations. Here, we present an empirical evaluation of this idea for $h = 2, 3$. We show results for four synthetic event pair sets in Figure 2.17. Two of those sets contain noises since in the corresponding cases, the correlation is hard to break, which means in those cases it is easy to detect the correlations. We can see that the results are as expected. The performance curves for $h = 3$ can keep high for a longer range of the number of reference nodes sampled from each V_v^h, compared to $h =$ 2. This is because 3-vicinities are usually much larger than 2-vicinities and 3-vicinities of event nodes tend to have more overlapped regions. Therefore, sampling a batch of reference nodes from 3-vicinities is less likely to be trapped in local correlations than from 2-vicinities. The results also indicate that we can sample a small number of reference nodes from each V_v^h for Importance sampling, without

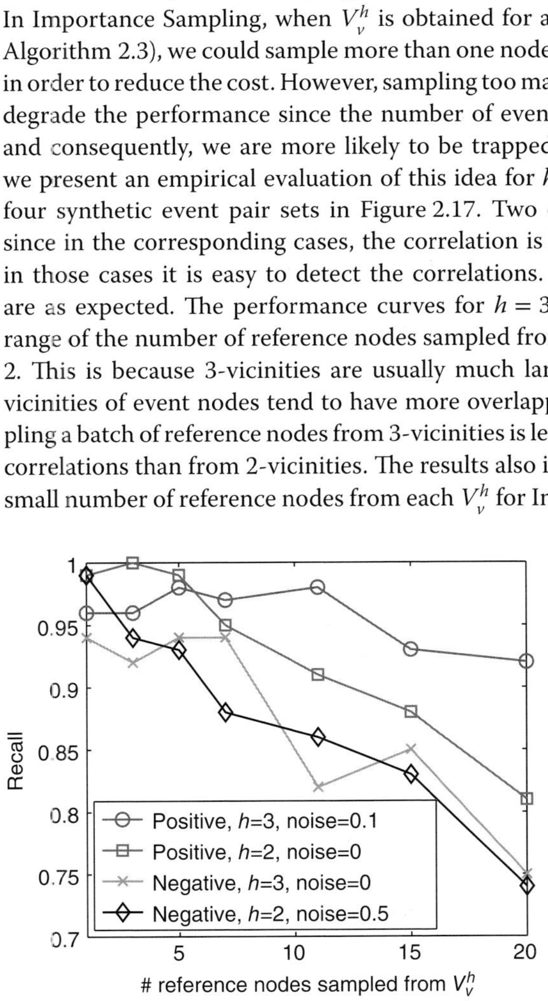

Figure 2.17 Performance of sampling different number of reference nodes from each V_v^h for Importance sampling. (See inset for color representation of this figure.)

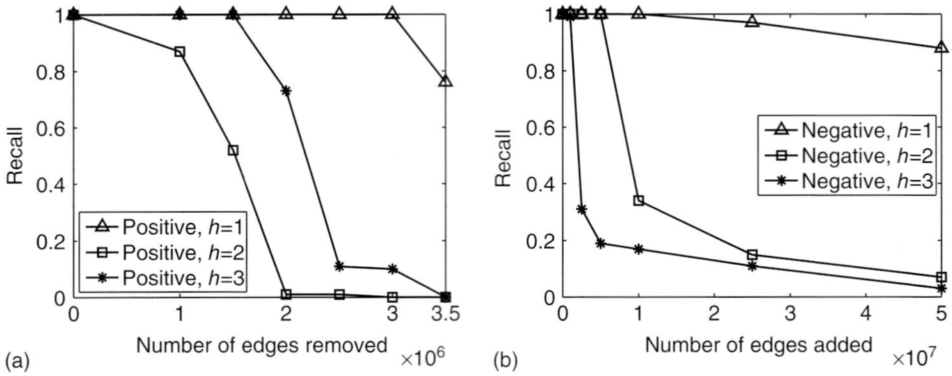

Figure 2.18 Impact of randomly removing or adding edges on the correlation results.

severely affecting its performance. In the following efficiency experiments, we set this number to 3 and 6 for $h = 2$ and $h = 3$, respectively.

2.4.4.5 Impact of Graph Density

We change the graph density to see the impact on the correlation results. Specifically, we alter the DBLP graph by randomly adding/removing edges and run Batch_BFS for the six-event pair sets (without noises) generated in Section 2.4.4.2. Figure 2.18 shows the results. We can see when removing the edges, the recall of positive pairs decreases, while adding edges leads to recall decline of negative pairs. In the remaining cases (e.g., negative pairs vs edge removal), the recall remains at 1. This is because removing the edges tends to increase distances among nodes, while adding edges makes nodes near one another. Figure 2.18a shows that 1-hop positive correlations are less influenced by edge removal, which is different from the observation in Section 2.4.4.3, that is, 1-hop positive correlations are easier to break. The reason is that in our correlation simulation model, 1-hop positive-event pairs tend to have more nodes with both events, due to the Gaussian distributed distances between event b nodes and corresponding event a nodes. Nodes with both events reflect a TC which is not influenced by the edge removal. However, TESC does not just measure the TCs. We will show in Section 2.4.4.7 there are real-event pairs which exhibit high positive TESC but are independent or even negatively correlated by the TC.

2.4.4.6 Efficiency and Scalability

We test efficiency and scalability of our TESC testing framework on Twitter graph. First, we investigate the running time of different reference node sampling algorithms with respect to the number of event nodes, that is, the size of $V_{a\cup b}$. In particular, we randomly pick nodes from the Twitter graph to form $V_{a\cup b}$ with sizes ranging from 1000 to 500 000. Then, each algorithm is run to generate sample reference nodes for these $V_{a\cup b}$'s in order to record its running time. The results are averaged over 50 test instances for each size of $V_{a\cup b}$. Figure 2.19 shows the results

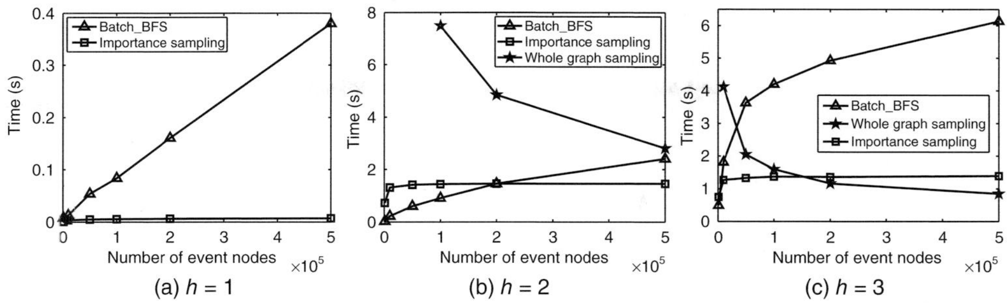

Figure 2.19 Running time of reference node sampling algorithms with increasing number of event nodes.

for the three vicinity levels. To keep the figures clear, we do not show the running time of Whole-graph sampling for some cases since its running time goes beyond 10 s. We can see that for different vicinity levels, the situations are quite different. Generally speaking, the running time of BFS increases significantly as $V_{a\cup b}$ grows, while that of Importance sampling hardly increases. This is consistent with our analysis in Section 2.4.3.4. The running time of Importance sampling increases a little, in that, the algorithm tends to choose event nodes with large V_v^h to peek in the sampling loop. By chance, there would be more and more event nodes with large sizes of V_v^h as $V_{a\cup b}$ grows. We can see the Importance sampling is definitely more efficient than Batch_BFS when $h = 1$. For $h = 2$ and 3, when the size of $V_{a\cup b}$ is small, we can use Batch_BFS; for large sizes of $V_{a\cup b}$, Importance sampling is a better choice. Whole-graph sampling is recommended only for $h = 3$ and for large sizes of $V_{a\cup b}$ (above 200k in the case of Twitter graph). To conclude, the results indicate our reference sampling algorithms are efficient and scalable, that is, we can process $V_{a\cup b}$ with 500K nodes on a graph with 20M nodes in 1.5 s.

Besides the reference node sampling, the TESC testing framework also needs to do one h-hop BFS search for each sample reference node to compute event densities and then calculate $z(a, b)$. Figure 2.20 shows that these two operations are

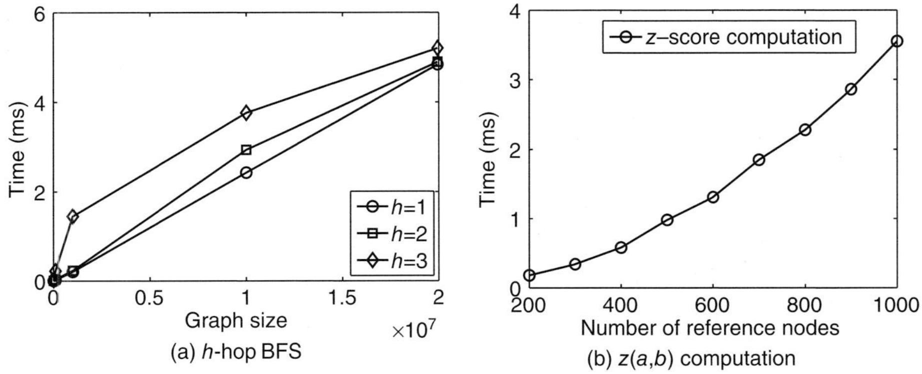

Figure 2.20 Running time of one h-hop BFS search and $z(a, b)$ computation.

also efficient and scalable. Figure 2.20a indicates that on a graph with 20 million nodes, one 3-hop BFS search needs only 5.2 ms, which is much faster than the state-of-art hitting time approximation algorithm (170 ms for 10 million nodes) (Guan *et al.*, 2011). Efficiency is the major reason that we choose this simple density measure, rather than more complicated proximity measures such as hitting time. On the other hand, although the measure computation has time complexity $O(n^2)$, we do not need to select too many reference nodes since the variance of $t(a,b)$ is upper bounded by $\frac{2}{n}(1-\tau(a,b)^2)$ (Kendall and Gibbons, 1990), regardless of N. Figure 2.20b shows we can compute $z(a,b)$ in 4 ms for 1000 reference nodes.

2.4.4.7 Real Events

We provide case studies of applying our TESC testing framework on real events occurring in real graphs. We use Batch_BFS for the reference node selection. As aforementioned in Section 2.4.4.3, low level positive correlations and high level negative correlations are of interest. Hence, we report typical highly correlated event pairs we found in DBLP and Intrusion datasets in terms of 1-hop positive TESC and 3-hop negative TESC respectively. We report z scores as the significance scores of the correlated event pairs. To give a notion of the correspondence between z scores and p values, a z score > 2.33 or < -2.33 indicates the corresponding p value < 0.01 for one-tailed significance testing. Before presenting the results, we would like to emphasize that our correlation findings are for specific networks and our measure detects the exhibition of correlation, but not its cause.

Tables 2.4 and 2.5 show the results for DBLP. For comparison, we also show correlation scores measured by treating nodes as isolated transactions. We use Kendall's τ_b (Agresti, 2007) to estimate the TC since τ_b can capture both the positive and the negative correlations. All scores in the tables are z scores. We can see that highly positively correlated keywords are semantically related and reflect hot research areas in different communities of computer science, while negatively correlated ones represent topics which are far away from each other. In DBLP, keyword pairs having positive TESC tend to also have positive TC. However, for the negative case, the results are not consistent. We can see in Table 2.5 many

Table 2.4 Five keyword pairs exhibiting high 1-hop positive correlation (DBLP). All scores are *z*-scores.

#	Pair	TESC			TC
		$h = 1$	$h = 2$	$h = 3$	
1	Texture versus image	6.22	19.85	30.58	172.7
2	Wireless versus sensor	5.99	23.09	32.12	463.7
3	Multicast versus network	4.21	18.37	26.66	123.2
4	Wireless versus network	2.06	17.41	27.90	198.2
5	Semantic versus RDF	1.72	16.02	24.94	120.3

Table 2.5 Five keyword pairs exhibiting high 3-hop negative correlation (DBLP). All scores are *z*-scores.

#	Pair	TESC			TC
		$h = 1$	$h = 2$	$h = 3$	
1	Texture versus Java	−23.63	−9.41	−6.40	4.33
2	GPU versus RDF	−24.47	−14.64	−6.31	1.24
3	SQL versus calibration	−21.29	−12.70	−5.45	−0.62
4	Hardware versus ontology	−22.31	−8.85	−5.01	3.38
5	Transaction versus camera	−22.20	−7.91	−4.26	4.85

Table 2.6 Five alert pairs exhibiting high 1-hop positive correlation (Intrusion). All scores are *z*-scores.

#	Pair	TESC ($h = 1$)	TC
1	Ping_Sweep versus SMB_Service_Sweep	13.64	1.91
2	Ping_Flood versus ICMP_Flood	12.53	5.87
3	Email_Command_Overflow versus Email_Pipe	12.15	−0.04
4	HTML_Hostname_Overflow versus HTML_NullChar_Evasion	9.08	0.59
5	Email_Error versus Email_Pipe	4.34	−3.52

pairs have positive TC. It means although some authors have used both the keywords, they are far away in the graph space, reflecting the fact that they represent quite different topics pursued by different communities in the coauthor social network.

Results for the Intrusion dataset are presented in Tables 2.6 and 2.7. Since the Intrusion graph contains several nodes with very high degrees (around 50k), its diameter is much lower than that of DBLP. In the Intrusion graph, 2-vicinity of a node tends to cover a large number of nodes. Therefore, for negative TESC we focus on $h = 2$. As shown in Table 2.6, positively correlated alerts reflect high-level intrusion activities. The first pair reflects pre-attack probes. The second one is related to ICMP DOS Attack. The third and fifth pairs indicate that the attacker is trying to gain root access of those hosts by vulnerabilities in email softwares and services. The fourth one is related to Internet Explorer's vulnerabilities. Notice that the third pair is nearly independent and the fifth pair is negatively correlated under TC. The reason could be that some attacking techniques consume bandwidth and there is a tradeoff between the number of hosts attacked and the number of techniques applied to one host. Attackers might choose to maximize coverage by alternating related intrusion techniques for hosts in a subnet, in order to increase the chance of success. Although these alerts represent related techniques, they do not exhibit positive TC. TESC can detect such positive structural correlations.

On the other hand, the highly negatively correlated alerts are those related to different attacking approaches, or in connection with different platforms. For

Table 2.7 Five alert pairs exhibiting high 2-hop negative correlation (Intrusion). All scores are z-scores.

#	Pair	TESC ($h = 2$)	TC
1	Audit_TFTP_Get_Filename versus LDAP_Auth_Failed	−31.30	−0.81
2	LDAP_Auth_Failed versus TFTP_Put	−31.12	−0.81
3	DPS_Magic_Number_DoS versus HTTP_Auth_TooLong	−30.96	−0.18
4	LDAP_BER_Sequence_Dos versus TFTP_Put	−30.30	−1.57
5	Email_Executable_Extension versus UDP_Service_Sweep	−26.93	−0.97

Table 2.8 Two rare alert pairs with positive 1-hop TESC which are not discovered by proximity pattern mining.

Pair (count)	z-score/p-value
HTTP_IE_Script_HRAlign_Overflow (16) versus HTTP_DotDotDot (29)	3.30/0.0005
HTTP_ISA_Rules_Engine_Bypass (81) versus HTTP_Script_Bypass (12)	2.52/0.0059

example, in the first pair of Table 2.7 LDAP_Auth_Failed is related to brute-force password guessing, while Audit_TFTP_Get_Filename is related to TFTP Attack which allows remote users to write files to the target system without any authentication; in the third pair, DPS_Magic_Number_DoS is exclusive for Microsoft Dynamics GP software, while HTTP_Auth_TooLong is for Netscape Enterprise Server software. These pairs also exhibit moderate negative TC.

We also compare our results with those produced by the proximity pattern mining problem (Khan, Yan, and Wu, 2010) for the positive case. Specifically, we set $minsup = 10/|V|$ for the pFP algorithm and $\alpha = 1$, $\epsilon = 0.12$ (Khan, Yan, and Wu, 2010). Then, we run the proximity pattern mining method on the Intrusion dataset. From the results, we find that most highly positively correlated pairs detected by TESC are also reported as proximity patterns or subsets of proximity patterns. However, some rare event pairs detected by TESC are not discovered by the proximity pattern mining method. Table 2.8 shows two such examples. Digits in parentheses are event sizes. The reason is that proximity pattern mining is intrinsically a frequent pattern mining problem (Khan, Yan, and Wu, 2010). It requires events to occur not only closely but also frequently closely on the graph. In TESC there is no such requirement and we could detect positively correlated rare event pairs.

2.4.5 Discussions

A straightforward measure for TESC could be to calculate the average distance between nodes of the two events. Measures of this kind try to capture the "distance" between the two events directly. However, for these direct measures, it is difficult to estimate their distributions in the null hypothesis (i.e., no correlation). An empirical approach is to use randomization: Perturbing events a and b independently in the graph with the observed sizes and internal structures, and calculating the empirical distribution of the measure. Unfortunately, it is hard to preserve each event's internal structure, thus making randomization not effective. Our approach avoids randomization by indirectly measuring the rank correlation between two events' densities in local neighborhoods of sampled reference nodes. Significance can be estimated by τ's nice property of being asymptotically normal under the null hypothesis. Our approach provides a systematic way to compute formal and rigorous statistical significance, rather than empirical one. Here, another rank correlation statistic, Spearman's ρ (Kendall and Gibbons, 1990), could also be used. We choose Kendall's τ since it can provide an intuitive interpretation and also facilitate the derivation of the efficient importance sampling method.

Another simple idea is that we first map nodes in a graph to a Euclidean space by preserving the structural properties and then apply existing techniques for spatial data. Nevertheless, (i) techniques for spatial data are not scalable; (ii) mapping introduces approximation errors. For example, researchers tried to approximate network distances using a coordinate system (Ng and Zhang, 2002; Zhao *et al.*, 2010). According to the recent work (Zhao *et al.*, 2010), one distance estimation costs 0.2 μs. Let us take the most recent method for spatial data (Smith, 2004) as an example. It requires estimating the distances between each reference point and all event points. Consequently, for 500K event points and 900 reference points, the total time cost is 90 s! Although we could build k-d tree indexes (Bentley, 1990) to improve efficiency, k-d tree only works well for low dimensional spaces. Reducing the dimensionality leads to a higher distance estimation error (Zhao *et al.*, 2010), indicating a tradeoff between accuracy and efficiency. Our method avoids these annoying issues and provides a scalable solution over the exact structure.

How to choose the sample size of reference nodes is a practical issue. While there is no theoretical criterion for choosing a proper sample size, in practice we can do the correlation/independence simulations on a graph (like in Section 2.4.4.2) and choose a large enough sample size so that the recall is above a user-defined threshold, for instance, 0.95. The recall is connected to the type I and type II errors in statistical tests for independence and correlation, respectively.

Our method can assess correlations in different vicinity levels, that is, h. Another scheme could be that we get rid of h by designing a weighted correlation measure where reference nodes closer to event nodes have higher weights. This is challenging since we cannot directly make use of τ's nice property of being asymptotically normal in the null case. Another possible extension is to consider event intensity on nodes, for example, the frequency by which an author used a keyword. We leave these possible extensions for future work.

2.5 Conclusions

In this chapter, we first studied the problem of measuring how strongly an event that took place in a graph is correlated to the graph structure, that is, SSC A novel measure was introduced to assess SSC. It can be used to derive statistical significance to test if an event is randomly distributed over a graph or not. We proposed using hitting time to instantiate our framework and derived a set of sampling and approximation algorithms so that the correlation score can be estimated very quickly in large-scale graphs. By comparing the score with the situation where the event is randomly distributed in the same network, our method is able to discover the events of nodes that are highly correlated with the graph structure. Our method is scalable and successfully applied to the coauthor DBLP network and a social network extracted from TaoBao.com, the largest online shopping network in China, with many exciting discoveries.

Next, we studied the problem of measuring TESC in graphs and proposed a novel measure and an efficient testing framework to address it. Given the occurrences of two events, we choose uniformly a sample of reference nodes from the vicinity of all event nodes and compute for each reference node the densities of the two events in its vicinity respectively. Then, we employ the Kendall's τ rank correlation measure to compute the average concordance of density changes for the two events, over all pairs of reference nodes. The correlation significance can then be assessed by τ's nice property of being asymptotically normal under the null hypothesis. We also proposed three different algorithms for efficiently sampling reference nodes. Finally, experiments on real-graph datasets with both synthetic and real events demonstrated that the proposed TESC testing framework was not only efficacious, but also efficient and scalable.

Acknowledgments

This research was sponsored in part by the U.S. Army Research Laboratory under cooperative agreements W911NF-09-2-0053 and W911NF-11-2-0086, and the National Science Foundation of China under Grant No. 61373118. The views expressed in this chapter are solely attributed to the authors and do not represent the opinions or policies of any of the funding agencies.

References

Agrawal, R., Imieliński, T., and Swami, A. (1993) Mining association rules between sets of items in large databases. *SIGMOD*, pp. 207–216.

Agresti, A. (2007) *An Introduction to Categorical Data Analysis*, Wiley-Blackwell.

Albert, R. and Barabási, A. (2002) Statistical mechanics of complex networks. *Reviews of Modern Physics*, **74**(1), 47–97.

Anagnostopoulos, A., Kumar, R., and Mahdian, M. (2008) Influence and correlation in social networks. SIGKDD, pp. 7–15.

Anselin, L. (1995) Local indicators of spatial association–LISA. *Geographical Analysis,* **27** (2), 93–115.

Arge, L., De Berg, M., Haverkort, H., and Yi, K. (2004) The priority R-tree: a practically efficient and worst-case optimal R-tree. *SIGMOD*, pp. 347–358.

Bao, H. and Chang, E.Y. (2010) Adheat: an influence-based diffusion model for propagating hints to match Ads. WWW 2010, pp. 71–80.

Bentley, J. (1990) K-d trees for semidynamic point sets. *Proceedings of the 6th Annual Symposium on Computational Geometry,* pp. 187–197.

Brand, M. (2005) A random walks perspective on maximizing satisfaction and profit. *SIAM International Conference on Data Mining.*

Brin, S. and Page, L. (1998) The anatomy of a large-scale hypertextual web search engine. *Proceedings of the 7th International Conference on World Wide Web*, pp. 107–117.

Brown, J.J. and Reingen, P.H. (1987) Social ties and word-of-mouth referral behavior. *Journal of Consumer Research,* **14**(3), 350–362.

Chen, W., Wang, C., and Wang, Y. (2010) Scalable influence maximization for prevalent viral marketing in large-scale social networks. *SIGKDD.*

Dehmer, M., Emmert-Streib, F., and Tripathi, S. (2013) Large-scale evaluation of molecular descriptors by means of clustering. *PLoS ONE,* **8** (12), e83956.

Diggle, P. and Cox, T. (1983) Some distance-based tests of independence for sparsely-sampled multivariate spatial point patterns. *International Statistical Review,* **51** (1), 11–23.

Ester, M., Ge, R., Gao, B.J., Hu, Z., and Ben-Moshe, B. (2006) Joint cluster analysis of attribute data and relationship data: the connected K-center problem. *SDM,* pp. 25–46.

Fréchet, M. and Shohat, J. (1931) A proof of the generalized second-limit theorem in the theory of probability. *Transactions of the American Mathematical Society,* **33** (2), 533–543.

Gilks, W. and Wild, P. (1992) Adaptive rejection sampling for Gibbs sampling. *Journal of the Royal Statistical Society Series C (Applied Statistics),* **41** (2), 337–348.

Gionis, A., Mannila, H., Mielikäinen, T., and Tsaparas, P. (2006) Assessing data mining results via swap randomization. *SIGKDD,* pp. 167–176.

Guan, Z., Wu, J., Zhang, Q., Singh, A., and Yan, X. (2011) Assessing and ranking structural correlations in graphs. *SIGMOD,* pp. 937–948.

Guan, Z., Yan, X., and Kaplan, L.M. (2012) Measuring two-event structural correlations on graphs. *Proceedings of the VLDB Endowment,* **5** (11), 1400–1411.

Han, J. and Kamber, M. (2006) *Data Mining: Concepts and Techniques,* Morgan Kaufmann.

Hesterberg, T. (1995) Weighted average importance sampling and defensive mixture distributions. *Technometrics,* **37** (2), 185–194.

Hoeffding, W. (1963) Probability inequalities for sums of bounded random variables. *Journal of the American Statistical Association,* **58** (301), 13–30.

Katz, L. (1953) A new status index derived from sociometric analysis. *Psychometrika,* **18** (1), 39–43.

Kautz, H., Selman, B., and Shah, M. (1997) Referral web: combining social networks and collaborative filtering. *Communications of the ACM,* **40** (3), 63–65.

Kempe, D., Kleinberg, J., and Tardos, E. (2003) Maximizing the spread of influence through a social network. *SIGKDD,* ACM, pp. 137–146.

Kendall, M. and Gibbons, J. (1990) *Rank Correlation Methods,* Edward Arnold.

Khan, A., Yan, X., and Wu, K. (2010) Towards proximity pattern mining in large graphs. *SIGMOD,* pp. 867–878.

La Fond, T. and Neville, J. (2010) Randomization tests for distinguishing social influence and homophily effects. WWW, pp. 601–610.

Lotwick, H. and Silverman, B. (1982) Methods for analysing spatial processes of several types of points. *Journal of the Royal Statistical Society Series B (Methodological),* **44** (3), 406–413.

Lovász, L. (1993) Random walks on graphs: a survey. *Bolyai Society Mathematical Studies,* **32** (2), 1–46.

Mei, Q., Zhou, D., and Church, K. (2008) Query suggestion using hitting time. *CIKM*.

Moser, F., Colak, R., Rafiey, A., and Ester, M. (2009) Mining cohesive patterns from graphs with feature vectors. *SDM*, pp. 593–604.

Ng, T. and Zhang, H. (2002) Predicting internet network distance with coordinates-based approaches. *INFOCOM*, pp. 170–179.

Nowell, D.L. and Kleinberg, J. (2003) The link prediction problem for social networks. *CIKM*, pp. 556–559.

Saigo, H., Nowozin, S., Kadowaki, T., Kudo, T., and Tsuda, K. (2009) gBoost: a mathematical programming approach to graph classification and regression. *Machine Learning*, **75** (1), 69–89.

Sarkar, P. and Moore, A. (2007) A tractable approach to finding closest truncated-commute-time neighbors in large graphs. *UAI*.

Sarkar, P., Moore, A.W., and Prakash, A. (2008) Fast incremental proximity search in large graphs. *ICML*, pp. 896–903.

Silva, A., Meira, W. Jr., and Zaki, M.J. (2010) Structural correlation pattern mining for large graphs. *Proceedinigs of the 8th Workshop on Mining and Learning with Graphs*, pp. 119–126.

Smith, T. (2004) A scale-sensitive test of attraction and repulsion between spatial point patterns. *Geographical Analysis*, **36** (4), 315–331.

Srikant, R. and Agrawal, R. (1995) Mining generalized association rules. *Future Generation Computer Systems*, **13** (2-3), 161–180.

Wasserman, L. (2004) *All of Statistics: A Concise Course in Statistical Inference*, Springer-Verlag.

Zhao, X., Sala, A., Wilson, C., Zheng, H., and Zhao, B. (2010) Orion: shortest path estimation for large social graphs. *WOSN*, pp. 1–9.

Zhou, Y., Cheng, H., and Yu, J. (2009) Graph clustering based on structural/attribute similarities. *Proceedings of the PVLDB*, **2** (1), 718–729.

3
Spectral Graph Theory and Structural Analysis of Complex Networks: An Introduction

Salissou Moutari and Ashraf Ahmed

3.1
Introduction

Recent developments of high-throughput technology in molecular biology have led to an overabundance of data from a variety of simple organisms (Mason and Verwoerd, 2007). The relationships between elements of these organisms are generally described as networks, in which the elements are the vertices (nodes) and the relationships are the edges (links). Typical biological networks at the molecular level include gene regulation networks, signal transduction networks, protein interaction networks, and metabolic networks. Although networks provide a framework to represent high-throughput data from biological experiments from next-generation sequencing technology, extraction of meaningful information from these data poses many scientific challenges, and is still experiencing intensive research effort.

Many research works, during the last two decades, highlighted the potential of graph theory as a natural mathematical tool for investigating and understanding structural properties of complex biological networks (see Mason and Verwoerd 2007 and the references therein for an overview). However, the main challenge posed by a discrete approach such as the graph theory appears to be the explosion of connectivities within the associated network. Although a significant progress has been made in investigating various network properties and characteristics including degree distributions, clustering coefficients, shortest connecting paths, etc., these quantities are still unable to describe satisfactorily the global properties of complex biological networks (Feizi *et al.*, 2013). Recent applications of spectral graph theory to complex networks provide a framework for quantitative modeling, enabling unprecedented insights into the organization and functioning of biological systems. Hence, spectral graph theory is poised to become a promising approach for extracting meaningful connectivities in complex networks such as those encountered in biological systems. Another more interesting aspect of the spectral graph analysis is that, in practice it reduces to an eigenvalue problem, for which various computationally efficient algorithms exist, see for example (Cullum and Willoughby, 2002; Golub and Van Loan, 2013; Saad, 2011) and the references

Computational Network Theory: Theoretical Foundations and Applications, First Edition.
Edited by Matthias Dehmer, Frank Emmert-Streib, and Stefan Pickl.

therein for an overview. However, although significant progress has been made in solving eigenvalue problems, some computational challenges still remain, in particular, for large-scale and non-linear eigenvalue problems (Mehrmann and Voss, 2005).

Our primary objective in this chapter is to provide some basic results from spectral theory relating graph properties and characteristics to their matricial representations, highlighting the potential of spectral graph theory in the analysis of complex graphs, such as biological networks. This chapter aims to be rather introductory and more advanced materials on the topic can be found for example in (Brouwer and Haemers, 2012; Cioaba, 2010; Cvetkovic, Rowkinson, and Simic, 2009; Cvetkovic, Doob, and Sachs, 1995; Hoory, Linial, and Wigderson, 2006; Krivelevich and Sudakov, 2006; van Damand and Haemers, 2009; Kannan and Vempala, 2008), and the references therein. The material discussed in the chapter can be divided naturally into four main sections. The first section of this chapter will primarily be concerned with some graph theoretical basic concepts and definitions, whereas the second section will recall some linear algebra results. The third section will be devoted to the presentation of some results on the characterization of graph properties using spectral theory. In the fourth section, we will discuss some computational challenges associated with spectral graph analysis, whereas the last section presents some concluding remarks.

3.2 Graph Theory: Some Basic Concepts

This section will introduce some basic definitions and fundamental concepts of graph theory as well as the terminologies and notations to be used throughout this chapter. For more detailed discussions regarding the topic, we refer the reader to classical textbooks, such as, (Bollobas, 1998; Berge, 1976; Godsil and Royle, 2001; Harary, 1969; West, 2001).

Graphs can be divided into two main classes: *directed graphs* and *undirected graphs*. Formally, a finite *directed graph* or a *digraph*, denoted $G = (V, E)$, consists of a set of vertices or nodes

$$V = V(G) = \{v_1, v_2, \dots, v_n\},$$

together with an edge set $E = E(G) = \{e_1, e_2, \dots, e_m\}$, where each edge $e_k = (v_i, v_j) \in E$ connects the starting node v_i with the terminal node v_j. When there is no direction associated with the edges, that is, for each edge, $e_k = (v_i, v_j) \in E$, v_i and v_j do not form ordered pairs; hence, either of them can be the starting or the terminal node, then the graph is called an *undirected graph*. Edges joining vertices to themselves are referred to as *loops*. When numerical values, called *weights* are assigned to the edges of a graph G, then G is termed a *weighted* graph.

Let $e_k = (v_i, v_j) \in E$ be an edge of a graph $G = (V, E)$. Then, the vertices v_i and v_j are said to be *adjacent vertices* or *neighbours*. The edge $e_k = (v_i, v_j)$ is said to be *adjacent* to v_i and v_j or *incident* with v_i and v_j. Two edges are said to be *adjacent edges* if they have a vertex in common. The set of neighbours of a vertex v, denoted $\mathcal{N}(v)$, is referred to as the *neighbourhood* of v. The number of neighbours of a vertex v, that is, the cardinality of $\mathcal{N}(v)$, denoted $\deg(v)$, is called the *degree* of the vertex v.

Definition 3.1 A graph G is called *simple* if any two vertices of G are connected by at most one edge.
If multiple edges are allowed between two vertices of a graph G, then G is called a *multigraph*.

Definition 3.2 Let $G_1 = (V_1, E_1)$ and $G_2 = (V_2, E_2)$ be two graphs. If $V_2 \subseteq V_1$ and $E_2 \subseteq E_1$, then G_2 is called a *subgraph* of G_1, and we write $G_2 \subseteq G_1$.
If $V_2 = V_1$ then G_2 is called a *spanning* subgraph of G_1.

Definition 3.3 An induced subgraph of a graph $G_1 = (V_1, E_1)$ is a subgraph $G_2 = (V_2, E_2) \subseteq G_1$ such that all edges of G_1 between pairs of vertices from V_2 are included in G_2.

Graph Isomorphism An *isomorphism* between graphs $G_1 = (V_1, E_1)$ and $G_2 = (V_2, E_2)$ is a bijective (one to one) mapping $f : V_1 \longrightarrow V_2$, which preserves adjacency, that is, for every pair of vertices $v_i, v_j \in V_1$, it holds that $e_k = (v_i, v_j)$ is an edge of G_1 if and only if $e_z = (f(v_i), f(v_j))$ is an edge of G_2.

Two graphs G_1 and G_2 are said to be *isomorphic* if there exists an isomorphism between them, and we write $G_1 \approx G_2$.

Modular Decomposition A *module* in a graph $G = (V, E)$ is a set of vertices, X, that have exactly the same set of neighbours in $V - X$, that is, for every pair of vertices v_i and v_j in X we have

$$\mathcal{N}(v_i) \cap (V - X) = \mathcal{N}(v_j) \cap (V - X).$$

3.2.1 Connectivity in Graphs

Graphs are generally used to model various interconnecting complex networks, including biological networks at the molecular level such as gene regulation networks, signal transduction networks, protein interaction networks, and metabolic networks. One is often interested in the susceptibility of these networks to be disrupted, that is, will these networks remain connected even after the failure of some small number of nodes (vertices) or links (edges)? This naturally leads to the study of their "connectivity."

Walk, Path, and Cycle Let $G = (V, E)$ be a graph. A *walk*, in G is a sequence $v_0, e_1, v_1, \ldots, v_k$ of graph vertices v_i and edges e_i such that for $1 \leq i \leq k$, the edge e_i has endpoints v_{i-1} and v_i. Then, we have the following:

- The walk is said to be *closed* if $v_0 = v_k$.
- A closed walk with at least three vertices is called a *cycle.*
- The *length* of a walk is defined by its number of edges.
- If all the vertices in a walk are distinct, then the walk is termed a *path.*

A *shortest path* between two vertices is a path with minimal length.
A graph is said to be *acyclic* if it has no cycles.
Let $G = (V, E)$ be a graph with $|V| = n$ and $|E| = m$. The *degree* of a vertex $v_i \in V$, also denoted $\deg(v_i)$, is defined by the number of edges incident to v_i. Thus, the sum of the degrees of the vertices of G is given by

$$\sum_{i=1}^{n} \deg(v_i) = 2m.$$

Distance The *distance* between two vertices of G, v_i and v_j, denoted $d_G(v_i, v_j)$ is the length of the *shortest walk* between v_i and v_j. Obviously, the distance satisfies the following metric properties:

- $d_G(v_i, v_j) \geq 0$ for all $v_i,\ v_j\ \in V$;
- $d_G(v_i, v_k) \leq d_G(v_i, v_j) + d_G(v_j, v_k)$ for all $v_i,\ v_j,\ v_k \in V$;
- If G is an undirected graph, then $d_G(v_i, v_j) = d_G(v_j, v_i)$ for all $v_i,\ v_j\ \in V$.

If there is no walk between two vertices, v_i and v_j, of a graph G, then

$$d_G(v_i, v_j) = \infty.$$

Connectivity A graph $G = (V, E)$ is said to be *connected* if for each pair of vertices $v_i, v_j \in V$, there is a walk between v_i and v_j.
A maximal connected subgraph of a graph G is called a *connected component* or simply a *component* of G.
Thus, a graph G is said to be connected if G consists of at most one connected component, whereas a disconnected graph has at least two components.
Let $G = (V, E)$ be a simple graph with $|V| = n$ and $|E| = m$, then G has at least $n - m$ components.

Edge and Vertex-Connectivity Let G be a connected graph. Then,

- G is said to be *k-edge-connected,* $k > 1$, if G remains connected after removal of any subset of s vertices, with $s \leq k - 1$.
- G is said to be *k-vertex-connected,* $k > 1$, if G remains connected after removal of any subset of s vertices, with $s \leq k - 1$.

A graph G is k-edge-connected if, and only if, there exists (at least) k edge-disjoint walks between any pair of vertices.
A graph G is k-vertex-connected if, and only if, there exists (at least) k internally disjoint walks between any pair of vertices.
Graphs that are 1-*vertex connected* or 1-*edge-connected* are simply called *connected* graphs.

The *edge-connectivity* of a graph G, denoted $\mu(G)$, is the minimum number of edges whose deletion disconnects G.
The *vertex-connectivity* of a graph G, denoted $\nu(G)$, is the minimum number of vertices whose deletion disconnects G.
If $\delta(G)$ denotes the minimum degree of G, then the following holds:

$$\nu(G) \leq \mu(G) \leq \delta(G).$$

Cut-Edge, Cut-Vertex, and Local Connectivity A *cut-edge* or *cut-vertex* of a graph G is an edge or vertex whose deletion increases the number of components in G.
The *local connectivity* of two nonadjacent vertices v_i and v_j in a graph is defined by the smallest number of vertices whose removal separates v_i and v_j.

Connectivity in Weighted Graphs Let $G = (V, E, w)$ be a simple connected weighted graph, without loops, and such that $w(e_k) > 0,\ \forall\ e_k \in E$. Then,

- the length of a *weighted walk* $\gamma = v_0, e_1, v_1, e_e, v_2, \ldots, e_n, v_n$ in G is defined by

  $$d_G^w(\gamma) = \sum i = 1^n w(e_i);$$

- the weighted distance between two vertices v_i, and v_j in G is defined by

  $$d_G^w(v_i, v_j) = \min_{\gamma \in \Gamma} d_G^w(\gamma),$$

  where Γ denotes the set of walks from v_i to v_j.

Eccentricity, Radius, Diameter, and Center Let $G = (V, E)$ be a simple connected graph, without loops. Then,

- the *eccentricity* of a vertex v_i of G, denoted $\varepsilon(v_i)$, is defined by

  $$\varepsilon(v_i) = \max_{v_j \in V} d_G(v_i, v_j);$$

- the *radius* of G, denoted $\mathrm{rad}(G)$, is the minimum eccentricity over all the vertices of G, that is,

  $$\mathrm{rad}(G) = \min_{v_i \in V} \varepsilon(v_i) = \min_{v_i \in V} \max_{v_j \in V} d_G(v_i, v_j);$$

- the *diameter* of G, denoted $\mathrm{diam}(G)$, is the maximum eccentricity over all the vertices of G, that is,

  $$\mathrm{diam}(G) = \max_{v_i \in V} \varepsilon(v_i) = \max_{v_i, v_j \in V} d_G(v_i, v_j);$$

- the *center* of G is the subset of vertices $\overline{V} \subseteq V$ such that

  $$\varepsilon(v_i) = \mathrm{rad}(G),\ \forall\ v_i \in \overline{V}.$$

3.2.2 Subgraphs and Special Graphs

Some special graph structures play an important role in the analysis of complex systems, for example, in biology, where some particular graphs are used to represent, for instance, the evolutionary relationships between species.

Tree Let G be an undirected connected graph, with $|V| = n$ and $|E| = m$. If G is acyclic, then G is called a *tree*, and the following definitions are equivalents:

- G is connected and has no cycles;
- G is connected and has $n - 1$ edges;
- G has no cycles and has $n - 1$ edges.

The vertices of a tree with degree 1 are called *leaves*, whereas the other vertices are called inner vertices.
A binary tree is a tree where each vertex has at most a degree of 3.
Let $G = (V, E, w)$ be a weighted graph with $|V| = n$. A *minimum spanning tree* of G is a set of $n - 1$ edges of minimum total weight that form a spanning tree of G.

Clique and Coclique A *clique* in a graph $G = (V, E)$ is a set $\hat{V} \subseteq V$ of pair-wise adjacent vertices.
The *clique number*, denoted $\omega(G)$, is defined by the number of vertices in the largest clique in G.
A *coclique* in a graph $G = (V, E)$ is a set $\overline{V} \subseteq V$ of pair-wise nonadjacent vertices. The number of vertices in the largest coclique of a graph G, denoted $\alpha(G)$,is called *independence number*.

Regular Graph A graph G is called *regular* of degree (or valency) k, when every vertex of G has precisely k neighbors or the degree of any vertex of G is k.
Let $|V|$ denote the cardinality of V, that is, the number of vertices in a graph G. If G is a k-regular graph with $k = |V| - 1$, then G is simply called a *regular* graph.
Let $\delta(G)$ and $\Delta(G)$ denote respectively the minimum degree and the maximum degree among the vertices of G. If G is regular of degree k, then

$$\delta(G) = \Delta(G) = k.$$

Complete Graph A graph G is called a *complete graph* if its vertices are pair-wise adjacent, that is, each pair of vertices of G is connected by an edge, and a complete graph on n vertices is generally denoted by K_n.

Bipartite Graph If the vertices of a graph $G = (V, E)$ can be partitioned into two non-empty sets V_1 and V_2 so that there is no edge joining two vertices in the same set, then G is called a *bipartite* graph or a *digraph*.

Line Graph Let $G_1 = (V_1, E_1)$ and $G_2 = (V_2, E_2)$ be two graphs. The graph G_2 is called the *line graph* of G_1 if the edges of G_1 form the vertices of G_2, such that two vertices of G_2 are adjacent if the corresponding edges in G_1 have a vertex in common.

3.3 Matrix Theory: Some Basic Concepts

Let A be a real symmetric matrix of order n, then A has n real eigenvalues and a corresponding orthogonal set of eigenvectors.

3.3.1 Trace and Determinant of a Matrix

Let A be an $n \times n$ real matrix. The *trace* of A, denoted tr(A), is the sum of the diagonal entries of A, that is

$$\mathrm{tr}(A) = \sum_{i=1}^{n} a_{ii}.$$

The *determinant* of A, denoted det(A), can be computed using the following recursive relation:

$$\det(A) = \begin{cases} a_{11}, & \text{if } n = 1, \\ \sum_{i=1}^{n} (-1)^{i+j} a_{ij} \det(M_{ij}), & \text{if } n > 1, \end{cases}$$

where, M_{ij} is the $(n-1) \times (n-1)$ matrix obtained by removing the ith row and the jth column of A.

Let A and B be two $n \times n$ matrices and k a real scalar. The following useful properties of the determinant should be noted:

1) $\det(AB) = \det(A)\det(B)$;
2) $\det(A^T) = \det(A)$;
3) $\det(kA) = k^n \det(A)$;
4) $\det(A) \neq 0$ if and only if A is nonsingular.

If A is an $n \times n$ upper triangular or lower triangular matrix then,

$$\det(A) = \prod_{i=1}^{n} a_{ii},$$

that is, the determinant of a triangular matrix is the product of the diagonal entries. Therefore, the most practical way to compute a determinant of a matrix A is to decompose it into a product of lower and upper triangular matrices.

3.3.2
Eigenvalues and Eigenvectors of a Matrix

Let A be an $n \times n$ matrix. The eigenvalues of A, denoted $\lambda_i(A)$ $i = 1, 2, \ldots, n$ or $\lambda = \{\lambda_i,\ i = 1, 2, \ldots, n\}$, are the n roots of the polynomial in λ of degree n defined by $\det(A - \lambda I)$, that is, the eigenvalues are solutions of

$$\det(A - \lambda I) = 0.$$

The polynomial $\det(A - \lambda I)$ is called the *characteristic polynomial.*
If A is a real matrix, then the eigenvalues of A are either real or pairs of complex conjugates. If A is a symmetric matrix, then all its eigenvalues are real.
The following properties of eigenvalues should be noted:

1) $\det(A) = \prod_{i=1}^{n} \lambda_i(A)$;
2) $\mathrm{tr}(A) = \sum_{i=1}^{n} a_{ii} = \sum_{i=1}^{n} \lambda_i(A)$.

Therefore, a squared matrix A is nonsingular if and only if all its eigenvalues are non-zero.

Definition 3.4 The *spectral radius* of a squared matrix A, denoted $\rho(A)$, is given by

$$\rho(A) = \max_{i=1,2,\ldots,n} |\lambda_i(A)|.$$

Definition 3.5 A non-null vector x such that

$$Ax = \lambda_i(A)x$$

is called the right eigenvector associated with the eigenvalue $\lambda_i(A)$.

For each eigenvalue $\lambda_i(A)$, its right eigenvector x is found by solving the system

$$(A - \lambda_i(A)I)x = 0.$$

Let A be an $n \times n$ real matrix. The following properties should be noted:

- if A is an upper triangular or lower triangular matrix, then its eigenvalues are given by its diagonal entries, that is,

$$\lambda_i(A) = a_{ii}, \quad \text{for } i = 1, 2, \ldots, n;$$

- if A is orthogonal, then $|\lambda_i(A)| = 1$, for all $i = 1, 2, \ldots, n$;
- if A is symmetric, then there exists an orthogonal matrix $Q \in \mathbb{R}^{n\times n}$ such that

$$Q^T AQ = D,$$

where, D is an $n \times n$ diagonal matrix whose diagonal entries are $\lambda_1(A), \lambda_2(A), \ldots, \lambda_n(A)$.

- if A is nonsingular, that is, $\lambda_i(A) \neq 0$ for $i = 1, \dots, n$, then

$$\lambda_i(A^{-1}) = \frac{1}{\lambda_i(A)}, \quad \text{for } i = 1, \dots, n.$$

Eigenvalues can be used to determine the definiteness of the symmetric matrices. Let A be an $n \times n$ symmetric real matrix and $\lambda_i(A)$, $i = 1, 2, \dots, n$ its associated eigenvalues. Then,

- A is said to be *positive definite* if and only if $\lambda_i(A) > 0$, for all $i = 1, \dots, n$;
- A is said to be *positive semi-definite* if and only if $\lambda_i(A) \geq 0$, for all $i = 1, \dots, n$;
- A is said to be *negative definite* if and only if $\lambda_i(A) < 0$, for all $i = 1, \dots, n$;
- A is said to be *negative semi-definite* if and only if $\lambda_i(A) \leq 0$, for all $i = 1, \dots, n$;
- A is said to be *indefinite* if and only if $\lambda_i(A) > 0$ for some i and $\lambda_j(A) < 0$ for some j.

Let A be a nonsingular $n \times n$ symmetric matrix. If A is *positive semi-definite* (respectively *negative semi-definite*), then A is *positive definite* (respectively *negative definite*).

Let A be an $m \times n$ matrix of with $m < n$. Then, the determinant of any $n \times n$ square submatrix of A is called the *major determinant* of A.

If A is an $n \times n$ matrix, then the system

$$Ax = 0$$

has a non-trivial solution $x \neq 0$ if and only if $\det(A) = 0$.

The set of vectors $x \in \mathbb{R}^n$ such that $Ax = 0$ forms a vector space called the null space of the matrix A, denoted $\ker(A)$. Moreover, we have

$$\text{rank}(A) + \text{rank}(\ker(A)) = n.$$

3.4 Graph Matrices

Various matrices are associated to graphs, and in the sequel, we will introduce some of them, including the adjacency matrix, the incidence matrix, the degree matrix, the diffusion matrix, the Laplacian matrix, the cut-set matrix, and the path matrix.

Definition 3.6 Let $G = (V, E)$ be an undirected graph. The degree or the *combinatorial degree* of a vertex v_i, denoted $\deg(v_i)$, is given by the cardinality of the neighborhood of v_i:

$$\deg(v_i) = |\mathcal{N}(v_i)|.$$

For a weighted graph $G = (V, E, w)$, the degree of a vertex v_i, also called the *weighted degree*, is given by the sum of the weights of the edges adjacent to v_i, that is,

$$\deg(v_i) = \sum_{v_j \in \mathcal{N}(v_i)} w(v_i, v_j).$$

3.4.1 Adjacency Matrix

Definition 3.7 Let $G = (V, E)$ be an undirected simple graph without loops, with $|V| = n$. Then, the adjacency matrix of G is the $n \times n$ matrix denoted $A = A(G)$, with entries a_{ij} such that:

$$a_{ij} = \begin{cases} 1 & \text{if } (v_i, v_j) \in E, \\ 0 & \text{otherwise.} \end{cases}$$

Thus, the adjacency matrix is a symmetric (0, 1)-matrix with zeros on its main diagonal.

If G is a multigraph, then the entries a_{ij} of the adjacency matrix A define the multiplicity of the edge joining the vertices v_i and v_j, in which case, A is a non-negative integer matrix.
For a weighted graph $G = (V, E, w)$, the entries of the adjacency matrix are given by

$$a_{ij} = \begin{cases} w(v_i, v_j) & \text{if } (v_i, v_j) \in E, \\ 0 & \text{otherwise.} \end{cases}$$

If G is a weighted graph, then the entry a_{ij} of the adjacency matrix A defines the weight of the edge joining the vertices v_i and v_j, in which case A is a real or complex matrix depending on the nature of the weights.
Let G_1 and G_2 be two simple graphs and let A_1 and A_2 denote their respective adjacency matrices. Then, the graphs G_1 and G_2 are isomorphic if and only if there exists a permutation matrix R such that

$$A(G_2) = R^{-1}A(G_1)R.$$

3.4.2 Incidence Matrix

Definition 3.8 Let $G = (V, E)$ be an undirected graph without loops, with $|V| = n$ and $|E| = m$. Then, the vertex-edge *incidence matrix* of G is the $n \times m$ matrix, denoted $Q = Q(G)$, with entries q_{ik} such that:

$$q_{ik} = \begin{cases} 1 & \text{if vertex } v_i \text{ is incident to edge } e_k, \\ 0 & \text{otherwise.} \end{cases}$$

Definition 3.9 Let $G = (V, E)$ be a directed graph without loops, with $|V| = n$ and $|E| = m$. Then, the *directed incidence matrix* of G is the $n \times m$ matrix, denoted $\overline{Q} = \overline{Q}(G)$, with entries $\overline{q}_{ik}$ such that:

$$\overline{q}_{ik} = \begin{cases} -1 & \text{if vertex } v_i \text{ is the terminal node of edge } e_k, \\ 1 & \text{if vertex } v_i \text{ is the starting node of edge } e_k, \\ 0 & \text{otherwise.} \end{cases}$$

3.4.3 Degree Matrix and Diffusion Matrix

Definition 3.10 Let $G = (V, E)$ be an undirected simple graph. Then, the *degree matrix* of G is the $n \times n$ diagonal matrix, denoted $D = D(G)$, with entries d_{ij} such that:

$$d_{ij} = \begin{cases} \deg(v_i) & \text{if } v_i = v_j, \\ 0 & \text{otherwise.} \end{cases}$$

Definition 3.11 Let $G = (V, E)$ be an undirected simple graph. Then, the *diffusion matrix* of a graph G, also referred to as the *walk matrix*, denoted $W = W(G)$, is defined by

$$W = AD^{-1},$$

where D and A are the degree matrix and adjacency matrix of G, respectively.

3.4.4 Laplace Matrix

Definition 3.12 Let $G = (V, E)$, with $|V| = n$, be an undirected simple graph without loops. The *Laplace matrix* of the graph G, also referred to as the *Laplacian*, the *Kirchhoff matrix*, or the *matrix of admittance*, is an $n \times n$ symmetric matrix, denoted $L = L(G)$, defined by

$$L = D - A,$$

where D and A are the degree matrix and the adjacency matrix of G, respectively. Thus, the entries l_{ij} of L are such that:

$$l_{ij} = \begin{cases} \deg(v_j) & \text{if } v_i = v_j, \\ -1 & \text{if } v_i \text{ and } v_j \text{ are adjacent,} \\ 0 & \text{otherwise.} \end{cases}$$

An important variant of the Laplace matrix, called *normalised Laplacian* and denoted $\overline{L} = \overline{L}(G)$, is defined by

$$\overline{L} = D^{-1/2} L D^{-1/2} = I - D^{-1/2} A D^{-1/2},$$

where I denotes the identity matrix, D and A are the degree matrix and adjacency matrix of G, respectively. The entries $\overline{l}_{ij}$ of $\overline{L}$ are such that:

$$\overline{l}_{ij} = \begin{cases} 1 & \text{if } v_i = v_j \text{ and } d_{v_j} \neq 0, \\ -\dfrac{1}{\sqrt{\deg(v_i)\deg(v_j)}} & \text{if } v_i \text{ and } v_j \text{ are adjacent,} \\ 0 & \text{otherwise.} \end{cases}$$

The matrix $|L| = D + A$ is called the *signless Laplace matrix* of G.

Some Properties of the Laplace Matrix The following important properties of the Laplace matrix should be noted:

- Let $\overline{Q}$ denote the incidence matrix of a directed graph G, then

$$L = \overline{QQ}^t.$$

Hence, L does not depend on the orientations.
- L is positive semidefinite;
- $\forall x \in \mathbb{R}^n,\ x^t Lx = \sum_{v_i \text{ and } v_j \text{ adjacent}} (x_i - x_j^2)$;
- $L = \alpha I - B$, where B denotes an $n \times n$ matrix with positive entries, I is an $n \times n$ identity matrix and $\alpha > \rho(B)$;
- $\min\limits_{i=1,2,\ldots,n} \lambda_i(L) = 0$;

3.4.5
Cut-Set Matrix

Definition 3.13 Let $G = (V, E)$ be a graph and let P be a partition $V = V_1 \cup V_2$ of the vertex of G. The set of edges whose end points are in different subsets of the partition, that is, edges from V_1 to V_2, is called a *cut* or a *cut-set.*

Let G be a graph with m edges and n cut-sets. The cut-set matrix of G, denoted C, is an $n \times m$ matrix with entries c_{ij} such that:

$$c_{ij} = \begin{cases} 1 & \text{if the } i\text{th cut-set contains the } j\text{th edge,} \\ 0 & \text{otherwise.} \end{cases}$$

3.4.6
Path Matrix

Let G be a graph with m edges, and let u and v be two vertices of G. Then, the *path matrix* for the vertices u and v denoted $P(u, v)$ is an $n \times m$ matrix, where n is the number of different paths between the two vertices, with entries p_{ij} such that:

$$p_{ij} = \begin{cases} 1 & \text{if the } j\text{th edge lies in the } i\text{th path} \\ 0 & \text{otherwise.} \end{cases}$$

The rows of the path matrix $P(u, v)$ correspond to the different paths between vertices u and v, whereas, the columns correspond to the different edges of the graph G.

3.5
Spectral Graph Theory: Some Basic Results

Spectral graph theory is a very well-developed mathematical field, which provided a framework to establish the fundamental relationships between graph theory and matrix algebra. The major advantage of using graph spectra in treating graphs lies in the relative short time for computing eigenvalues and eigenvectors of several graph matrices. Spectral graph parameters encompass various information on the graph's local and global structure, which are usually obtained using exponential algorithms. In the subsequent sections, we will use spectral theory to characterize some structural properties of some graphs through their matricial representations. More advanced results on spectral graph theory can be found for example in (Biggs, 1993; Brouwer et al., 1989; Chung, 1997; Godsil, 1993; Mohar and Poljak, 1993; Spielman, 2012).

Definition 3.14 Let G be a graph and A its associated adjacency matrix. Then, the *ordinary spectrum* of G is defined by the spectrum of the matrix A, that is, the set of eigenvalues of A together with their multiplicities.
Let G be an undirected graph without loops and L its associated Laplace matrix. Then, the *Laplace spectrum* of G is defined by the spectrum of L.

Definition 3.15 Graphs with the same spectrum of an associated matrix M (e.g., the adjacency matrix, the Laplace matrix, the signless Laplace matrix) are called *cospectral* graphs with respect to M or *M-cospectral* graphs.
Graphs with the same ordinary spectrum are called *cospectral* or *isospectral.*

Definition 3.16 Let $G = (V, E)$ be a graph with $|V| = n$. Then, the *characteristic polynomial* of G, denoted $p(G, x)$ is defined by

$$p(G, x) = \det(xI - A) = x^n + c_1 x^{n-1} + \dots + c_n,$$

where A is the adjacency matrix of G and I the $n \times n$ identity matrix.
Furthermore, we have the following facts:

- $c_1 = \mathrm{tr}(A) = 0$;
- $-c_2 = |E|$.

Definition 3.17 The Euclidean distance between the eigenvalue sequences of two graphs on the same number of vertices is called the *spectral distance* of graphs.

3.5.1
Spectral Characterization of Graph Connectivity

Let G be a graph with connected components G_i $(1 \leq i \leq k)$. Then, the ordinary (respectively Laplace) spectrum of G is the union of the ordinary (respectively Laplace) spectra of G_i (and the multiplicities are added).

Definition 3.18 Let $0 = \lambda_1(L) \leq \lambda_2(L) \leq \dots \leq \lambda_n(L)$ be the eigenvalues of the Laplace matrix, L, associated to a graph G with at least two vertices . The second smallest Laplace eigenvalue, $\lambda_2(L)$, is called the *algebraic connectivity* of G or the *Fiedler value* of G, and its corresponding eigenvector is referred to as the *Fiedler vector*.

The Fiedler value provides good bounds on the connectedness of a graph. When a graph is disconnected, then its Fiedler value is null, that is, $\lambda_2(L) = 0$.

Theorem 3.1 *If G is a connected graph, then*

$$rank(C) = rankA = rank(G)$$

where C and A are the cut-matrix and adjacency matrix associated to G, respectively.

3.5.1.1 Spectral Theory and Walks

In this section, we will present some results relating spectral theory and walks and distance parameters in graphs.

Let A denote the adjacency matrix of a graph $G = (V, E)$ with $|V| = n$, then the following holds:

- for $k = 0, 1, \ldots$, the entry a_{ij} of the matrix A^k is the number of walks of length k from vertex v_i to vertex v_j;
- $a_{ii}^2 = \deg(v_i)$;
- the number of edges of G is given by

$$|E| = \frac{1}{2}\sum_{i=1}^{n} \deg(v_i) = \frac{1}{2}\mathrm{tr}(A^2).$$

- if A has r distinct eigenvalues, then

$$\mathrm{diam}(G) \leq r - 1,$$

  where diam(G) denotes the diameter of graph G.

Let L denote the Laplace matrix of G and $0 = \lambda_1(L) \leq \lambda_2(L) \leq \cdots \leq \lambda_n(L)$ the corresponding Laplace eigenvalues. Then, the following holds:

$$\frac{4}{n\lambda_2(L)} \leq \mathrm{diam}(G) \leq \left\lceil \frac{\cosh^{-1}(n-1)}{\cosh^{-1}\left(\frac{\lambda_n(L)+\lambda_2(L)}{\lambda_n(L)-\lambda_2(L)}\right)} \right\rceil + 1.$$

Let $\overline{\rho}(G)$ denote the average distance between distinct vertices of G, then

$$\frac{1}{n-1}\left(\frac{2}{\lambda_2(L)} + \frac{n-2}{2}\right) \leq \overline{\rho}(G) \leq \frac{n}{n-1}\left\lceil \frac{\Delta(G) + \lambda_2(L)}{4\lambda_2(L)} \ln(n-1) \right\rceil,$$

where $\Delta(G)$ denotes the maximum degree among the vertices of G.

Theorem 3.2 *Let $G = (V, E)$, with $|V| = n$, be a simple graph different from a complete graph K_n, and let L be its associated Laplace matrix and $\lambda_2(L)$ the second smallest eigenvalue of L. Let $v(G)$ and $\mu(G)$ denote the vertex- and edge-connectivity of G, respectively. Then, the following hold*

- $\lambda_2(L) \leq v(G) \leq \mu(G)$ *and*
- $\lambda_2(L) \geq 2\mu(G)(1 - \cos(\pi/n))$.

Theorem 3.3 *Let L denote the Laplace matrix associated to a graph $G = (V, E)$ with $|V| = n$, $|E| = m$ and having c connected components. Then,*

$$rank(L) = n - c.$$

Theorem 3.4 *The multiplicity of the eigenvalue* $\lambda_1(L) = 0$ *of an undirected graph G equals the number of connected components in G.*

Maximum Cut Problem Let $G = (V, E, w)$ be a weighted graph with $|V| = n$ and Laplace matrix L. Then, the *maximum cut* problem (or *max-cut* problem in short) is to find a cut-edge $S \subset E$ for which the weight $\sum_{e \in S} w(e)$ is maximum.

Let $mc(G)$ denote the value of the maximum cut in a weighted graph G. Then, the following holds:

$$mc(G) \leq \frac{n}{4} \max_i(\lambda_i(L)).$$

3.5.2
Spectral Characteristics of some Special Graphs and Subgraphs

3.5.2.1 Tree

Theorem 3.5 *Let* $0 = \lambda_1(L) \leq \lambda_2(L) \leq \cdots \leq \lambda_n(L)$ *be the eigenvalues of the Laplace matrix, L, associated to a graph G. Then, the number, N, of spanning trees of G is given by,*

$$N = \frac{\prod_{i=2}^{n} \lambda_i(L)}{n}.$$

Let $G = (V, E)$, with $|V| = n$, be a tree and $\lambda_2(L) \leq \cdots \leq \lambda_n(L)$ its corresponding non-zero Laplace eigenvalues. Then, the average distance between distinct vertices of G, denoted $\overline{\rho}(G)$, is defined by

$$\overline{\rho}(G) = \frac{2}{n-1} \sum_{i=2}^{n} \frac{1}{\lambda_i(L)}.$$

3.5.2.2 Bipartite Graph

The adjacency matrix of a bipartite graph has the form $A = \begin{pmatrix} 0 & U \\ U^T & 0 \end{pmatrix}$. Therefore, a graph is bipartite if and only if its spectrum is symmetric about 0.

Theorem 3.6 *A graph G is bipartite if and only if its Laplace spectrum and its signless Laplace spectrum are equal.*

Theorem 3.7 *The number of bipartite connected components of an undirected graph G is defined by the multiplicity of the* 0 *signless Laplace eigenvalue of G.*

Theorem 3.8 *Let G be a graph with adjacency matrix A. Then, G is bipartite if and only if for each eigenvalue* $\lambda_i(A)$ *of G,* $-\lambda_i(A)$ *is also an eigenvalue of G, with the same multiplicity.*

Theorem 3.9 *Let G be a graph with adjacency matrix A. Then, G is bipartite if and only if*

$$\min_i(\lambda_i(A)) = -\max_i(\lambda_i(A)).$$

3.5.2.3 Complete Graph

Theorem 3.10 *Let $G = (V, E)$ be a complete graph with adjacency matrix A. Let $\lambda_{\min}(A)$ be the minimal eigenvalue of A, then,*

$$\lambda_{\min}(A) = -1.$$

3.5.2.4 Regular Graph

Let $G = (V, E)$ be a regular graph with valency r and $|V| = n$. Let A and L denote the adjacency matrix and the Laplace matrix of G, respectively. Then,

$$L = rI - A,$$

where I denotes the identity matrix.
Furthermore, if G has ordinary eigenvalues $r = \lambda_1(A) \geq \lambda_2(A) \geq \cdots \geq \lambda_n(A)$ and Laplace eigenvalues $0 = \lambda_1(L) \geq \lambda_2(L) \geq \cdots \geq \lambda_n(L)$, then,

- $\lambda_i(A) = r - \lambda_i(L), \forall\, i = 1, \ldots, n;$
- $\lambda_i(L) \leq r, \forall\, i = 1, \ldots, n.$

Theorem 3.11 *Let $G = (V, E)$ be a regular graph, with $|V| = n$, and let A and D denote the adjacency matrix and degree matrix of G, respectively. Let $\lambda_{\max}(A)$ be the maximal eigenvalue of A, then,*

$$\frac{1}{n}\sum_{i=1}^{n} d_{ii} = \lambda_{\max}(A) = \max_{i} d_{ii},$$

where d_{ii} are the diagonal entries of the degree matrix D.

Theorem 3.12 *Let G be a connected regular graph with valency r and $\lambda_2(L)$ its second smallest Laplace eigenvalue. Then,*

$$diam(G) \leq 2\left\lceil \frac{2r - \lambda_2(L)}{4(r - \lambda_2(L))}\ln(n-1)\right\rceil.$$

Theorem 3.13 *Let G be an undirected regular graph with valency r. Then, r is the largest eigenvalue of G, and its multiplicity equals the number of connected components in G.*

3.5.2.5 Line Graph

Theorem 3.14 *Let G_1 be a graph, and G_2 its associated line graph with adjacency matrix A_2. Then*

$$\lambda_i(A_2) \geq -2, \quad \forall\, i.$$

Let $G_1 = (V_1, E_1)$ be a graph, with $|E_1| = m$. Let $\lambda_1(|L_1|) \geq \cdots \geq \lambda_k(|L_1|)$ be the positive signless Laplace eigenvalues of G_1. Then, the eigenvalues of G_2, the line graph of G_1, are defined by

- $\lambda_i(|L_2|) = \lambda_i(|L_1|) - 2$, for $i = 1, \ldots, k$ and
- $\lambda_i(|L_2|) = -2$, if $k < i < m$.

3.5.3 Spectral Theory and Graph Colouring

The *chromatic number* of a graph G, denoted $\chi(G)$ is the minimum number of colors needed to color the vertices of G so that no two adjacent vertices have the same color.
Let A denote the adjacency matrix of a simple graph G. Then,

$$1 + \frac{\max_i(\lambda_i(A))}{\min_i(\lambda_i(A))} \leq \chi(G) \leq 1 + \max_i(\lambda_i(A)).$$

3.5.4 Spectral Theory and Graph Drawing

Let $G = (V, E)$ be a connected graph, with $|V| = 100$. Let A, D and L denote the adjacency matrix, the degree matrix, and the Laplace matrix of G, respectively. Let $u_2(L)$, $u_3(L)$ and $u_4(L)$ denote the eigenvectors associated with the second, third, and fourth smallest eigenvalues of L. Then, the graph G can be drawn by generating 100 randomly chosen points in the unit square, as depicted in Figure 3.1. However, drawing the graph G by taking advantage of its spectral characteristics, in particular the Laplace eigenvectors $u_2(L)$, $u_3(L)$ and $u_4(L)$ provides a more clearer structure of G, as illustrated in Figures 3.2–3.4.

3.6 Computational Challenges for Spectral Graph Analysis

In practice, spectral graph analysis reduces to an eigenvalue problem. Therefore, the basic computational challenges associated with the spectral graph analysis are those inherent to eigenvalue problems. Since large-scale eigenvalue problems are generally solved using numerical methods, then the main questions that naturally arise are about the accuracy and stability of these methods, that is, how well the numerically estimated eigenvalues approximate the true eigenvalues. In this section, we will highlight some key features of the most commonly used algorithms to solve large-scale eigenvalue problems and discuss their limitations.

3.6.1 Krylov Subspace Methods

Krylov subspace methods, such as Arnoldi algorithm (Arnoldi, 1951), and Lanczos algorithm (Lanczos, 1952) and their enhanced versions, see for example (Golub and Van Loan, 2013; Saad, 2011; Cullum and Willoughby, 2002), for an overview, are among the most popular numerical approaches used to solve eigenvalue problems. However, for certain type of matrices, Krylov subspace methods require the construction of a high-degree polynomial in order to produce an accurate approximation of the eigenpairs (eigenvalues and eigenvectors). Thus,

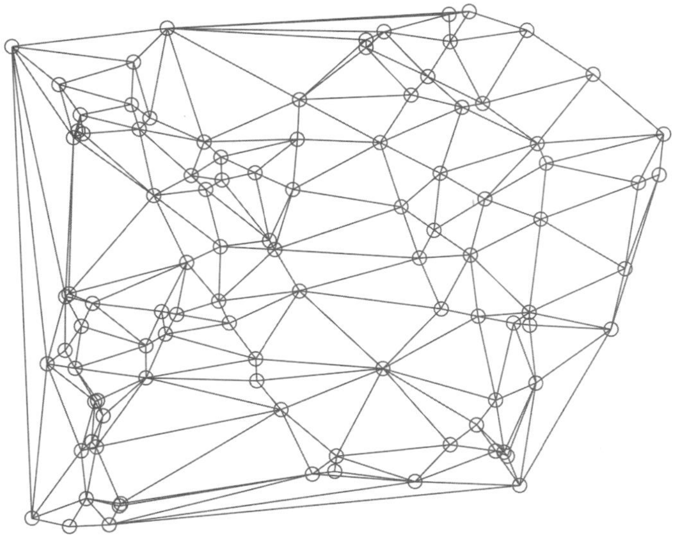

Figure 3.1 The graph G, drawn using its adjacency matrix A and randomly chosen positions for the vertices. (See inset for color representation of this figure.)

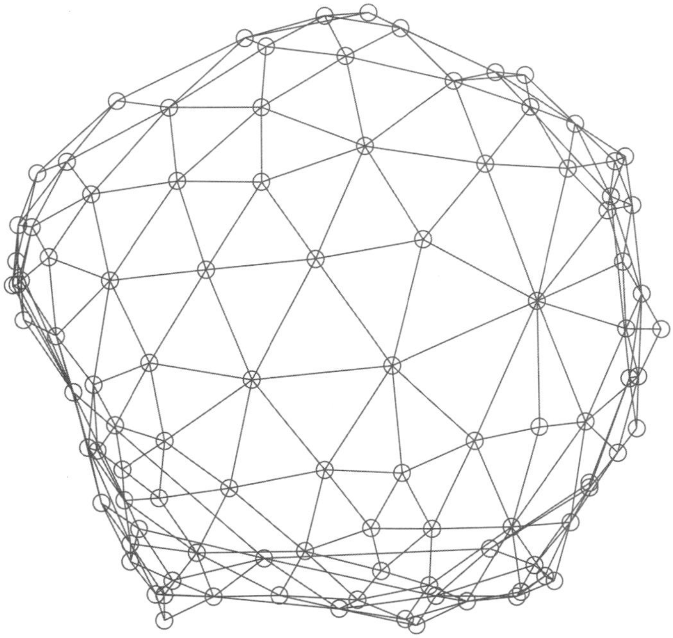

Figure 3.2 The graph G, drawn using its adjacency matrix A and its two Laplace eigenvectors $u_2(L)$ and $u_3(L)$. (See inset for color representation of this figure.)

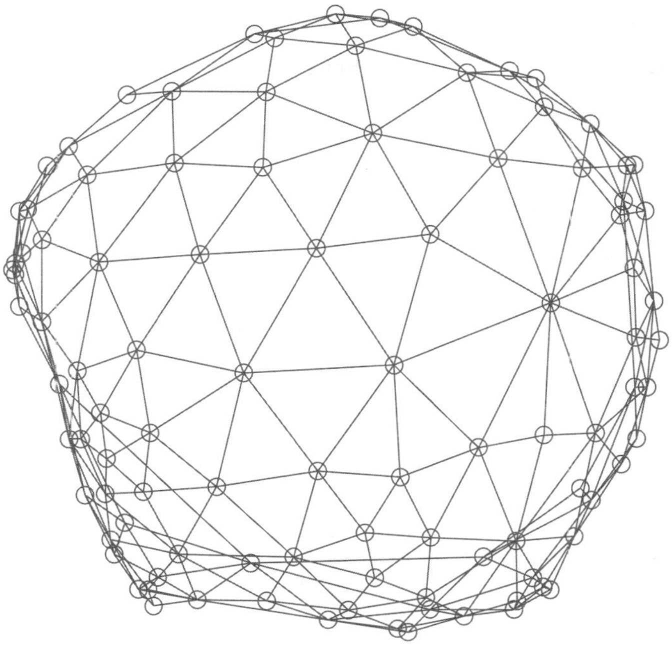

Figure 3.3 The graph G, redrawn using its walk matrix $W = AD^{-1}$ and its two Laplace eigenvectors $u_2(L)$ and $u_3(L)$. (See inset for color representation of this figure.)

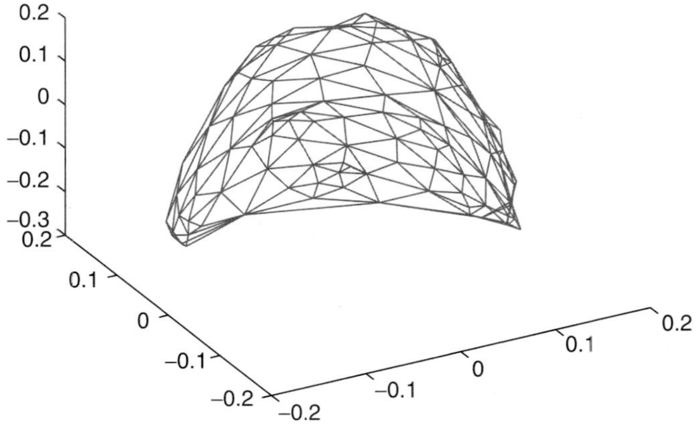

Figure 3.4 The graph G, drawn using its adjacency matrix A and its three Laplace eigenvectors $u_2(L)$, $u_3(L)$, and $u_4(L)$. (See inset for color representation of this figure.)

the application of these methods requires a preconditioning of the matrix to guarantee their convergence.

3.6.2
Constrained Optimization Approach

In order to overcome the difficulties associated with the preconditioning required by Krylov subspace methods, alternative approaches based on constrained optimization are used to solve eigenvalue problems. Let A be a symmetric matrix, then the eigenvalue problem can be formulated as,

$$\begin{aligned} \min_{x} \quad & x^T A x \\ & x^T x = 1. \end{aligned} \tag{3.1}$$

Using an appropriate change of variable, which is equivalent to the preconditioning, the optimization problem (Eq. (3.1)) can be solved using, for example, the locally optimal block preconditioned conjugate gradient algorithm proposed in Knyazev (2001). However, when the approximate solution to Eq. (3.1), denoted $\hat{x}$ is close to the exact optimal solution, that is, the exact eigenvector, then minimizing the objective-function, in Eq. (3.1), is no longer productive. In which case, instead of Eq. (3.1), the following formulation is adopted to refine the eigenvector approximation while maintaining the approximate eigenvalue, which is already very accurate:

$$\begin{aligned} \min_{x} \quad & \|A\hat{x} - \xi\hat{x}\| \\ & \|\hat{x}\| = 1, \\ & \hat{x} \in \Omega, \end{aligned} \tag{3.2}$$

where Ω is an orthonormal basis for a subspace, which contains $\hat{x}$, and ξ is a highly accurate approximation to the desired eigenvalue.

Such a refinement of eigenvector has proven its effectiveness in solving large-scale eigenvalue problems for various applications, see for example Wood and Zunger (1985), Jia (1997), and Stewart (2001) for an overview.

3.7
Conclusion

This chapter introduced some basics of spectral graph theory, and its applications to characterize some graph properties, including graph connectivity, graph modularity, and graph drawing. More precisely, the chapter focused on the most commonly used types of matrices in spectral graph theory, namely, the adjacency matrix, the Laplace matrix, the cut matrix, and presented some new and old basic results, which used these matrices not only to analyze and obtain some topological information of graphs but also to identify some classes of special graphs and subgraphs. Furthermore, we have discussed some computational challenges, which are associated with the spectral graph analysis.

Many research works have pointed out the potential of spectral graph theory as a rigorous framework to infer and analyze global properties of complex networks; (Banerjee and Jost, 2009; Brouwer and Haemers, 2012; Cioaba, 2010; Cvetkovic, Rowkinson, and Simic, 2009; Hoory, Linial, and Wigderson, 2006; Krivelevich and Sudakov, 2006; van Damand and Haemers, 2009; Kannan and Vempala, 2008) including the deconvolution of indirect information flows and the inference of direct interactions over networks (Feizi *et al.*, 2013). This chapter intends to be primer to those interested in studying spectral graph theory, and it aims to illustrate how basic tools from matrix algebra and graph theory can be leveraged to get information about some topological structures of complex networks, as well as the underlying computational challenges.

References

Arnoldi, W.E. (1951) The principle of minimized iterations in the solution of the matrix eigenvalue problem. *Quarterly of Applied Mathematics*, **9**, 17–29.

Banerjee, A. and Jost, J. (2009) Graph spectra as a systematic tool in computational biology. *Discrete Applied Mathematics*, **157**, 2425–2431.

Berge, C. (1976) *Graphs and Hypergraphs*, North-Holland, Amsterdam.

Biggs, N. (1993) *Algebraic Graph Theory*, 2nd edn, Cambridge University Press, Cambridge.

Bollobàs, B. (1998) *Modern Graph Theory*, Springer-Verlag, New York.

Brouwer, A.E., Cohen, A.M., and Neumaier, A. (1989) *Distance-Regular Graphs*, Springer-Verlag, Berlin.

Brouwer, A. and Haemers, W. (2012) *Spectra of Graphs*, Springer-Verlag, Berlin.

Chung, F.R.K. (1997) *Spectral Graph Theory*, Published for the Conference Board of the Mathematical Sciences, Washington, DC.

Cioaba, S. (2010) Some applications of eigenvalues of graphs, in *Structural Analysis of Complex Networks* (ed. M. Dehmer), Birkhäuser.

Cullum, J.K. and Willoughby, R.A. (1995) *Lanczos Algorithms for Large Symmetric Eigenvalue Computations*, Birkhäuser.

Cvetkovic, D.M., Doob, M., and Sachs, H. (1995) *Spectra of Graphs*, 3rd edn, Johann Ambrosius Barth, Heidelberg.

Cvetkovic, D.M., Rowkinson, M.P., and Simic, S. (2009) *An Introduction to the Theory of Graph Spectra of Graphs*, London Mathematical Society - Student Texts, Cambridge University Press.

van Dam, E. and Haemers, W. (2009) Developments on spectral characterization of graphs. *Discrete Mathematics*, **309**, 576–586.

Feizi, S., Marbach, D., Médard, M., and Kellis, M. (2013) Network deconvolution as a general method to distinguish direct dependencies in networks. *Nature Biotechnology*, **31** (8), 726–733.

Godsil, C.D. (1993) *Algebraic Combinatorics*, Chapman & Hall, New York.

Godsil, C.D. and Royle, G. (2001) *Algebraic Graph Theory*, Graduate Texts in Mathematics, Springer.

Golub, G.H. and Van Loan, C.F. (2013) *Matrix Computations*, 4th edn, The Johns Hopkins University Press.

Harary, F. (1969) *Graph Theory*, Addison-Wesley, Reading, MA.

Hoory, S., Linial, N., and Wigderson, A. (2006) Expander graphs and their applications. *Bulletin American Mathematical Society*, **43** (4), 439–561.

Jia, Z. (1997) Refined iterative algorithm based on Arnoldi's process for large unsymmetric eigenproblems. *Linear Algebra and its Applications*, **259**, 1–23.

Kannan, R. and Vempala, S. (2008) Spectral algorithms. *Foundations and Trends in Theoretical Computer Science*, **4** (3), 132–288.

Knyazev, A.V. (2001) Toward the optimal preconditioned eigensolver: locally optimal block preconditioned conjugate gradient method. *SIAM Journal of Scientific Computing*, **23**, 517–541.

Krivelevich, M. and Sudakov, B. (2006) Pseudo-random graphs, in *More Sets, Graphs and Numbers*, Bolyai Society Mathematical Studies, Vol. 15, Springer, pp. 199–262.

Lanczos, C. (1952) Solution of systems of linear equation*s by minimized iterations. *Journal of Research of the National Bureau of Standards*, **49**, 33–53.

Mason, O. and Verwoerd, M. (2007) Graph theory and networks in biology. *IET Systems Biology*, **1**, 89–119.

Mehrmann, V. and Voss, H. (2005) Nonlinear eigenvalue problems: a challenge for modern eigenvalue methods. *GAMM Mitteilungen*, **27**, 121–152.

Mohar, B. and Poljak, S. (1993) Eigenvalues in combinatorial optimization, in *Combinatorial and Graph-Theoretical Problem in Linear Algebra* (ed. M. Boyle), Springer, New York.

Saad, Y. (2011) *Numerical Methods for Large Eigenvalue Problems*, SIAM - Classics in Applied Mathematics. Revised edn., SIAM.

Spielman, D. (2012) Spectral graph theory, in *Combinatorial Scientific Computing* (eds U. Naumann and O. Schenk), Chapman and Hall/CRC.

Stewart, G.W. (2001) *Matrix Algorithms: Eigensystems*, Vol. II, SIAM, Philadelphia, PA.

West, D. (2001) *Introduction to Graph Theory*, 2nd edn, Prentice Hall.

Wood, D.M. and Zunger, A. (1985) A new method for diagonalising large matrices. *Journal of Physics A: Mathematical and General*, **18**, 1343–1359.

4 Contagion in Interbank Networks

Grzegorz Hałaj and Christoffer Kok

4.1 Introduction

Financial intermediation serves the purpose of reallocating funds from the net savers to the net borrowers of the economy. Without financial intermediation, economic entities with excess funds would have difficulties finding and providing financing to other economic agents in need of funds. As a result of its role in intermediating between savers and borrowers, the financial system consists of a large number of interlinkages. Financial transactions thus create links both between financial intermediaries (i.e., banks and other financial institutions) and the nonfinancial sectors of the economy (i.e., households, firms, government entities), and also among financial intermediaries inside the financial system. While it is generally acknowledged that the financial system – at least in normal times – helps smooth the wheels of the economy by making the resource allocation more efficient (as financial intermediaries are specialised in and have comparative advantages in reallocating savings), this intermediation process can be disrupted when, for example, key financial institutions get into trouble or when macroeconomic and asset price shocks affect the financial system as a whole. Such adverse shocks can lead to cascading contagion effects throughout the financial system due to the many, and often highly complex, interrelationships that exist between the actors in the financial system. Disruptions to the provision of financial services can in turn have serious implications for the real economy.

For theses reasons, there has been increasing interest among economists and policy makers to understand and measure the risks posed by the complex system

1) Grzegorz Hałaj. European Central Bank, Kaiserstrasse 29, D-60311 Frankfurt am Main, Deutschland
DISCLAIMER: The chapter should not be reported as representing the views of the European Central Bank (ECB). The views expressed are those of the authors and do not necessarily reflect those of the ECB.
The authors are indebted to I. Alves, S. Dubecq, J. Henry, G. Rünstler who provided valuable comments and to C. Minoiu and VS Subrahmanian who discussed the paper about the endogenous networks during INET conference in Ancona.

Computational Network Theory: Theoretical Foundations and Applications, First Edition.
Edited by Matthias Dehmer, Frank Emmert-Streib, and Stefan Pickl.

of financial interrelations characterising the modern economy. In order to be able to identify, assess, and possibly address the potential contagion risks existing within the financial system, network-based models have proved particularly useful. In this light, and especially triggered by the financial crisis erupting in 2007, an extensive literature on the contagion analysis using network theory and modeling has emerged (see also Section 0.2 later for a survey of some of the most recent studies). The approaches to analyzing financial networks have often relied on network-based tools developed in other sciences, such as biology, physics, and medicine. Compared to many other sciences, network applications in finance are constrained by the considerable computational challenges related to the fact that sufficiently granular data on financial networks are often not available. Moreover, also in contrast to some other sciences, it is difficult to construct realistic counterfactual simulations that are able to fully capture the multi-layered and dynamic complexity characterising financial networks in the real world.

In order to shed light on some of the key computational issues in financial networks, in this chapter, we demonstrate two recent applications related to the interbank networks. The illustrations serve to highlight approaches to overcome the computational issues related to the limited data availability and the highly complex dynamic interactions underlying the financial interrelations (here exemplified with applications for the interbank market). We furthermore demonstrate how regulatory measures can be employed to contain contagion risk embedded in the interbank market structures, to the extent that prudential actions can be shown – using network-based models – to be effective in pulling the interbank network structures in a direction that makes them more resilient, such network applications can be useful for policy purposes; for example, to help calibrating prudential policy measures.

The interbank market has been one of the key topics for researchers using network-based approaches to study financial contagion. The market for interbank transactions where banks directly interact with each other is an obvious place to start when looking for network links between financial institutions. Another reason for the interest in network-based analysis of the interbank contagion results from the fact that the interbank market was one of the main victims of the financial crisis erupting in 2007. The crisis led to a general loss of trust among market participants and resulted in severe interbank market disruptions. Moreover, failures of some key market players triggered concerns about risks of interbank contagion, whereby even small initial shocks could have potentially detrimental effects on the overall system. As a result of these concerns, and also reflecting a broader aim of making the financial sector more resilient, in recent years financial regulators have introduced various measures that aim at mitigating (and better reflecting) the risks inherent through the bilateral links between banks in the interbank network. These international reform initiatives range inter alia from limits on large counterparty exposures, higher capital requirements on counterparty exposures, and requirements to settle standardised OTC derivatives contracts via central counterparty clearing (CCP) houses. While it seems plausible that these initiatives should help alleviate contagion risks in the interbank market, there is still only little research aiming to quantify and understand the effects of these reforms on network structures and the contagion risk that might emerge from these structures.

For this purpose, the chapter presents two models to derive the interbank networks that are determined by certain characteristics of the banks' balance sheets, the structure of which is assumed to be an outcome of banks' risk-adjusted return optimization of their assets and liabilities. The first model takes the random simulation approach to analysing contagion within the interbank networks, which is related to the so-called *Stochastic block modeling* of networks, as for instance suggested by Lu and Zhou (2010), whereby link prediction algorithms are used to produce the missing links between the agents (nodes) in a given network. The second model of the interbank network combines bank balance sheet optimization with the random network generation technique presented in Hałaj and Kok (2013b). This allows us to study the endogenous network formation based on the optimizing bank behavior.

The models can thus help to understand the foundations of topology of the interbank network. It, furthermore, provides a tool for analyzing the sensitivity of the interbank structures to the heterogeneity of the banks (in terms of size of balance sheet, capital position, general profitability of non-interbank assets, counterparty credit risk), and to changes of the market and bank-specific risk parameters. Such parameter changes could, for example, be due to the regulatory policy actions (e.g., pertaining to capital buffers as well as the size and diversity of interbank exposures) aiming at mitigating systemic risk within the interbank system. The framework developed in this paper can, therefore, be used to conduct a normative analysis of macro- and micro-prudential policies geared towards more resilient interbank market structures.

4.2 Research Context

The recent crisis events have highlighted the systemic risks to the financial system of individual bank failures via the interlinkages that exist between banks; especially in the unsecured interbank market. Particular attention has been paid to the potential counterparty risks banks are exposed to via their bilateral interbank exposures.[2] This, in turn has led to a flurry of academic research to help understand, measure, and assess the impact of contagion within the network of banks and other institutions that constitute the financial system. In addition, a number of policy initiatives have been introduced in recent years to counter the potential contagion risks of the interlinked banking networks; especially exemplified by the additional capital requirements on globally systemic institutions (G-SIBs).

2) See Rochet and Tirole (1996), Allen and Gale (2000), and Freixas, Parigi, and Rochet (2000) for some early prominent examples.

The academic literature analyzing financial contagion has followed different strands. One area of research has focused on capturing contagion using financial market data. Kodres and Pritsker (2002) provides a theoretical model, whereby in an environment of shared macro-economic risks and asymmetric information, asset price contagion can occur even under the assumption of rational expectations. On the empirical side, some early studies attempted to capture contagion using event studies to detect the impact of bank failures on stock (or debt) prices of other banks in the system.[3] The evidence from these studies was, however, rather mixed. This may be due to the fact that stock price reactions typically observed during normal periods do not capture well the non-linear and more extreme asset price movements typically observed during periods of systemic events where large-scale contagion effects could be expected. In this light, some more recent market data studies have applied extreme-value theory to better capture such extraordinary events.[4] In a similar vein, Polson and Scott (2011) apply an explosive volatility model to capture stock market contagion measured by excess cross-sectional correlations. Stock market and CDS spread correlations were investigated with network-based techniques by Emmert-Streib and Drehmer (2010), and Peltonen, Scheicher, and Vuillemey (2013). Other studies have tried to capture the conditional spillover probabilities at the tail of the distribution by using quantile regressions.[5] Diebold and Yilmaz (2011) proposes in turn to use variance decompositions as connectedness measures to construct networks among financial institutions based on market data.

A different strand of the literature has been based on balance sheet exposures (such as interbank exposures and bank capital) with the aim of conducting counterfactual simulations of the potential effects on the network of exposures if one or more financial institutions encounter problems. This may overcome some of the deficiencies of the market data-based literature, such as the fact that asset prices can be subject to periods of significant mis-pricing, which may distort the signals retrieved from the analysis. The starting point to analyze bank contagion risks and interconnectedness on the basis of balance sheet data is having reliable information on the interbank networks. One can view a financial exposure or liability within a network as a relationship (or edge) of an institution (node) vis-à-vis another, whereby the relationship portrays a potential channel of shock transmission among the institutions. Mutual exposures of financial intermediaries are generally beneficial as they allow for a more efficient allocation of financial assets and liabilities, and are a sign of better diversified

3) See, e.g., Aharony and Swary (1983), Peavy and Hempel (1988), Docking, Hirschey, and Jones (1997), Slovin, Sushka, and Polonchek (1999), Cooperman, Lee, and Wolfe (1992), Smirlock and Kaufold (1987), Musumeci and Sinkey (1990), Wall and Peterson (1990), and Kho, Lee, and Stulz (2000).
4) See, e.g., Longin and Solnik (2001), Hartmann, Straetmans, and de Vries (2004), Hartmann, Straetmans, and de Vries (2005), Gropp, Duca, and Vesala (2009).
5) See, e.g., Cappiello, Gerard, and Manganelli (2005), Engle and Manganelli (2004), White, Kim, and Manganelli (2010), and Adrian and Brunnermeier (2011).

financial institutions.[6] At the same time, when large shocks hit the financial system, financial networks – especially if exposures are concentrated among a few main players – can act as an accelerator of the shock's initial impact by propagating it throughout the financial system via network links. As emphasized by Allen and Gale (2000), the underlying structure of the network determines how vulnerable it is to contagion.[7] For example, Allen and Gale (2000) emphasize the contagion risk prevailing in complete networks, that is, those having $\binom{N}{2}$ linkages, where N is the number of nodes.[8] It is, furthermore, emphasized in the literature that in the presence of asymmetric information about the quality of counterparties and of the underlying collateral, adverse selection problems may arise which can render the interbank networks dysfunctional in periods of distress.[9]

In most of the financial applications of the network theory, the network structures are reconstructed from a very limited information sets. Usually, the missing linkages are created using entropy maximizing techniques (Elsinger, Lehar, and Summer, 2006; Mistrulli, 2011) that may smooth out the real contagion effects in the system.[10]

The financial contagion literature is furthermore related to the complex network analysis in other academic fields (medicine and physics in particular). It thus relates to the so-called "robust-yet-fragile" network characterization, by which networks are found to be resilient to most shocks but can be susceptible to pernicious contagion effects when specific nodes are targeted.[11] Recent models of the interbank market that incorporates this knife-edge character of the financial networks include (Nier *et al.*, 2007; Iori *et al.*, 2008; Georg, 2011).

The content of this chapter is also related to research on the network formation, which was only recently pursued in finance. Understanding the emergence process of the interbank networks can be critical to control and mitigate these risks. Endogenous networks (and their dynamics) are a difficult problem since the behavior of the agents (banks in particular) is very complex. In other areas of social studies, the network formation was addressed by means of network-game techniques (Jackson and Wolinsky, 1996). In financial networks, researchers also applied recently game theoretical tools (Acemoglu, Ozdaglar, and Tahbaz-Salehi,

6) For example, interbank connections may produce co-insurance against liquidity shocks and may enhance peer monitoring; see e.g., Bhattacharya and Gale (1987), Flannery (1996), Rochet and Tirole (1996), and Freixas, Parigi, and Rochet (2000).
7) See also Battiston *et al.* (2009), Gai, Haldane, and Kapadia (2011), and Battiston *et al.* (2012). Nier *et al.* (2007) and Allen and Babus (2009) provides surveys of the recent literature.
8) Brusco and Castiglionesi (2007) in contrast highlights that in the presence of moral hazard among banks, in the sense that liquidity co-insurance via the interbank market entails higher risk-taking, more complete networks may in fact prove to be more, not less, contagious.
9) See, e.g., Flannery (1996), Ferguson *et al.* (2007), Heider, Hoerova, and Holthausen (2009), and Morris and Shin (2012).
10) An alternative interesting approach was taken by Anand, Craig, and von Peter (2013) who reconstructed networks of minimal potential density given the marginal constraints of the total interbank lending and borrowing volumes. For applications in other research disciplines, see also the example Tighe, van Eerd, and Vlugt (2008) and Tsuda and Noble (2004).
11) See, e.g., Albert, Jeong, and Barabáasi (2000), Barabási and Albert (1999), and Doyle *et al.* (2005).

2013; Babus and Kondor, 2013; Bluhm, Faia, and Krahnen, 2013; Cohen-Cole, Patacchini, and Zenou, 2011; Gofman, 2013) or portfolio optimization (Georg, 2011).[12] For instance, Acemoglu, Ozdaglar, and Tahbaz-Salehi (2013) shows that the equilibrium networks generated via a game on a space of interbank lending contracts posted by banks, can be socially inefficient since financial agents "do not internalize the consequences of their actions on the rest of the network."[13] In Cohen-Cole, Patacchini, and Zenou (2011) banks respond optimally to shocks to incentives to lend. Bluhm, Faia, and Krahnen (2013) approach to modeling endogenous interbank market is closely related to ours. However, the main distinctions from our approach are: risk neutrality of banks, riskiness of the interbank assets introduced to the model only via capital constraints, and not including funding risk as having a potential impact on the interbank structure. Castiglionesi and Lavarro (2011) presented a model with endogenous network formation in a setting with micro-founded banking behavior.[14] These advances notwithstanding, owing to the complexity of the equilibrium-based studies of network formation, agent-based modeling of financial networks is one promising avenue that can be followed (Grasselli, 2013; Markose, 2012). Furthermore, a few recent studies likewise highlight the importance of considering the various dimensions of interbank linkages (direct and indirect) for capturing the true contagion risk. In other words, these papers argue that a "holistic" modeling approach to the multiple layers of interbank relations is warranted to truly capture the contagion risks embedded in the financial system.[15]

This paper adds to this strand of the literature by taking a model of portfolio optimizing banks to a firm-level dataset of European banks, which in turn allows us to study within an endogenous network setting the impact of plausible internal limit systems based on credit valuation adjustments (CVAs) accounting for counterparty credit risk (Deloitte and Solum Financial Partners, 2013) and various regulatory policy measures on interbank contagion risk. In addition to the asset-liability optimizing behavior that we impose on the agents (i.e., the banks), our network formation model also incorporates sequential game theoretical elements. If the portfolio optimization of interbank investment and interbank funding does not lead to a full matching of interbank assets and liabilities, banks will engage in a bargaining game while taking into account deviations in their optimal levels of diversification of investment and funding risks (Rochet and Tirole, 1996).[16] The sequence of portfolio optimization and matching games is repeated until the full allocation of interbank assets at the aggregate level has been reached. The outlined mechanism is also related to the studies on matching in the loan

12) Some earlier contributions incorporating multi-agent network models, albeit with fixed network and static balance sheet assumptions, include (Iori, Jafaray, and Padilla, 2006; Nier *et al.*, 2007).
13) See also Gai and Kariv (2003) for an earlier contribution.
14) Other studies in this direction include (Babus, 2011; Castiglionesi and Wagner, 2013).
15) See, e.g., Gomez-Gardeñes *et al.* (2013), Foti *et al.* (2013) and Montagna and Kok (2013).
16) While not explicitly taken into account in this paper, this is related to the literature on interbank lending where due to asymmetric information, banks are not able to perfectly monitor their peers. Such information asymmetries may be reinforced by adverse shocks as, for example, experienced during the recent financial crisis, see (Heider, Hoerova, and Holthausen, 2009).

market (see, e.g., Chen and Song (2013) and Fox (2010)). Furthermore, to further reduce mismatches between banks' funding needs and the available interbank credit emerging from the portfolio optimizing choices, we introduce an interbank loan pricing mechanism that is related to models of money market price formation (Eisenschmidt and Tapking, 2009; Ewerhart *et al.*, 2004; Hamilton, 1998). Importantly, as argued by Afonso, Kovner, and Schoar (2011), such pricing mechanisms can be expected to be more sensitive to the borrower characteristics (and risks) during periods of stress. The model presented here would be able to account for such effects.

The chapter is structured as follows: Section 4.3 presents the model of network formation under optimizing bank behavior. In Section 4.4, some topology results from the network simulations are presented and it is illustrated how the model can be applied for studying various macro-prudential policy measures aimed at strengthening the resilience of the financial system. Section 4.6 concludes.

4.3 Models

We present two models of interbank contagion focusing on the derivation of interbank topologies from the usually very incomplete and aggregate set of data on interbank lending and borrowing. We develop our approaches under the belief that the real interbank network structures are richer than the usually assumed entropy-maximizing topologies, and that they are very much influenced by banks' active and optimizing behavior. The behavioral aspects of the interbank market are as much important for the topology of the network as some statistical properties that can be inferred from the data (e.g., incompleteness, core-periphery features, centrality, etc.).[17]

In both approaches, we try to reconstruct the interbank network based on aggregate balance sheet structures of banks and some general information on possible interbank lending relationship patterns. We assume that there are N banks in the system. Aggregate interbank borrowing and lending volumes are known for all the banks; each bank i lends a_i interbank placements and borrows l_i interbank deposits. For simplicity, all interbank deposits are assumed to be of the same maturity. This usual assumption allows for abstracting from some intertemporal effects on the interbank market (e.g., related to market liquidity). Each bank i holds capital e_i protecting it against most of the probable losses it may incur in the nearest future. The risk of the losses is quantified by the risk weighted assets (RWA_i) which is a common regulatory measure for all banks in the financial system. Lending relationship is understood as a probability measuring likelihood

17) *Notation for the rest of the chapter*: $\mathbb{N}$ stands for set $\{1, 2, \dots, N\}$, "$*$" denotes entry-wise multiplication, that is, $[x_1, \dots, x_N] * [y_1, \dots, y_n] := [x_1 y_1, \dots, x_N y_N]$, "$^\top$" is transposition operator and – for matrix X – $X_{\cdot j}$ denotes jth column of X and $X_{i\cdot}$ denotes ith row of X, $\#C$ – number of elements in a set C, $\mathbb{I}_A$ denotes indicator function of a set A.

of a placement extended by one bank to another. The second, behavioral model assumes additionally that risk and return characteristics of the interbank lending and borrowing are available. These risk and return parameters are related to the interbank interest rate levels and volatility and a default risk of an interbank borrower. All these pieces of information can be found in public domain; in banks' financial reports, in regulatory disclosures, or in commercial databases. The main goal of the two model suits is to derive an $N \times N$ matrix $L = [L_{ij}]$ of bilateral interbank exposures, whereby L_{ij} denotes a placement of bank j in bank i. The subsequent interbank contagion analysis, illustrated in Section 4.4, is based on the structure of the matrix L.

4.3.1 Simulated Networks

A standing assumption of many interbank contagion models estimating the linkages on the interbank market imposes an entropy-maximizing structure of an interbank network. This may lead to an inaccurate measuring of risk transmission of financial problems of some banks to other participants of the market. In contrast, the approach that we take relaxes a constraint of having just one particular interbank structure to study and analyzes all possible structures permitted by the aggregate data on interbank lending and borrowing. Therefore, it can be thought of as more a *simulation* than *estimation* technique.

The model of the simulated interbank networks relies on a random generator of various possible structures of interbank networks in which we apply a shock to one bank (or a set of banks) that is subsequently transmitted within this interbank system. The network is generated based on the banks' balance sheet data, on their total interbank placements. and deposits and on the assessment of the banks' geographical breakdown of activities. Notably, we do not have data on the individual banks' bilateral exposures, which are instead derived based on their total interbank placements and deposits. All in all, the proposed approach to interbank contagion analysis helps to overcome the usual deficiency of data on bilateral exposures.

The model is parameterized for large 89 banks, mostly from euro area countries. These are banks included in the EU-wide stress tests conducted by the European Banking Authority (EBA), but the data used to parameterize the model are taken from Bureau van Dijk's Bankscope and the banks' financial reports. The resilience of the resulting simulated interbank structures to defaults of banks is analyzed in a contagion framework *à la* (Eisenberg and Noe, 2001). The framework allows for measuring of a propagation of a default shock across the network of interbank liabilities. The default shock is simply meant to be a given bank's default on all its interbank payments. It then spreads across the banking system, transmitted by the interbank network of the simulated bilateral exposures.

There are three main building blocks of the model. First, the probability map that a bank, in a given country, makes an interbank placement to a bank in another (or the same) country was proposed; second, an iterative procedure to generate interbank networks by randomly picking a link between banks and accepting it with probability taken from the probability map. Finally, the algorithm of clearing payments proposed by Eisenberg and Noe (2001) on the interbank market in two versions was applied: without and (modified) with a "fire sales" mechanism.

4.3.1.1 **Probability Map**

Bank-by-bank bilateral interbank exposures are not readily available. For that reason, to define the probability structure of the interbank linkages (a probability map P^{geo}), as a starting point, the EBA disclosures on the geographical breakdown of individual banks' activities (here, measured by the geographical breakdown of exposures at default) were employed.[18] The probabilities were defined at the country level, that is, the exposures were aggregated within a country and the fraction of these exposures towards banks in a given country was calculated. These fractions were assumed to be probabilities that a bank, in a given country, makes an interbank placement to a bank in another (or the same) country.

The probability map based on the EBA disclosures is an arbitrary choice contingent on the very limited availability of data about interbank market structures. An idea of the market fragmentation along the national borders, while treating separately the internationally active banks, seems to be justified. Nevertheless, the results (the structure of the network and the contagion spreading) are dependent on the particular probability structure (geographical proximity matters). In results Section 4.4, we perform some sensitivity analysis of the systemic importance of banks if the probability map is distorted.

4.3.1.2 **Interbank Network**

The network is generated randomly based on the probability map P^{geo}. A possible interbank network (realization from a distribution of networks given by the probability map) is generated in the following way. A pair of banks (i, j) is randomly drawn (all pairs have equal probability) and the pair is kept as an edge (link) in the interbank network with a probability given by the probability map. It means that bank j lends to bank i. If the drawn link is kept as an interbank exposure, then the random number is generated (from the uniform distribution on $[0, 1]$) indicating what percentage of reported interbank liabilities (l_i) of the first bank in the pair comes from the second bank in the pair (the amount is appropriately truncated to account for the reported interbank assets (a_j) of the second bank). If not kept, then the next pair is drawn (and accepted with a corresponding probability or not). Ultimately, the stock of interbank liabilities and assets is reduced by the volume of the assigned placement. The procedure is repeated until no more interbank liabilities are left to be assigned as placements from one bank to another.

Analysing many different interbank structures instead of just one specific (either observed at the reporting date or – if not available – estimated, e.g., by means of entropy measure) accounts for a very dynamic, unstable nature of the interbank structures confirmed by many studies (Gabrieli, 2011; Garratt, Mahadeva, and

18) The bank level exposure data were downloaded from the EBA website: *http://www.eba.europa.eu*.

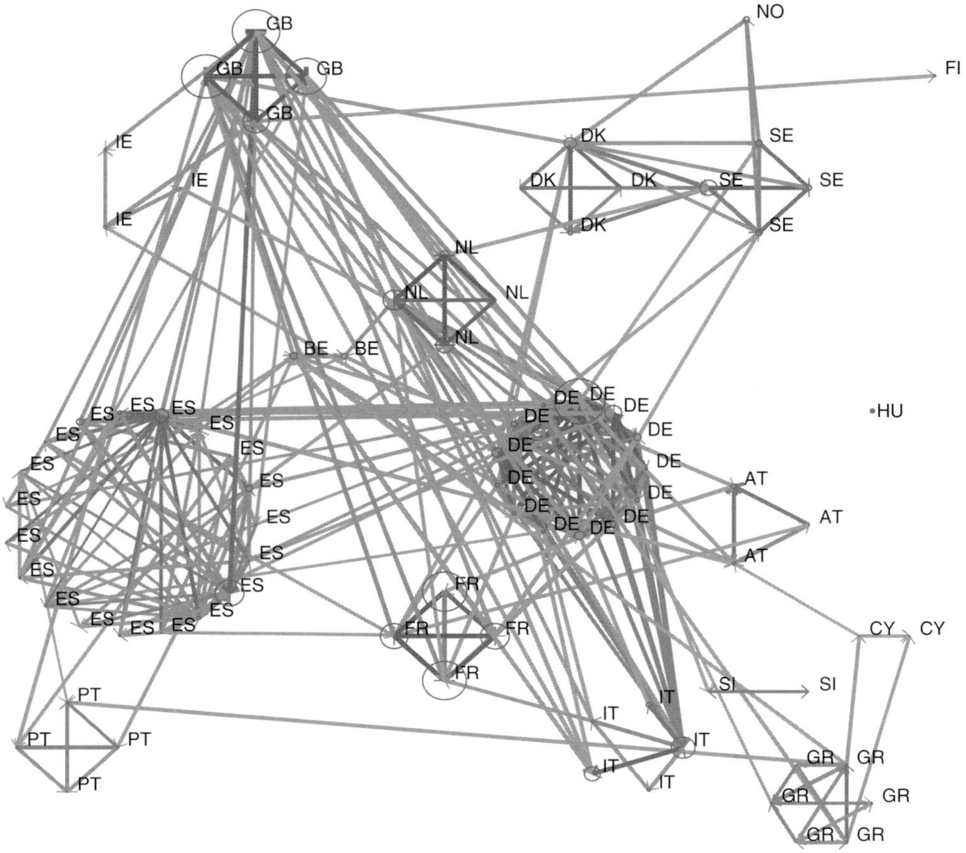

Figure 4.1 A generated interbank network. (See inset for color representation of this figure.) Note: an arrow between bank A and B indicates an interbank deposit of bank B placed in bank A; the width of an arrow reflects the size of the exposure; the lighter the green color of an arrow, the lower the probability that the arrow joins a given pair of banks. Source: own calculations.

Svirydzenka, 2011). We construct 20 000 structures for the purpose of our contagion analysis. The way in which linkages are drawn may still be an issue for the distribution of the whole network. It may underestimate the probability of networks in which nodes have many linkages of similar size. However, the algorithm does not exclude such configurations, which are typical for the real interbank networks with money centers.

Figure 4.1 illustrates one realization from the whole distribution of network structures for the EU banking sector generated using the random network modeling approach. The width of the arrows indicates the size of exposures (logarithmic scale) and the coloring scale (from light to dark green) denotes the probability (inferred from the interbank probability map) that a given bank grants an interbank deposit to the other bank. Most of the connections are between banks from the same country but the connectivity between the biggest domestic banking systems is also quite high (the German, Spanish, and British banking systems, in particular).

The very general characteristics of the network and of the role played by the particular nodes can be performed by means of some standard network statistics. The simulated network approach gives the whole distribution of measures that, further statistically analyzed, may indicate some candidate banks to be systemically important. We looked at three centrality measures in the result section – that is, degree, closeness, and betweenness, which inform about network activity, independence of nodes, and nodes' control of activity in the network, respectively.[19)]

4.3.1.3 Contagion Mechanism

Our assessment of the size of the interbank contagion is inspired by the so-called interbank clearing payments vector, derived by Eisenberg and Noe (2001) and which we define in our modification by a vector p^* solving the following equation,

$$p^* = \min\{\max\{e - a + l + \pi^\top p^*, 0\}, l\} \tag{4.1}$$

where π is a matrix of the relative interbank exposures with π_{ij} entry defined as bank j interbank exposure toward bank i divided by the total interbank liabilities of bank i. The expression $e - a + l$ can be interpreted as banks' own funding sources adjusted by the net interbank exposures. The interbank liabilities l are a proxy for a buffer set aside in the assets assuming that banks keep some liquid sources to cover the potential outflow of the interbank funding. Any decline in this buffer can be introduced via capital e shock. The ultimate interbank payments are derived as the equilibrium of flows in the interbank network. The contagious default on the interbank deposits is detected by comparing l_i and p^* – if the difference is >0, then it means that bank i defaults on its interbank payments. The loss for the interbank creditors is calculated as

$$loss = \pi^\top (l - p^*)$$

The applied clearing payments vector procedure does not require any assumption about the size of interbank loss incurred at default of a counterparty. The loss given default (LGD) is endogenous and can be expressed as a loss ratio L^*:

$$L_i^* = \frac{a_i - (\pi^\top p^*)_i}{a_i},$$

where $(v)_i$ denotes ith component of vector v. In order to compare the interbank losses in a standardised way across the banking system, we calculate an impact of the losses on a capital adequacy measure (CAR) defined as the Core Tier 1 capital divided by the RWA. Consequently, the CAR reduction of bank i as a result of the

19) Further interesting reading about the application of network measures can be found in von Goetz (2007).

losses incurred on the interbank exposures is defined as,

$$\Delta\mathrm{CAR}_i = 100 \cdot \left(\frac{\mathrm{CT1}_i - loss_i}{\mathrm{RWA}_i} - \mathrm{CAR}_i \right)$$

The equilibrium payments vector is calculated in an iterative (sequential) procedure. Namely, let us define a function $F : [0, l_1] \times \cdots \times [0, l_N] \rightarrow [0, l_1] \times \cdots \times [0, l_N]$ as

$$F(p) = \min\{\max\{e - a + l + \pi^\top p, 0\}, l\} \tag{4.2}$$

The value of F for a given p can be interpreted as the vector of the interbank payment, given the banks receive back as much as $\pi^\top p$ of their interbank assets. It can be shown that a sequence (p^n) defined as $p^0 = l$ and $p^n = F(p^{n-1})$ converges to the clearing payments vector p^*.

In an event-driven concept of contagion, it is interesting to decompose the first and second round effects of contagion. First, we introduce a notion of a triggering bank, that is, a bank that initially defaults on their interbank deposits (due to some exogenous shock not encompassed by the model). Second, we define the first round effects as those related purely to the default of banks on their interbank payments given

- default of a triggering bank or a group of triggering banks on all its interbank deposits,
- all other banks declaring to pay back all their interbank debts.

Third, the default of other banks following bank-triggers' inability to pay back their interbank debts would be classified as second round contagion effects if,

- they would pay back all their debts, if all non-triggering banks, which are their debtors, returned their debts,
- they are not capable of paying back part of their interbank deposits in the clearing payments equilibrium.

Formal definition of the decomposition of contagion effects are presented in Hałaj and Kok (2013a).

4.3.1.4 Fire sales of Illiquid Portfolio

The concept of the sequence (p^n) is helpful in introducing the "fire sales" mechanism to the interbank equilibrium. In order to meet their obligations, banks may need to shed part of their securities portfolio; the less interbank assets they receive back, the higher is the liquidation need. This may adversely affect the mark-to-market valuation of their securities portfolios and further depress their capacity to pay back their interbank creditors. Consequently, this mechanism may lead to a spiral effect of fire sales of securities (as, for example, suggested in recent papers by Geanakoplos (2009) and Brunnermeier (2009).

Banks may respond in different ways to the losses on the interbank exposures depending on their strategies and goals. In order to cover the resultant liquidity shortfall, they may simply shed some assets. However, the sell-off may be much more severe for banks targeting their capital ratio, as in Cifuentes, Ferrucci, and Shin (2005), leverage ratio (see also Adrian and Shin (2010)). In the latter case, the usually double digit ratio "x" of assets to capital would translate into securities disposal of "$x * $ loss". We account for both cases in our modeling framework of the "fire sales". Again, details of fire sales mechanism are presented in Hałaj and Kok (2013b).

4.3.2
Systemic Probability Index

Simulations of a large number of interbank networks is very costly in terms of the computation time. The main goal of the section is to define a simplified measure of systemic fragility in the system derived from the assumptions of the simulated networks. We have four general objectives:

- building an index (called systemic probability index (SPI)) measuring the contagion risk stemming from the interbank structure rather than the risk related to an external shock;
- taking into account the whole range of possible interbank structures accounting for the probability map introduced in Section 4.3.1.1;
- designing it in such a way that it is easy and fast to compute for large interbank systems, at least substantially reducing the time of Monte Carlo simulations;
- being consistent with the simulation as far as the most systemically important banks are concerned.

SPI reflects the likelihood of the contagion spreading across the banking system after a default of a given bank on its interbank debt. Therefore, it is a bank-specific measure, depending on the distribution of the interbank deposits and placements among banks and on the probability map of linkages. The rest of the section is devoted to describing the sufficient assumptions underlying the construction of an index satisfying the four general objectives.

Our starting point was to use a probability structure based on the simulated interbank networks to construct a measure of how likely, how broad, and how fast is the interbank contagion spreading after a given bank defaults on all its interbank payments. Let us suppose that a node I defaults on its interbank payments. What is the probability that node j defaults? In short, it is a probability that losses of bank j incurred on its interbank exposures against I surpass bank's j capital. Formally, for a loss ratio $\eta_I \in [0,1]$, this probability can be expressed as:

$$p^{(1)}_{Ij} = \mathbf{P}\left(G_{Ij}\pi_{Ij}\eta_I l_I > e_j\right),$$

where G_{ij} is a random variable taking values from the set $\{0,1\}$, whereas value 1 occurs with probability P^{geo}_{ij}. By introducing G, we mimic the randomness of the simulated networks as far as the accept-reject algorithm to establish links between banks concerned. The expected payment $\pi_{Ij}l_I$ is the liability of bank I toward j. The adjusted value $\eta_I\pi_{Ij}l_I$ is the corresponding LGD of bank I. Finally, G_{Ij} informs about the probability that the link between I and j exists. Therefore, G introduces the probabilistic nature of the interbank structure. The relative exposure π_{Ij} can formally be characterised by the joint probability of values of all elements of the matrix π.

What is the impact of a default at round k on the probability of default at round $k+1$? More precisely, what is the relationship between probability of default at k and $k+1$? Let us assume that the default at k means that the whole volume of debt is not returned back by the defaulted bank to its creditors. Thus,

$$P_{Ij}^{(k+1)} = \mathbf{P}\left(\sum_{i=1}^{N} G_{ij}\pi_{ij} \cdot \eta_i P_{Ii}^{(k)} l_i > e_j\right) \tag{4.3}$$

is the probability of default of bank j at time $k+1$ given that the probabilities of default of banks 1, 2,... N at time k are $P_{I1}^{(k)}, P_{I2}^{(k)}, \ldots P_{IN}^{(k)}$, respectively.

Hałaj and Kok (2013c) study the properties of an approximation to $P^{(k)}$ assuming the joint normal distribution of exposures between banks in pairs drawn from distribution P^{geo}.

A vector measure $P_{I.}^{(k)}$ should be aggregated across the banking system to obtain a scalar and comparable measure of bank's default impact on the interbank system, that is, Systemic Probability Index. For instance, we weigh the individual indices at their limits by banks' total assets. that is:

$$\mathrm{SPI}_I = \frac{\sum_{j=1}^{N} \mathrm{TA}_j P_{Ij}^{(\infty)}}{\sum_{j=1}^{N} \mathrm{TA}_j}. \tag{4.4}$$

Consequently, the SPI_I measures an aggregate risk of a contagion impact of bank I's default.

4.3.3
Endogenous Networks

The simulated network approach to the interbank contagion analysis presented in the previous section relaxes the very limiting assumption of entropy-maximizing structures but still leaves aside all important behavioral aspects of interbank network. In reality, banks actively manage counterparty risk in their interbank portfolios and respond very quickly to any signs of potential deterioration of credit quality of their interbank debtors. They also assess the funding risk of their interbank borrowing portfolio to minimize the risk of losing a large chunk of their funding sources in case some of their interbank peers refuse to roll-over funding. The interbank network described in this section is an outcome of a sequential game played by the banks trying to invest on the interbank market and to borrow interbank funding. Banks optimize their interbank assets taking into account the risk and regulatory constraints, as well as the demand for the interbank funding and propose their preferred portfolio allocation. For what concerns the funding side, banks define their most acceptable structure of funding sources with the objective to limit refinancing (roll-over) risk. Banks meet in a bargaining game in which the supply and demand for interbank lending is determined. In order

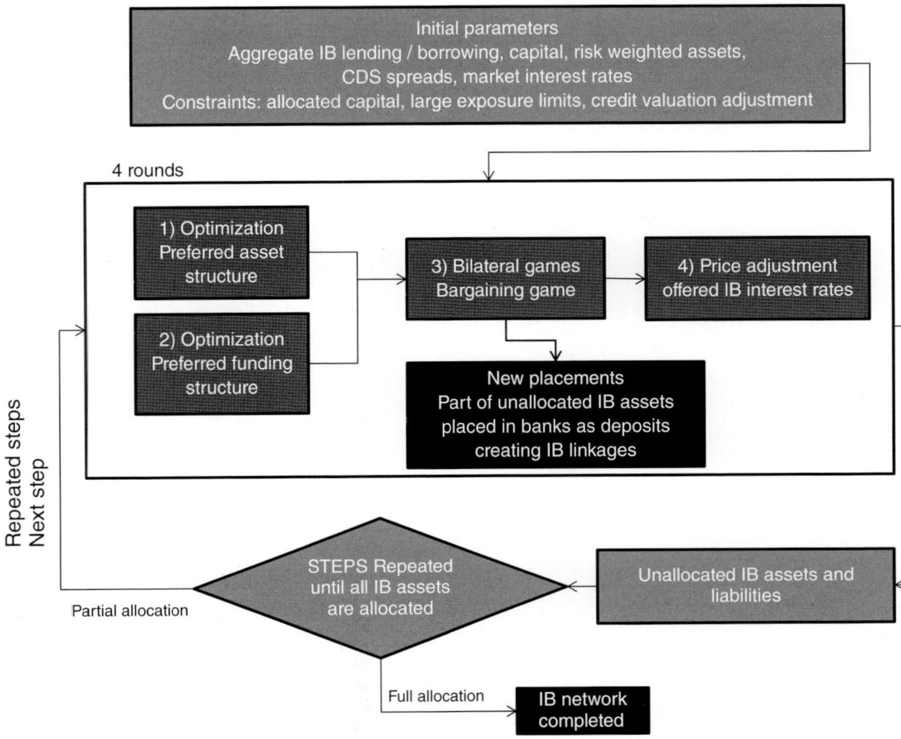

Figure 4.2 The sequential four-round procedure of the interbank formation.

to account for the quite complex aspects of the interbank market formation, we propose a sequential optimization process, each step of which consists of four distinctive rounds (see the block scheme in Figure 4.2).

There are three main general assumptions of the model:

1) Banks know their aggregate interbank lending and borrowing, as well as those of other banks in the system.
2) Banks optimize the structure of their interbank assets, that is, their allocation across counterparties.
3) Banks prefer diversified funding sources in terms of roll-over risk (i.e., liquidity risk related to the replacement of the maturing interbank deposits).

This first, rather strong assumption has its motivation in the stable overtime fraction of interbank assets to total liabilities, confirmed empirically in a sample of 89 largest EU banks.[20] In theory, part of those assets and liabilities, in particular with the shortest maturities, can be volatile since it reacts to volatile banks' liquidity needs. However, the interbank investment portfolio and interbank funding portfolio may be much more stable since their volumes should result from a

20) A standard deviation of quarterly ratios of interbank assets or interbank liabilities to total assets amounts on average to 2.5%.

general asset-liability planning within the asset and liability modeling (ALM) process defining, inter alia a target for product mix of assets and funding sources and income parameters.

The second assumption follows the standard portfolio choice theory. Optimization of counterparty structure in the interbank portfolio is constrained by regulatory liquidity and capital rules, and the relationship lending; banks are assumed to optimize their portfolio in a set of counterparties with whom they built up relationship lending. The third assumption refers to the set of banks' counterparties. Based on the lending relationship, each bank has a subgroup of partners on the interbank market with whom it is likely to trade. It is reasonable to assume that banks try to minimize funding risk in their subgroups. Notably, there is some empirical evidence (Bräuning and Fecht, 2012) that the relationship lending may impact the pricing of interbank loans and consequently, also the funding structure. All in all, the decision about the funding structure is a function of diversification needs and build-up of relationship lending.

In the *first round*, banks specify the preferred allocation of interbank assets by maximizing the risk-adjusted return from the interbank portfolio. Banks are assumed to be risk averse which follows the approach taken in the capital management practice, where RAROC[21] and RARORAC are common practice in banking for managing accepted levels of exposure and are standard part of ALM indicators (Adam, 2008).[22]

In the optimization process of the interbank assets, each bank first draws a sample of banks according to a pre-defined probability map P^{geo} describing the likelihood of connections between banks. The probability map has already been introduced in the Section 4.3.1.1 using the geographical breakdown of banks' exposures, disclosed during the EBA 2011 capital exercise. Second, they make offers of interbank placements trying to maximize the return adjusted by investment risk. Banks take into account:

- expected interest income;
- risk related to interest rate volatility and potential default of counterparts, and correlation among risks;
- internal risk limits for capital allocated to the interbank portfolio, based on the CVA concept[23] and regulatory constraints in the form of large exposure limits specifying the maximum size of an exposure in relation to the capital base;
- exogenous volume of total interbank lending.

21) Risk-Adjusted Return on Capital and Risk-Adjusted Return on Risk-Adjusted Capital.
22) The literature of the banking theory is not equivocal about risk averseness of banks. The traditional strand of literature builds on the assumption of risk neutrality (Baltensperger, 1980; Boyd and Nicoló, 2005; Pelizzon and Schaefer, 2005). In contrast, Howard and Matheson (1972), Danielsson, Jorgensen, and de Vries (2002), and Cuoco and Liu (2006) models banks' decisions as risk sensitive.
23) This CVA element is not to be mistaken with the CVA capital charge on changes in the credit spread of counterparties on OTC derivative transactions. However, the line of calculation is similar. Some banks use CVA internally to render the exposure limits, sensitive to the counterparty risk, in a consistent, model-based way (Deloitte and Solum Financial Partners, 2013).

Notably, the structure rather than the aggregate volume of lending is optimized. The aggregate interbank lending and borrowing of banks in the model is exogenous.

Obviously, the recipients of the interbank funding can have their own preferences regarding funding sources. Therefore, in the *second round* of the model, after the individual banks' optimization of interbank assets, banks calculate their optimal funding structure, choosing among banks that offered funding in the first round. They decide about the preferred structure based on the funding risk of the resulting interbank funding portfolios.

The offers of interbank placements may not fully meet the funding needs of banks. In the *third round* we, therefore, assume that banks negotiate the ultimate volume of the interbank deposit. We model these negotiations by means of a *bargaining game* in which banks may be more or less willing to deviate from their optimization-based preferred asset-liability structures. Notably, also at this round, banks take into account their risk and budget constraints.

Since interbank asset and interbank funding optimization followed by the game may not result in complete allocation of the predefined interbank assets and in full satisfaction of all the interbank funding needs, the prices on the interbank market may be adjusted. In the *fourth round* banks with an open funding gap are assumed to propose a new interest rate for the new interbank investors depending on the relative size of the gap to their total interbank funding needs. Implicitly, we do not model the role of the central bank which normally stands ready to provide liquidity.

The four consecutive rounds are repeated with a new drawing of banks to be included into subsamples of banks with which each bank prefers to trade. Consequently, each bank enlarges the group of banks considered to be their counterparties on the interbank market and proposes a new preferred structure of the interbank assets and liabilities for the unallocated part in the previous step. In this way, the interbank assets and liabilities are incrementally allocated among banks.

A modeling of the network formation process in a sequential way is obviously stylised as in reality, banks are likely to conduct many of the steps described here in a simultaneous fashion. At the same time, the step-by-step approach is a convenient way of presenting the complex mechanisms that determine the formation of interbank linkages, which may realize in a very short time-span, even only several ticks long.

The following sections describe in detail how the endogenous networks are derived. Some important notations used thereafter are introduced later.

4.3.3.1 **Banks**

First, a description of the banks' balance sheet structures, interbank assets and liabilities in particular, is warranted. Each institution i aims to invest a_i volume of interbank assets and collect l_i of interbank liabilities. These pre-defined volumes are dependent on various exogenous parameters. For instance, individual banks' aggregate interbank lending and borrowing can be an outcome of ALM.[24] The interest rates paid by interbank deposits depend on:

- some reference market interest rates r^m (e.g., the 3-month offered interbank rate in country m),
- a credit risk spread (s_i) reflecting the credit risk of a given bank i,
- a liquidity premium q_i referring to the general market liquidity conditions and bank i's access to the interbank market,[25]
- LGD related to the exposure, denoted λ.

The LGD is assumed to be equal for all banks and exposures and amounting to 40%. All interbank assets and liabilities have the same maturity.

The credit spread s_i is translated into a bank-specific interest rate paid by bank i to its interbank creditors$-r_i$. It is based on the notion of equivalence of the expected returns from interbank investment to a specific bank and from investing into the reference rate r^m,

$$r^m + q_i \equiv r_i p_i \lambda + (1 - p_i) r_i, \tag{4.5}$$

where p_i denotes marginal probability of default on the interbank placement, extended to bank i and is calculated as

$$p_i : = s_i / \lambda,$$

Interest rate r_i can be interpreted as a rate that realizes the expected return of r^m given the default risk captured by the spread s_i.[26] We use a very basic approximation of the default probability p_i derived from the spread s_i but still we are able to gauge differences in default risk among banks.

Moreover, the cost–or return from the interbank placement perspective–is risky. The riskiness is described by a vector $\sigma : = [\sigma_1 \cdots \sigma_N]^\top$ of standard deviations of historical (computed) rates r_i and 4.3.1 correlation matrix Q of these rates calculated from Eq. (4.5) taking into account time series of interbank rates and CDS spreads. The riskiness stems from the volatility of the market rates and variability of default probabilities. Likewise, correlation is related to:

- the common reference market rate for banks-debtors in one country or co-movement of reference rates between countries;
- to the correlation of banks' default risk.[27]

The probability map P^{geo} described in Section 4.3.1.2 plays a role in defining the lending relationship also in the endogenous formation of the interbank network

24) Georg (2011) or Hałaj (2013) developed frameworks based on the portfolio theory to optimize the structure of investments and funding sources that could be followed.
25) We assume for simplicity that $q \equiv 0$ while indicating how liquidity can be captured in the framework.
26) Currency risk related to the cross-border lending between countries with different currencies is not addressed in the model.
27) Reason: banks operate on similar markets, have portfolios of clients whose credit quality depends on similar factors, their capital base is similarly eroded by the deteriorating market conditions, and so on.

structure. Banks are assumed to trade on the interbank market according to their established customer relationship. The relationship is captured by the banks' geographical proximity as well as the international profile of the bank. It is assumed that banks are more likely to trade with each other if they operate on the same market. The probability map is used to sample banks with which a given bank intends to trade.

Banks are also characterised by several other parameters not related to the interbank market but important in our framework from the risk absorption capacity perspective.

- banks' risk aversion is measured by $\kappa \geq 0$;
- capital e_i; and capital allocated to the interbank exposures e_i^I (e.g., economic capital budgeted for treasury management of the liquidity desk);
- RWA_i – similarly, RWA_i^I risk-weighted assets calculated for the interbank exposures. This may depend on the composition of the portfolio, that is, exposure to risk of different counterparts;
- CVA assumed to impact the economic capital and, consequently, the potential for interbank lending. For simplicity, we assume that an interbank exposure of volume L_{ij} requires $\gamma_i L_{ij}$ to be deducted from capital e_j^I, for γ_i being bank specific CVA factor, to account for the market-based assessment of the credit risk related with bank i. The parameter γ can also be viewed as a risk sensitive add-on to the risk weights (RWs) applied to the interbank exposures.

All the aforementioned balance sheet parameters are used in the following sections to define the banks' optimal investment and funding programs.

4.3.3.2 First Round – Optimization of Interbank Assets

Each bank is assumed to construct its optimal composition of the interbank portfolio given the market parameters, risk tolerance, diversification needs (also of a regulatory nature), and capital constraints (risk constraints including the CVAs introduced within Basel III).

The formation of the interbank network is modeled in a sequential way. We set the initial values of banks' assets and liabilities to be matched on the interbank market at the steps $k = 1, 2, 3...$ and of a structure of the interbank network, that is, for $k = 0$

$$\overline{l}^0 = l,$$

$$\overline{a}^0 = a,$$

$$\overline{L}^0 = N \times N \text{ matrix of } 0.$$

Vectors $\overline{a}^k, \overline{l}^k$ denote banks' aggregate interbank lending and borrowing which is still not allocated among banks before step k. A matrix $\overline{L}^k$ denotes the structure of linkages on the interbank market created up to the step k of the algorithm. Additionally, for notational convenience, we denote $\overline{B}_j^0 = \emptyset$ the initial empty set of banks in which a given bank j intends to invest.

At step k, bank j draws a sample of banks $B_j^k \subset \overline{\mathbb{N}}/\{j\}$. Each counterparty i of the bank j is accepted with probability P_{ij}^{geo}. Banks from the set B_j^k are assumed to enlarge the set of investment opportunities of bank j, that is, $\overline{B}_j^k = \overline{B}_j^{k-1} \cup B_j^k$. At step k, the bank considers (optimally) extending interbank placements to bank borrowers $\overline{B}_j^k$.

Bank j maximizes the following functional representing a risk-adjusted return from the interbank investment:

$$J(L_{1j}^k, \dots, L_{Nj}^k) = \sum_{i|i\neq j} r_i^k L_{ij}^k - \kappa_j(\sigma * L_{\cdot j}^k)^\top Q(\sigma * L_{\cdot j}^k), \tag{4.6}$$

where $r^1 \equiv r$ and rates r^k in steps $k \geq 2$ of the endogenous network algorithm can vary according to the adjustments related to the funding needs of banks that have problems with finding enough interbank funding sources (Section 4.3.3.5). The vector of risk measures σ was defined in Section 4.3.3.1. The interest rates r^k paid by the interbank deposits are the transaction rates defined by Eq. (4.5) and the risk – both related to market interest rate risk and default risk – is captured by the covariance $(\sigma * L_{\cdot j}^k)^\top Q(\sigma * L_{\cdot j}^k)$.

The maximization of the functional (4.6) is subject to some feasibility and capital constraints.

1) budget constraint – $\sum_{j|j\neq i} L_{ij}^k = \overline{a}_j^k$ and $L_{jj}^k = 0$, whereby $\overline{a}_i^0$ is exogenously determined;
2) counterpart's size constraint – $L_{ij}^k \leq \overline{l}_i^k$;
3) capital constraint – $\sum_{i|i\neq j} \omega_i(\overline{L}_{ij}^k + L_{ij}^k) \leq e_j^I - \gamma \cdot (\overline{L}_{\cdot j}^k + L_{\cdot j}^k)$ or equivalently $\sum_{i|i\neq j}(\omega_i + \gamma_i)(\overline{L}_{ij}^k + L_{ij}^k) \leq e_j^I$;
4) (optionally) large exposure (LE) limit constraint – $(\overline{L}_{ij}^k + L_{ij}^k) \leq \chi e_j$.

Given the risk constraints and the general riskiness of the interbank lending, it may not be possible for a bank i to place exactly $\overline{a}_i^k$ interbank deposits in total in step k. Therefore, the budget constraint may not be plausible – as a consequence, the bank i should consider lending less.[28] We apply a compromising iterative procedure, whereby we assume that banks solve the problem with the budget constraint $\overline{a}_i^k$ replaced with $\overline{a}_i^k - \Delta\overline{a}_i^k$, for some (small enough and positive) $\Delta\overline{a}_i^k$. The optimization is solved for the smallest possible k_i. The procedure can be interpreted as banks' gradual adjustments of the total interbank assets until the risk requirements are satisfied.

4.3.3.3 Second Round – Accepting Placements According to Funding Needs

The funding side of the interbank market is assumed to accept placements according to their preferred funding structure. Banks try to diversify their interbank funding sources.

28) In an extreme case, also the large exposure limits constraints may prove to be too severe. The system is not solvable if there exists a pair (k, j) such that $\chi \sum_{j|j\neq k} e_j < l_k$, which means that bank k is not able to find the predefined volume l_k of the interbank funding.

Each bank i aims at minimizing the funding risk. It is assumed that a default of a creditor results in an inability to roll over funding which means materialization of the funding risk. The risk of bank i accepting funding from j is measured by X_j, which is a random variable taking value 0 or 1: 0 with probability p_j inferred from the credit default spread s_j (Section 4.3.3.1) and 1 with probability $1 - p_j$. Probability p_j is a realization of a random variable $\tilde{p}_j$ implicitly defined by the following conditions. For a uniformly distributed u_j on the interval [0, 1], independent of $\tilde{p}_j$ and u_i for $i \neq j$, X_j has the following concise representation:

$$X_j = \mathbb{1}_{\{u_j > \tilde{p}_j\}}$$

For a vector of deposits $[L^k_{i1}, \dots, L^k_{iN}]$ it is quantified by $F : \mathbb{R}^N_+ \to \mathbb{R}$ defined:

$$F(L^k_{i1}, \dots, L^k_{iN}) = \kappa^F [L^k_{i1} \ \cdots \ L^k_{iN}] \overline{D}^2_X [L^k_{i1} \ \cdots \ L^k_{iN}]^\top, \tag{4.7}$$

where κ^F is funding risk aversion parameter. $\overline{D}^2_X$ denotes the covariance matrix of $[X_1, \dots, X_N]$, which implicitly defines the first two moments and covariance of random variables $\tilde{p}_1, \dots, \tilde{p}_N$.

Minimization of the funding risk for bank i means solving the following program:

$$\text{minimize } F(y) \text{ on } \underbrace{\mathbb{R} \cup \{0\} \times \cdots \times \mathbb{R} \cup \{0\}}_{N}$$

subject to

- relationship lending rule: $j \notin \overline{B}^k_j \Rightarrow y_j = 0$;
- budget constraint:

$$\sum_j y_j = \overline{l}^k_i,$$

- limit on the cost of funding: $(\overline{L}^k_{i\cdot} + L^k_{i\cdot})r \leq r^l_i$. Banks are willing to pay on their interbank funding rates on average r^l_i. This internal limit is related to the expected profitability of assets. The monitoring of such limiting values are critical for banks' income management processes. Typically, limits are implied by budgeting/funding transfer pricing (FTP) systems (see Adam (2008) for definitions and applications). In order to deactivate this option for a bank i, r^l_i needs to be set to a very large number.

The minimising vector is denoted $L^{F,k}_i$. The optimization of the funding portfolio is performed by all the banks in the system simultaneously.

4.3.3.4 Third Round–Bargaining Game

The structure of the interbank asset viewed from investment portfolio perspective may be, as is usually the case, different from the one derived from the interbank funding structure. In those instances, banks may need to somewhat deviate from their optimized interbank asset-liability structure and therefore, enter into negotiations with other banks in a similar situation. In order to address the issue about banks' willingness to accept a counteroffer to the optimization-based placement, we consider each pair of banks entering a type of a bargaining game with utilities (or disutilities) reflecting a possible acceptable deviation from the optimal allocation of portfolios. The game is performed simultaneously by all pairs of banks. The disutility – which is assumed to be of a linear type – is measured by a change of the optimized functional to a change in the exposure between the preferred volumes.

More specifically, the proposed games give one possible solution to the following question: What may happen if at step k, bank j offers a placement of $L_{ij}^{I,k}$ in bank i and bank i would optimally fund itself by a deposit $L_{ij}^{F,k}$ from bank j, which is substantially different in volume from the offered one? Perhaps, the banks would not reject completely the offer since it may be costly to engage in finding a completely new counterparty. By doing that, they may encounter risk of failing to timely allocate funds or replenish funding since the interbank market is not granular. Instead, we assume that these 2 banks would enter negotiations to find a compromising volume. We model this process in a bargaining game framework. Banks have their disutilities to deviate from the optimization-based volumes. The more sensitive their satisfaction is to the changes in the individually optimal volumes, the less willing they are to concede. We assume that each pair of banks play the bargaining game at each step of the sequential problem in isolation taking into account their risk constraints. This is a key assumption bringing the framework to a tractable one.

4.3.3.5 **Fourth Round – Price Adjustments**

Both the individual optimization and the bargaining game at round k may not lead to the full allocation of the interbank assets and there may still be some banks striving for interbank funding. By construction of the bargaining game, there are no banks with excess funding sources. In order to increase the chance of supplementing the interbank funding in the next step, banks with interbank funding deficiency adjust their offered interest rate. The adjustment depends on the uncovered funding gap. Let us assume that the market is characterised by a price elasticity parameter α, which translated the funding position into the new offered price. If at the step $k+1$ the gap amounts to $g_i^{k+1} : = l_i - \sum_j \overline{L}_{ij}^{k+1}$, then the offered rate $r_i^{k+1} = r_i^k \exp(\alpha g_i^{k+1}/l_i)$.

The initially drawn sample of banks $\overline{B}_{ij}^{1}$ may not guarantee a full allocation of interbank assets across the interbank market. There are various reasons for that: Some samples may be too small, consisting of banks that are not large enough to accept deposits or not willing to accept all offered deposits, given their preferred interbank funding structure. Therefore, at each step the samples are enlarged by randomly drawing additional banks (again with the probability p^{geo}) and the four steps are repeated until full allocation (or in practice, numerically close to full allocation) is achieved.

4.4 Results

4.4.1 Data

The models of simulated networks and endogenous interbank formation were applied to the EU banking system. The dataset regarding balance sheet structures of banks has the following structure:

- a sample of banks being a subset of 2011 EBA stress testing exercise disclosures $-N = 80$;
- Bankscope van Dijk's data on individual banks' balance sheet aggregates of total assets (TA_i), interbank borrowing and lending, customer loans (L_i), securities holding (S_i) and capital position (e_i);
- RWA of banks in the sample broken down (if available) by total customer loans, securities, and interbank lending. These pieces of information are used to proxy the allocation of capital to the interbank exposures assuming the Basel II 20% RW for the interbank lending and calculating the average RWs for customer loans and securities in the sample, denoted RW^L and RW^S, respectively. The allocated capital e^I is approximated in the following way:

$$e_i^I = \frac{20\% a_i}{20\% a_i + \text{RW}^L L_i + \text{RW}^S S_i} e_i$$

  The averages of RW of customer loans and securities instead of the bank by bank weight were necessitated by gaps in the dataset with respect to the portfolio breakdown of RWAs;
- The geographical breakdown of banks' aggregate exposures allow for parametrization of the probability map P^{geo}.

The straightforward caveat of the approximation of e^I is that the averaging of RW^L and RW^S across banks may lead to excessively stringent capital constraints for some of the banks. The compromising procedure of replacing of total interbank assets a_i with $a_i - k_i \Delta a_i$ accounts for that as well.

Additionally, CDS spreads (s) – for individual banks if available, otherwise country-specific – and 3-month money market rates for EU countries (r^m) were used to approximate the bank-specific interbank rates and their riskiness measured by the standard deviation of rates. Some projected paths of the CDS spread under the baseline economic scenario were applied to calculate the CVA of the interbank exposures.[29)]

The estimation of the correlations Q and Q^X is followed by the testing of the statistical significance of all the entries. Insignificant ones (at the probability level of 5%) are replaced by zeros. Three years of data with monthly frequency are used for the estimation.

29) The projected series of bank individual CDS spreads were kindly provided to us by M. Gross and calculated according to a method developed in Gross and Kok (2013).

4.4.2
Simulated Networks

The very first conclusion about how reasonable is the simulated network approach rather than approaches focusing just on one particular network structure can be inferred from the topological properties of the simulated networks. For that purpose, we calculate the distribution of the betweenness measures for all nodes in the 20 000 simulated networks and compare those with the entropy-maximizing network (using the efficient RAS algorithm (Mistrulli, 2011)) and the average network (described by the sum of all the simulated relative exposure matrices π divided by 20 000). The results are shown in Figure 4.3.[30] The complex shape of the resulting distributions suggests that the two calculated special networks are far from approximating the set of simulated networks.

Summing up, the simulated networks allow for analyzing much richer structures related to the probability map of the geographical breakdown of banks' activities than just the usually available (or estimated) one period snapshots. Otherwise, some very useful pieces of information about probabilities of the interbank links are ignored. Simulation of many feasible network structures is helpful in studying the tail contagion risk related to the variety of possible formation of the interbank structures.

Against this background, we now turn to discuss the contagion results based on our simulated networks. First, to illustrate the outcome of the network simulation, we compute – for each simulated network – the average Capital Adequacy Ratio reduction in the banking system (i.e., average ΔCAR_i) in the event of one bank failing on its interbank liabilities. Figure 4.4 shows the distribution of average CAR reductions across all the simulated networks; with and without "fire sale" losses. It is observed that for the large majority of simulated networks, the average solvency implications are relatively muted. In other words, contagious bank default is a tail-risk phenomenon. Broadly speaking, in 99% of the scenarios, the CAR reduction is negligible, while only in 1% point of the network realizations, the CAR reduction surpasses 0.2% point. This suggests that the interbank network structures are overall fairly robust against idiosyncratic shocks to the system, which thus serves the purpose of diversifying risks among the banks. This notwithstanding, we also observe substantial non-linear effects in terms of contagion as for some, albeit limited in number, network structures the impact on overall banking sector capitalization turns out to be much larger than for the vast majority of the networks.

It is furthermore noticeable from Figure 4.4 that inclusion of a "fire sale" mechanism increases the CAR reduction related to the interbank contagion. It is, however, also observed that the additional contagion impact compared to the case without any fire sales is relatively limited. Notably, the liquidation of the assets to retain a specific target leverage ratio implied by the contagious bank defaults

30) In some case, we present results only for the internationally active banks since banks from this group trigger the interbank contagion.

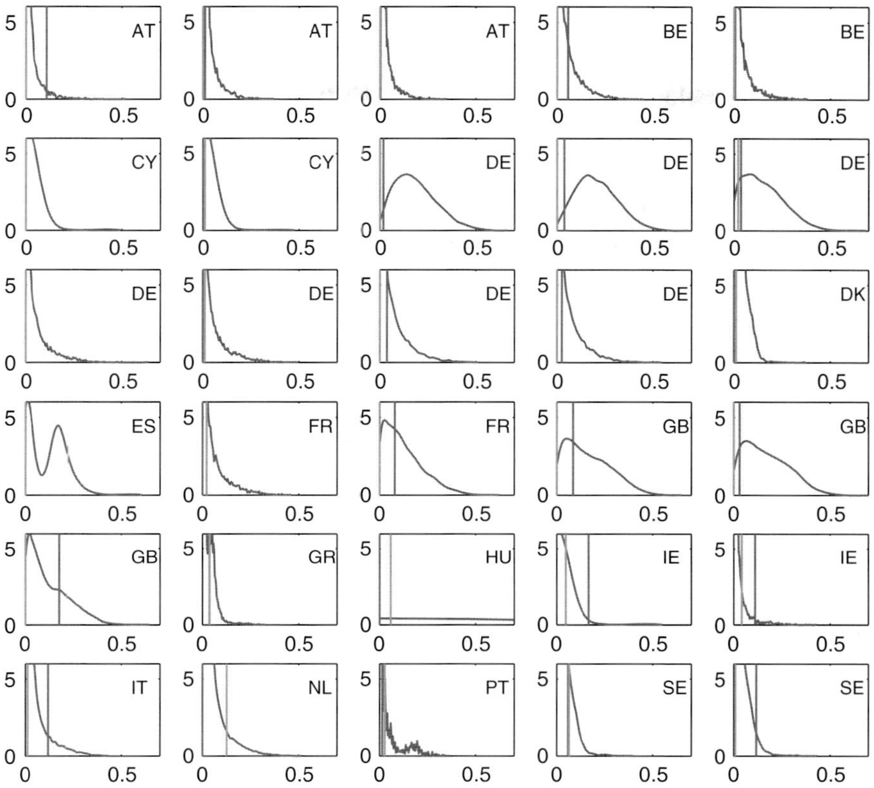

Figure 4.3 Betweenness-centrality measures: distribution on the simulated networks versus the average network. (See inset for color representation of this figure.) Note: Blue line: distribution on the simulated networks; red (vertical) line: measure for the average simulated network: green (vertical) line: measure for he entropy maximising network. Only internationally active banks are presented. Source: own calculations.

is triggered at substantially lower percentiles of the distribution of the simulated networks. This finding is consistent with theoretical predictions about the potential for substantial and long-lasting spill-over effects when financial intermediaries aim at controlling their leverage metrics.[31]

Figure 4.5 shows the distribution of individual banks' CAR reduction. The figure confirms that contagion in general is a tail-risk phenomenon across all banks. However, for some banks, contagion can be initiated in a substantial number of the simulated network structures. This indicates that some nodes in the network are more important than others for producing contagious effects.

We can also decompose the CAR reductions into first-round and second-round contagion effects; as proposed in Eq. 4.3 (Figure 4.5). We observe that, while the first round or direct effects are clearly dominating the overall impact across all

31) See, e.g., Adrian and Shin (2010), Geanakoplos (2009), and Brunnermeier (2009).

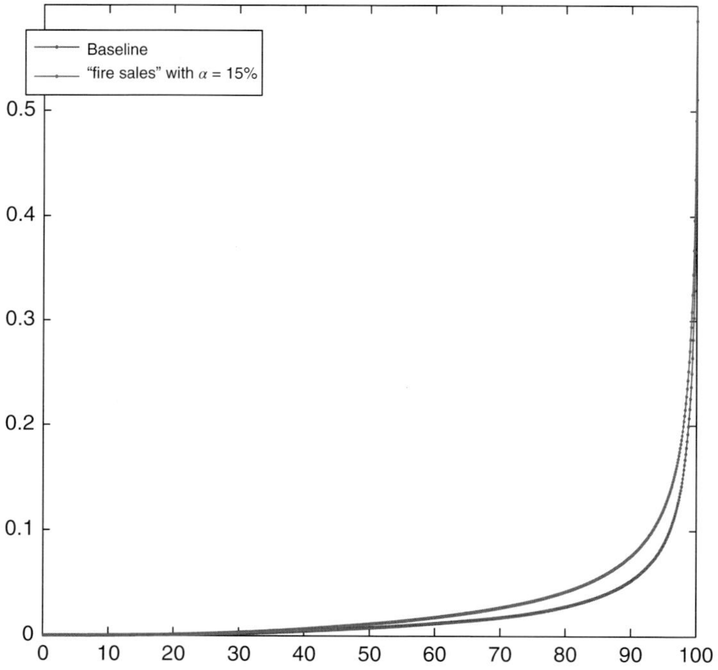

Figure 4.4 Distribution of the average CAR reduction (in p.p.). (See inset for color representation of this figure.) Source: own calculations.

banks, at least for some banks, also, the second-round shock propagation adds to the overall losses in the system. This illustrates that when analyzing interbank contagion, one needs to look beyond the direct bilateral exposures between the banks in the network, but also needs to consider potential knock-on effects once the first-round impact of bank defaults has been accumulated.

So far, in our simulated networks, we did not restrict the size of exposures a bank is allowed to hold against another bank. However, in practice, banks are constrained by the so-called "large exposure limits".[32] To account for such regulations, we impose two conditions:

1) the sum of all exposures that (individually) exceed 10% of the capital should not exceed 800% of the capital;
2) each exposure should not exceed 25% of the total regulatory capital.

As expected, this has the effect of substantially reducing the overall contagion impact across the networks compared to the situation without any limitations to counterparty exposures. Large exposure limits are integrated into the behavioral model of endogenous networks and their role in curbing the contagion losses is further investigated.

32) See Article 111 of Directive 2006/48/EC that introduces the limits.

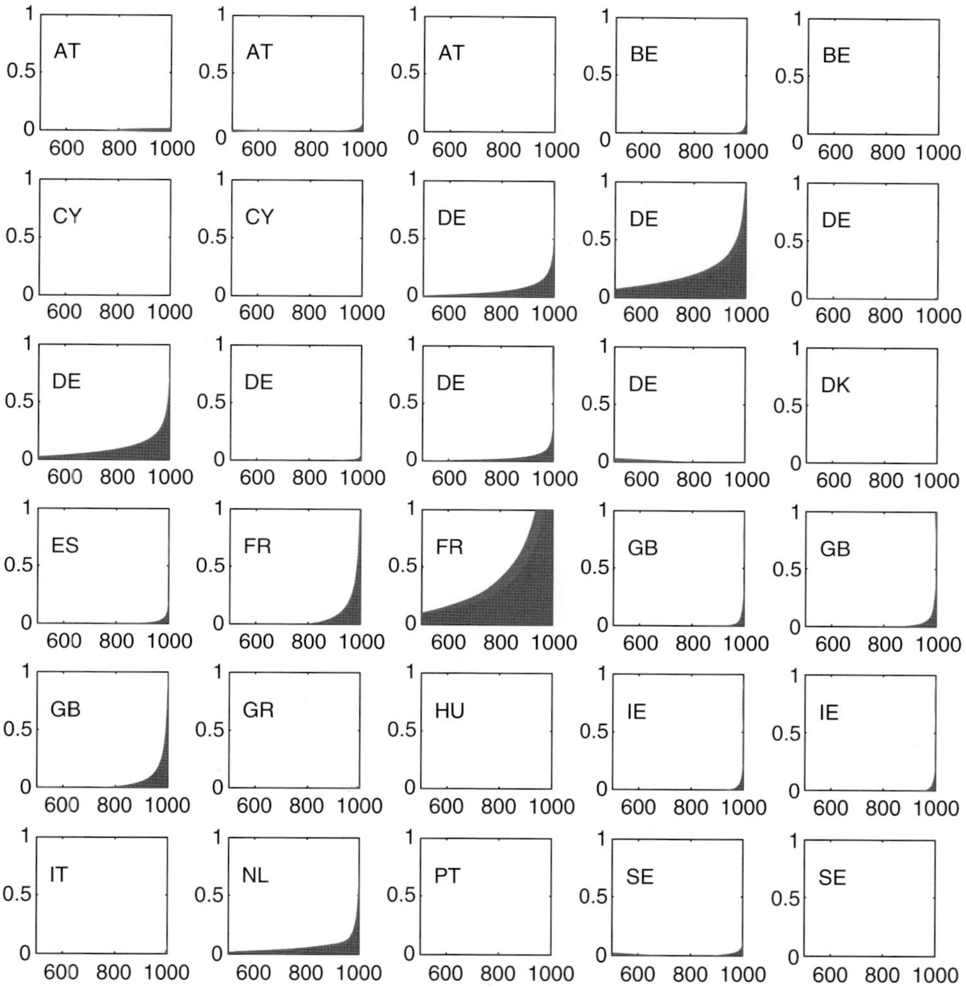

Figure 4.5 Decomposition of the distribution of individual banks' CAR reduction into first-and second-round contagion (in p.p.). (See inset for color representation of this figure.) Note: blue area–aggregate effect of first-round contagion; red area–second-round contagion. Only internationally active banks are presented. Source: own calculations.

4.4.3
Structure of Endogenous Interbank Networks

Since in the endogenous interbank networks model an interbank network *emerges* from the interactions of banks-agents, the resulting topology is far from being obvious. Some common statistical measures can help in understanding the structure at large. In general, the interbank networks are not complete. On an average, bank nodes have a degree of not more than 0.20 but the dispersion among nodes is substantial with some nodes having a degree of 0.30, while others only have a

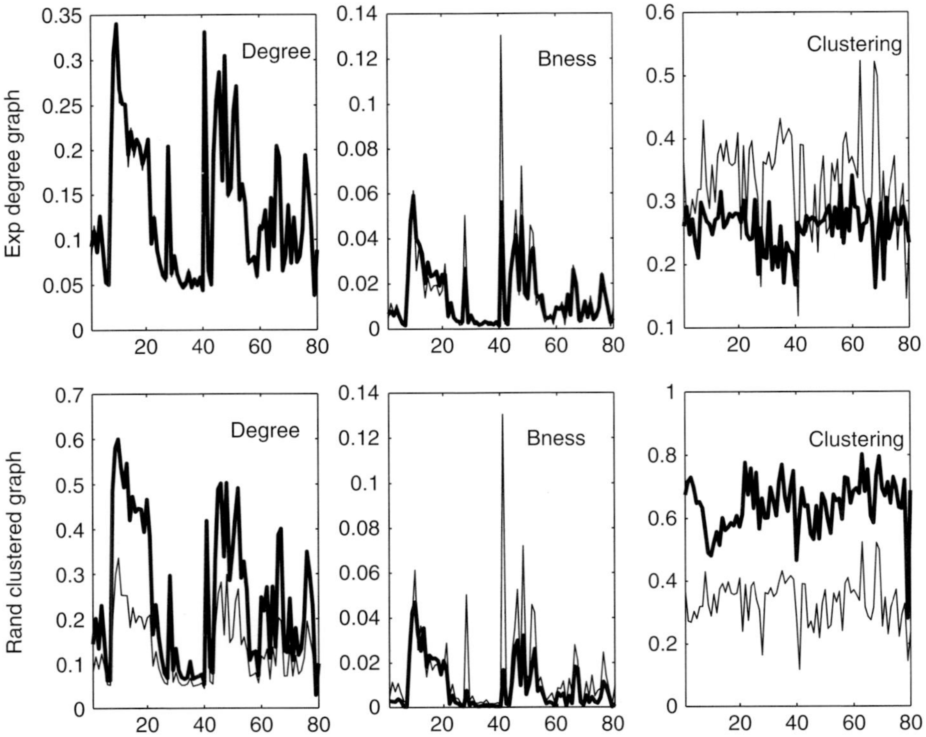

Figure 4.6 Endogenous networks versus random graphs generated with parameters inherited from the endogenous ones. Note: *x*-axis: banks in the sample. *y*-axis: statistical measure of topological properties. Blue-wide lines: referring to endogenous networks (average in a random sample of 100 networks). Red-thin lines: referring to random graphs (top row: random degree graphs; bottom row: randomly clustered graph. *NetworkX* library in Python was used to generate and analyze the random graphs.). Source: own calculations.

degree of 0.05 (Figure 4.6). The heterogeneity is more visible for centrality measures that aim at gauging the importance of a node as a hub in the system. These measures are deemed particularly important for capturing the risk of contagion, by detecting the nodes that may be most prone to spreading contagion across the system. For instance, it is observed that betweenness centrality is several times higher for some particular nodes. Some studies focus on core / periphery properties which mean that there is a subset of nodes in the system that is fully connected, whereas other nodes are only connected to that subset. There are various algorithms selecting the core and they may lead to a fuzzy classification – some nodes are "almost" core or "almost" periphery. In case of our endogenous networks, we have not found any significant classification of the nodes to the core and periphery (using the (Borgatti and Everett, 1999) approach). This is probably due to the fact that we capture global, internationally active bank hubs and connected subsystems of domestic banks. Overall, these findings suggest that the endogenous networks algorithm generates interbank structures that are not easy to be classified in a simple way by just a few topological parameters.

A usual approach to get a deeper understanding of the network structure is to compare it with graphs that possess well-studied statistical properties. The simplest benchmark random graph is generated in Erdös-Renyi (E-R) model constructing an edge between two given nodes with a given probability p, independent of all other pairs of nodes. Since we operate with a probability map assigning different probabilities to links between different banks, it is straightforward to imagine that the E-R approximation of endogenous networks should fail. A more promising method in terms of accuracy of approximations is based on the detailed information about degree and clustering of our generated endogenous networks. An *expected degree* graph (Chung and Lu, 2002) is the first example. In this model, links between nodes i and j are drawn with probability $deg_i deg_j / \sum_k deg_k$, where deg_i is a degree of a node i. The second type of potentially useful graphs is a *random clustered* graph model (Newman, 2009). Given a degree sequence of all nodes and a triangle sequence of nodes.[33] The random clustered algorithm chooses linkages uniformly from a joint distribution of possible set of triangles complying with the given degree sequence. In this way, the algorithm potentially has a better control not only of the degree distribution but also of clustering coefficients which are important indicators of contagion transmission channels.

The results of the comparison of the endogenous and random graphs are shown in Figure 4.6. Random graphs are constructed in such a way that for a given endogenous interbank network,

- the expected degree graph is generated using the degree sequence of nodes in the network;
- the random clustered graph is generated with a sequence of pairs consisting of a degree and triangle degree of all nodes.

We analysz 200 realizations of endogenous networks and 200 corresponding random networks. The generated expected degree networks are almost identical to the endogenous networks with respect to the degree distribution. It is not surprising, given that the degree of nodes is the only parameter of the expected degree graph algorithm. However, betweenness centrality, measuring direct and indirect connectivity of a given node with all other nodes in the system, proves to be less consistent. Some nodes of the endogenous networks are substantially more important in terms of centrality. The differences between endogenous and expected degree networks are even more striking for clustering measures gauging the concentration of the linkages. The random clustered graphs do not perform better, even though their parameters have more degrees of freedom. The algorithm of random clustered networks preserves the ranking of the nodes in terms of degree measures, but produces graphs with nodes possessing many more links than in the corresponding endogenous networks. The resulting clustering coefficients are

33) A triangle degree of a node is the number of triangles containing a given node. A triangle sequence is the sequence of the triangle degrees of all the nodes in a graph.

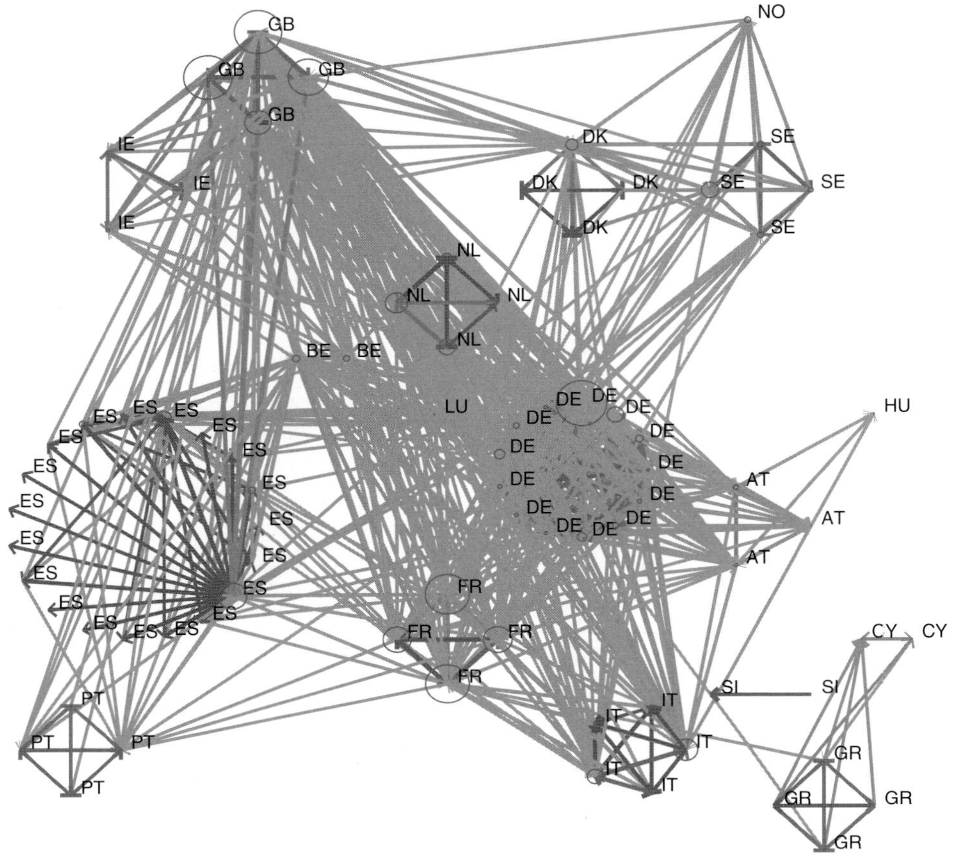

Figure 4.7 Incompleteness of the interbank network structure with the estimated correlation of risks. (See inset for color representation of this figure.) The darker the line, the higher the probability that the link exists. The circles around the nodes indicate bank sizes (proportional to log of total assets). Source: own calculations.

in general higher as well. All in all, the complex topology of the endogenously modeled EU interbank network implies that random graphs may oversimplify their structure. This notwithstanding, random graphs offer a valuable benchmarking tool for understanding the relationship between various topological properties of the networks.

One particular interbank structure estimated in the developed endogenous network model is presented in Figure 4.7. It is incomplete with the largest banks being the most interconnected nodes, which is in line with the findings elsewhere in the literature.[34)]

34) For a few representative country-specific studies, using real-time overnight transactions data or large exposure data as well as entropy approaches, see, e.g., Furfine (2003), Upper and Worms (2004), Boss *et al.* (2004), van Lelyveld and Liedorp (2006), Soramaki *et al.* (2007), and Degryse and Nguyen (2007).

Topological properties of the networks provide only partial insight into potential channels of contagion. Much more accurate methods rely on simulations of the flow of shocks across the system and are illustrated and applied in the next section devoted to contagion initiated by some adverse scenarios of solvency in the banking system.

4.5 Stress Testing Applications

The proposed approach to model the interbank networks opens many potential ways to study contagion risk in adverse market conditions and the effectiveness of various policy instruments in curbing contagion risk on the interbank market.

The most straightforward application of the simulated networks approach assumes the default of some banks on their interbank payments due to some adverse macroeconomic scenario inducing severe losses to vulnerable banks and driving them to insolvency. The defaults are linked to some adverse macroeconomic scenarios impacting the capital position of banks. The scenarios comprise developments of lending and borrowing interest rates, credit risk in banks' loan portfolios and securities valuation. In the stress-testing applications, the scenarios are particularly severe to verify the resilience of the bank, in particular to assess their potential to remain solvent even in a serious economic and financial downturn. Following the stress-testing scenarios, some banks fall into insolvency meaning that the losses related to the scenario substantially erode their capital. The insolvency is measured by a normalizing ratio of the capital to the RWA which weigh asset classes by their riskiness. For that purpose, we use the standard Core Tier 1 ratio (CT1 ratio), as described in Section 4.3.1.3. Insolvency of a bank usually means serious financial problems to it (for instance activation of various restructuring programs or liquidation) and it is reasonable to assume that it may trigger defaults of interbank payments.

We illustrate the contagion measures in the simulated interbank networks by picking four different examples of adverse macroeconomic scenarios in EU and analyze the propagation of contagion in the analyses sample of EU banks. The initial defaults of banks on their interbank liabilities are assumed for those institutions with CT1 ratio falling below 6%. Usually, banks tend to keep the ratios clear above regulatory 8% and a drop below 6% means a serious financial disorder for these banks. Then, the transmission of contagion losses across the system and measured by the clearing payments vector p^*. Analyzing 20 000 randomly generated networks gives the whole distribution of contagion losses ΔCAR, which are synthetically presented on Figure 4.8. The 4 box-whisker plots show that in most of the simulated interbank structures, the size of the contagion is rather limited (for 75% of the networks reduction of the analyzed capital ratio stays below 100 bps). However, in a number of isolated cases (i.e., for particular network structures) depending on the scenario, the CAR reduction reaches 400 bps. This is an evidence of a very non-linear feature of the resilience of the interbank networks

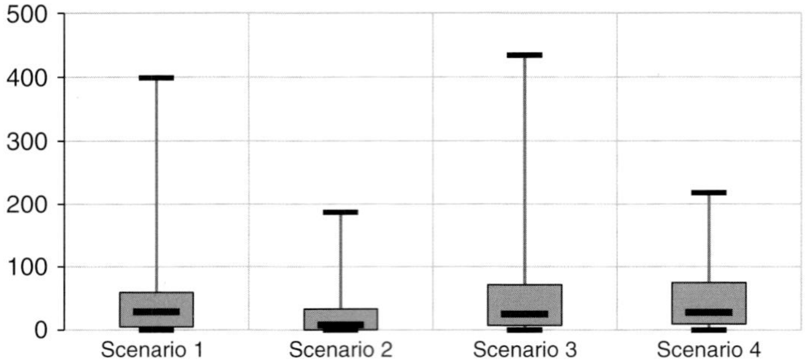

Figure 4.8 Worst-case bp reduction in CT1 ratio due to interbank contagion–cross-country dispersion. Note: *y*-axis: basis point CT1 ratio reduction; interquartile range represents 25th and 75th percentiles of cross-country contagion effects under the most severe of the simulated networks. Source: own calculations.

to contagion risk. This has important policy implications. The identification of systemically important banks should not be based on one particular estimated network topologies (either by means of entropy maximization or density minimization methods) but should rather try to capture many theoretically possible structures.

The adverse macroeconomic scenario can impact the network structure itself. The endogenous network formation allows for passing through the parameters of the network model some macroeconomic and financial shocks. The methodology applied to analyze the impact on the network structures under different macroeconomic conditions is as follows:

- We first compute each bank's total interbank investment and funding needs under a baseline macroeconomic scenario. The framework developed by Hałaj (2013) is used to translate the macroeconomic shock into the restructuring of the banks' assets. The outcome of the model is the change of volume of broad balance sheet categories: Customer loan portfolio, securities portfolio, and interbank lending portfolio. The relative change of the volume of the interbank assets of bank j is used to scale the volume of the interbank funding of j.
- Second, we construct the interbank network applying the method proposed in Section 4.4.3 under baseline scenario parameters and total interbank lending and borrowing in various regimes of Large Exposure limits and CVA.
- Finally, we impose an adverse macroeconomic shock to banks' capital position and subsequently, run a contagion model of banks' defaults.[35] The clearing payments vector approach is used to measure the contagion effects (Eisenberg and Noe, 2001; Hałaj and Kok, 2013b).

Figure 4.9 illustrates the impact of having different LE limit thresholds in the context of an adverse shock. Specifically, the y-axis illustrates the difference between networks formed under a 15% LE limit and under the standard 25% LE

35) To insure robustness of the results, a couple of adverse scenarios were applied.

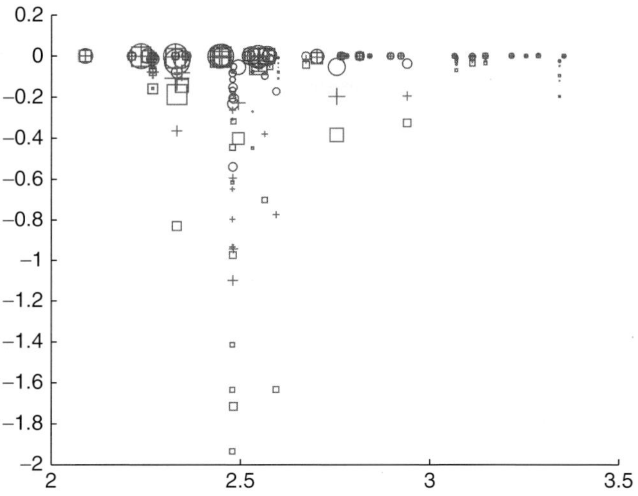

Figure 4.9 Counterparty credit quality and the impact of LE limits on the losses incurred due to contagion. (See inset for color representation of this figure.) Note: x-axis: CDS spread (in bps). y-axis: difference of CAR after adverse stress testing shock estimated for CVA (∘), 3× estim. CVA (+), 5× estim. CVA (□) versus no CVA regime (in pp, negative number means that introduction of CVA charge decreases contagion losses). No CVA adjustment (i.e., $\gamma \equiv 0$). The size of a circle is proportional to a bank's total assets. Source: own calculations.

limit in terms of the capital loss following an adverse shock. A positive value implies that contagion losses rise when lowering the LE limit. On the x-axis, we plot the banks according to the size of their riskiness (measured in terms of their CDS spreads). It is observed that more stringent LE limits overall tend to lower the contagion risk. Interestingly, this effect is especially pronounced for the group of banks perceived (by the markets) to be the soundest. In other words, the forced reduction of counterparty concentration risk that would be implied by a lowering of the LE limits would seem to particularly benefit the safest part of the banking system, whereas the more vulnerable segments are found to be less affected by changes in the LE limits.

A similar analysis was carried out also for what concerns the CVA based add-on to the RWs applied to the interbank exposures, see Figure 4.10. The Art. 458 of the CRDIV/CRR EU Directive[36] defines a macro-prudential instrument of specific RWs attached to exposures to financial institutions. A rather broad formulation allows for a risk-sensitive implementation of the instrument based on the CVA concept. In the context of our model, the parameter γ plays a role of that special add-on. As was the case for the large exposure limits, the impact of the CVA on contagion risk is mainly affecting the sounder banks. In the case of the CVA based

36) Directive 2013/36/EU of the European Parliament and of the Council of 26 June 2013 on access to the activity of credit institutions and the prudential supervision of credit institutions and investment firms, amending Directive 2002/87/EC and repealing Directives 2006/48/EC and 2006/49/EC.

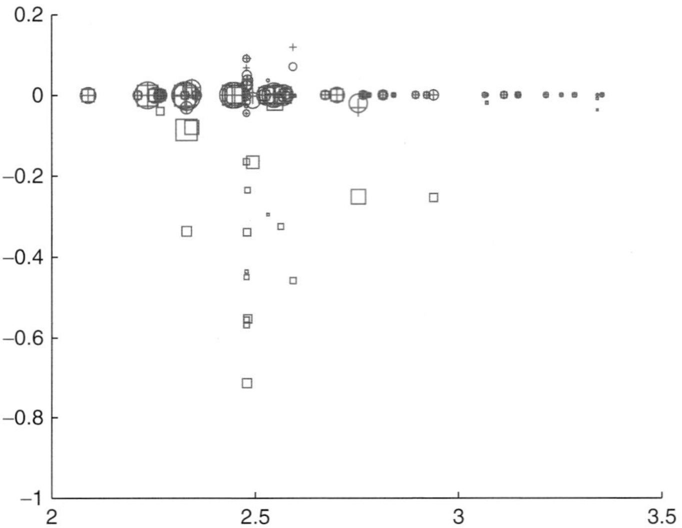

Figure 4.10 Counterparty credit quality and the impact of CVA capital charge on the losses incurred due to contagion. (See inset for color representation of this figure.) *x*-axis: CDS spread (in bps). *y*-axis: difference of CAR after adverse stress testing shock between no CVA regime calculations and CVA regime with: estimated CVA (∘), 3× estimated CVA (∘) and 5× estimated CVA (∘), in pp, positive number means that introduction of CVA charge increases contagion losses. LE= 25%. The size of a circle is proportional to a bank's total assets. Source: own calculations.

add-on, the direction of changes is, however, not unambiguous as we observe that depending on the bank, contagion risk can both increase and decrease when increasing the CVA charge.

4.6 Conclusions

We described methods to analyze the contagion risk in the interbank networks, which are the recent extensions of the vividly developing research and policy area in finance. Both abstract from the snapshot perspective of the network structures taken in most contagion studies.

In the first approach, contagion risk in the banking system is assessed based on the simulated networks concept. The tool allows for generating many possible interbank structures and for analyzing distribution of clearing payments vector *á la* (Eisenberg and Noe, 2001). Since the simulation of the random networks is computationally costly, we propose a simplified measure of contagion risk related to the network structure of interbank exposures; the so-called SPI.

The simulations that we perform confirm that contagion is heterogenous across the banking system and strongly non-linear. We have found that there are banks that pose much higher contagion risk to the banking system than other banks. At the same time, a small fraction of possible network structures may spread relatively sizable contagion losses across the system, thus highlighting the non-linear nature of shock propagation effects. Contagion is very much a tail-risk problem.

Our simulated networks' approach allows for comparison of the tail-risk networks. Although, all of the simulated structures on average can transmit contagion of only very limited size, the impact of bank triggers on the system may substantially differ in extreme cases. This is both confirmed by the simulations of contagion losses and by the Systemic Probability Index.

In the second approach presented in the chapter, we try to deviate from the standard, mechanistic cascading mechanism employed in the traditional interbank contagion models. Instead, we have developed an agent-based model that is able to account for banks' dynamic, endogenous responses both in the formation process of the interbank network and in response to contagious shocks.

Behavioral responses of banks may change the topology of the interbank market. For instance, high correlation of risks may imply more complete and interconnected structures. However, it does not mean *per se* that the contagion risk always decreases or increases. It depends on the size and distribution of shocks potentially triggering contagion spreading. Notably, behavioral aspects of interbank network formation should be taken into account when designing a regulatory framework for mitigating the contagion risk.

Apart from analysing network formation processes and the implications for interbank contagion risks in a setting where banks are dynamically optimising their actions, our model approach can also be used to assess the impact of different regulatory and macro-prudential policy instruments on the structure of interbank networks and their implied contagion risk. In this light, we presented a few policy experiments related to the effects of large exposure limits on counterparty credit risks. These macro-prudential policy instruments were analyzed both in the context of network formation and in terms of their ability to mitigate interbank contagion risks under adverse circumstances.

All in all, while the reported results obviously hinges on the specific characteristics of the banks included in the network system and on the specific adverse scenarios considered, the overriding conclusion from these policy experiments is that macro-prudential policies can make a significant difference through their impact on the network formation and ultimately on the risk of interbank contagion to adverse shocks. From this perspective, the modelling approach presented in this paper can be employed for conducting impact assessments of selected macro-prudential policy instruments and in this way help inform the calibration of such tools.

References

Acemoglu, D., Ozdaglar, A., and Tahbaz-Salehi, A. (2013) *Systemic risk and stability in financial networks.* Working Paper 18727, National Bureau of Economic Research, January 2013. *http://www.nber.org/papers/w18727* (accessed 17 December 2014).

Adam, A. (2008) *Handbook of Asset and Liability Management: From Models to Optimal Return Strategies*, The Wiley Finance Series, John Wiley & Sons, Ltd.

Adrian, T. and Brunnermeier, M. (2011) *CoVaR*. Working Paper 17454, NBER.

Adrian, T. and Shin, H.S. (2010) Financial intermediaries and monetary economics, in *Handbook of Monetary Economics* (eds B. Friedman and M. Woodford), North-Holland, New York.

Afonso, G., Kovner, A., and Schoar, A. (2011) Stressed, not frozen: the federal funds market in the financial crisis. *Journal of Finance*, **66** (4), 1109–1139.

Aharony, J. and Swary, V. (1983) Contagion effects of bank failures: evidence from capital markets. *Journal of Business*, **56** (3), 305–317.

Albert, R., Jeong, H., and Barabási, A.-L. (2000) Error and attach tolerance of complex networks. *Nature*, **406**, 378–382.

Allen, F. and Babus, A. (2009) Networks in finance, in *The Network Challenge: Strategy, Profit, and Risk in an Interlinked World* (eds P. Kleindorfer and J. Wind), Wharton School Publishing.

Allen, F. and Gale, D. (2000) Financial contagion. *Journal of Political Economy*, **108** (1), 1–33.

Anand, K., Craig, B., and von Peter, G. (2013) Fill in the blanks: network structure and interbank contagion. Presentation at INET 2013, Ancona.

Babus, A. (2011) *Endogenous intermediation in the over the counter markets*. Imperial College London. working paper.

Babus, A. and Kondor, P. (2013) *Trading and information diffusion in OTC markets*. CEPR Discussion Papers 9271, C.E.P.R. Discussion Papers, January.

Baltensperger, E. (1980) Alternative approaches to the theory of the banking firm. *Journal of Monetary Economics*, **6** (1), 1–37.

Barabási, A.-L. and Albert, R. (1999) Emergence of scaling in random networks. *Science*, **268**, 509–512.

Battiston, S., Gatti, D.D., Gallegat, M., Greenwald, B., and Stiglitz, J. (2009) *Liaisons dangereuses: increasing connectivity, risk sharing, and systemic risk*. NBER Working Paper, (1561).

Battiston, S., Gatti, D.D., Gallegati, M., Greenwald, B., and Stiglitz, J.E. (2012) Default cascades: when does risk diversification increase stability? *Journal of Financial Stability*, **8** (3), 138–149.

Bhattacharya, S. and Gale, D. (1987) Preference shocks, liquidity and central bank policy, in *New Approaches to Monetary Economics* (eds W. Barnett and K. Singleton), Cambridge University Press, New York.

Bluhm, M., Faia, E., and Krahnen, P. (2013) *Endogenous banks' networks, cascades and systemic risk*. mimeo, University Frankfurt, March 2013. *http://www.wiwi.uni-frankfurt.de/profs/faia/* (accessed 17 December 2014).

Borgatti, S.P. and Everett, M.G. (1999) Models of core/periphery structure. *Quantitative Finance*, **21**, 375–395.

Boss, M., Elsinger, H., Thurner, S., and Summer, M. (2004) Network topology of the interbank market. *Quantitative Finance*, **4**, 1–8.

Boyd, J.H. and De Nicoló, G. (2005) The theory of bank risk taking and competition revisited. *Journal of Finance*, **60** (3), 1329–1343.

Bräuning, F. and Fecht, F. (2012) *Relationship lending in the interbank market and the price of liquidity*. Discussion Papers 22/2012, Deutsche Bundesbank, Research Centre.

Brunnermeier, M. (2009) Deciphering the liquidity and credit crunch 2007-8. *Journal of Economic Perspectives*, **23** (1), 77–100.

Brusco, S. and Castiglionesi, F. (2007) Liquidity coinsurance, moral hazard and financial contagion. *Journal of Finance*, **65** (5), 2275–2302.

Cappiello, L., Gerard, B., and Manganelli, S. (2005) *Measuring comovements by regression quantiles*. Working Paper 501, ECB.

Castiglionesi, F. and Lavarro, N. (2011) *Fragile Financial Networks*, Mimeo, Tilburg University.

Castiglionesi, F. and Wagner, W. (2013) On the efficiency of bilateral interbank insurance. *Journal of Financial Intermediation*, forthcoming.

Chen, J. and Song, K. (2013) Two-sided matching in the loan market. *International Journal of Industrial Organization*, **31**, 145–152.

Chung, F. and Lu, L. (2002) Connected components in random graphs with given expected degree sequences. *Annals of Combinatorics*, **6** (2), 125–145, doi: 10.1007/pl00012580.

Cifuentes, R., Ferrucci, G., and Shin, H.S. (2005) Liquidity risk and contagion. *Journal of the European Economic Association*, **3** (2/3), 556–566.

Cohen-Cole, E., Patacchini, E., and Zenou, Y. (2011) *Systemic risk and network formation in the interbank market*. CEPR Discussion Papers 8332, C.E.P.R. Discussion Papers, April 2011.

Cooperman, E., Lee, W., and Wolfe, G. (1992) The 1985 Ohio thrift crisis, FSLIC's solvency, and rate contagion for retail CDs. *Journal of Finance*, **47** (3), 919–941.

Cuoco, D. and Liu, H. (2006) An analysis of VaR-based capital requirements. *Journal of Financial Intermediation*, **15** (3), 362–394.

Danielsson, J., Jorgensen, B.N., and de Vries, C.G. (2002) Incentives for effective risk management. *Journal of Banking & Finance*, **26** (7), 1407–1425.

Degryse, H. and Nguyen, G. (2007) Interbank exposures: an empirical examination of systemic risk in the Belgian banking system. *International Journal of Central Banking*, **3** (2), 123–171.

Deloitte and Solum Financial Partners (2013) *Counterpary risk and CVA survey*. Survey February, Deloitte and Solum Financial Partners.

Diebold, F.X. and Yilmaz, K. (2011) *On the network topology of variance decompositions: measuring the connectedness of financial firms*. Working paper.

Docking, D., Hirschey, M., and Jones, V. (1997) Information and contagion effects of bank loan-loss reserve announcements. *Journal of Financial Economics*, **43** (2), 219–239.

Doyle, J.C., Alderson, D., Li, L., Low, S.H., Roughan, M., Shalunov, S., Tanaka, R., and Willinger, W. (2005) The "robust yet fragile" nature of the Internet. *Proceedings of the National Academy of Sciences of the United States of America*, **102** (40), 14123–14475.

Eisenberg, L. and Noe, T.H. (2001) Systemic risk in financial systems. *Management Science*, **47** (2), 236–249.

Eisenschmidt, J. and Tapking, J. (2009) *Liquidity risk premia in unsecured interbank money markets*. Working Paper 1025, European Central Bank.

Elsinger, H., Lehar, A., and Summer, M. (2006) Using market information for banking system risk assessment. *International Journal of Central Banking*, **2** (1), 1-29.

Emmert-Streib, F. and Drehmer, M. (2010) Identifying critical financial networks of DJIA: towards network based index. *Complexity*, **16** (1), 24–33.

Engle, R.F. and Manganelli, S. (2004) CaViar: conditional autoregressive Value at Risk by regression quantile. *Journal of Business and Economic Statistics*, **22** (4), 367–381.

Ewerhart, C., Cassola, N., Ejerskov, S., and Valla, N. (2004) *Liquidity, information and the overnight rate*. Working Paper 378, ECB.

Ferguson, R., Hartmann, P., Panetta, F., and Portes, R. (2007) International financial stability. Geneva Report on the World Economy 9, CEPR.

Flannery, M. (1996) Financial crises, payment system problems, and discount window lending. *Journal of Money, Credit and Banking*, Part II, **28**, 804–824.

Foti, N., Caccioli, F., Farmer, J.D., and Rockmore, D. (2013) *How interbank lending amplifies overlapping portfolio contagion: a case study of the Austrian banking network*. arXiv: 1306.3704v1 [q-fin.GN].

Fox, J. (2010) *Estimating Matching Games with Transfers*, University of Michigan. Working paper,

Freixas, X., Parigi, B.M., and Rochet, J.-C. (2000) Systemic risk, interbank relations, and liquidity provisions. *Journal of Money, Credit and Banking*, **32** (3), 611–638.

Furfine, C. (2003) Interbank exposures: quantifying the risk of contagion. *Journal of Money, Credit and Banking*, **35** (1), 111–638.

Gabrieli, S. (2011) *The microstructure of the money market before and after the financial crisis: a network perspective.* Research Paper 181, CEIS.

Gai, P., Haldane, A., and Kapadia, S. (2011) Complexity, concentration and contagion. *Journal of Monetary Economics,* **58** (5), 453–470.

Gai, D. and Kariv, S. (2003) Bayesian learning in social networks. *Games and Economic Behaviour,* **45**, 329–346.

Garratt, R.J., Mahadeva, L., and Svirydzenka, K. (2011) *Mapping systemic risk in the international banking network,* Working Paper 413, Bank of England.

Geanakoplos, J. (2009) The leverage cycle, in *NBER Macroeconomics Annual,* Vol. 24 (eds K. Rogoff, D. Acemoglu, and M. Woodford). Published in April 2010 by University of Chicago Press.

Georg, C.-P. (2011) *The effect of the interbank network structure on contagion and common shocks.* Discussion Paper Series 2: Banking and Financial Studies 2011, 12, Deutsche Bundesbank, Research Centre, 12.

von Goetz, P. (2007) *International banking centres: a network perspective.* BIS Quarterly Review.

Gofman, M. (2013) *Efficiency and Stability of a Financial Architecture with Too Interconnected to Fail Institutions* mimeo, University of Wisconsin-Madison.

Gomez-Gardeñes, J., Pérez-Vicente, C.J., Moreno, Y., Gomez, S., Diaz-Guilera, A., and Arenas, A. (2013) *Diffusion dynamics on multiplex networks. arXiv: 1207.2788 [physics.soc-ph].*

Grasselli, M. (2013) An agent-based computational model for bank formation and interbank networks, in *Handbook on Systemic Risk* (eds J.-P. Fouque and J. Langsam), Cambridge University Press.

Gropp, R., Lo Duca, M., and Vesala, J. (2009) Cross-border contagion risk in Europe. *International Journal of Central Banking,* **5** (1), 97–139.

Gross, M. and Kok, C. (2013) *Measuring contagion potential among sovereigns and banks using a mixed-cross-section GVAR.* mimeo (forthcoming in the ECB working paper series), European Central Bank, March 2013.

Hałaj, G. (2013) *Optimal asset structure of a bank–bank reactions to stressful market conditions,* (ECB Working Paper Series) 1533, European Central Bank, April 2013.

Hałaj, G. and Kok, Ch. (2013a) *Assessing interbank contagion using simulated networks.* ECB Working Paper Series 1506, European Central Bank, January 2013a.

Hałaj, G. and Kok, Ch. (2013b) Assessing interbank contagion using simulated networks. *Computational Management Science,* **10** (2–3), 157–186.

Hałaj, G. and Kok, Ch. (2013c) *Modeling emergence of the interbank networks.* ECB Working Paper Series (forthcoming) XX, European Central Bank.

Hamilton, J.D. (1998) Measuring the liquidity effect. *American Economic Review,* **87** (1), 80–97.

Hartmann, P., Straetmans, S., and de Vries, C. (2004) Asset market linkages in crisis periods. *Review of Economics and Statistics,* **86** (1), 313–326.

Hartmann, P., Straetmans, S., and de Vries, C. (2005) *Banking system stability: a cross-Atlantic perspective.* Working Paper 11698, NBER.

Heider, F., Hoerova, M., and Holthausen, C. (2009) *Liquidity hoarding and interbank market spreads.* Working Paper 1107, ECB.

Howard, R.A. and Matheson, J.E. (1972) Risk-sensitive Markov decision processes. *Management Science,* **18** (7), 356–369. ISSN: 00251909.

Iori, G., De Masi, G., Precup, O.V., Gabbi, G., and Caldarelli, G. (2008) A network analysis of the Italian overnight money market. *Journal of Economic Dynamics and Control,* **32** (1), 259–278.

Iori, G., Jafaray, S., and Padilla, F. (2006) Systemic risk on the interbank market. *Journal of Economic Behavior and Organization,* **61**, 525–542.

Jackson, M.O. and Wolinsky, A. (1996) A strategic model of social and economic networks. *Journal of Economic Theory,* **71** (1), 44–74, *http://ideas.repec.org/a/eee/jetheo/v71y1996i1p44-74.html.*

Kho, B., Lee, D., and Stulz, R. (2000) US banks, crises and bailouts: from Mexico to LTCM. *American Economic Review,* **90** (2), 28–31.

Kodres, L.E. and Pritsker, M. (2002) A rational expectations model of financial contagion. *Journal of Finance*, **57** (2), 769–799.

van Lelyveld, I. and Liedorp, F. (2006) Interbank contagion in the Dutch banking sector: a sensitivity analysis. *International Journal of Central Banking*, **2** (2), 99-133.

Longin, F. and Solnik, B. (2001) Extreme correlation of international equity markets. *Journal of Finance*, **56** (2), 649–676.

Lu, L. and Zhou, T. (2010) Link prediction in complex networks: a survey. CoRR, abs/1010.0725.

Markose, S. (2012) *Systemic risk from global financial derivatives: a network analysis of contagion and its mitigation with superspreader tax*. Working Paper 12/282, IMF.

Mistrulli, P.E. (2011) Assessing financial contagion in the interbank market: maximum entropy versus observed interbank lending patterns. *Journal of Banking & Finance*, **35**, 1114–1127.

Montagna, M. and Kok, C. (2013) *Multi-layered interbank model for assessing systemic risk*. Kiel Working Papers 1873, Kiel Institute for the World Economy.

Morris, S. and Shin, H.S. (2012) Contagious adverse selection. *American Economic Journal: Macroeconomics*, **4** (1), 1–21.

Musumeci, J. and Sinkey, J.F. Jr. (1990) The international debt crisis, investor contagion, and bank security returns in 1987. *Journal of Money, Credit and Banking*, **22**, 209–220.

Newman, M.E.J. (2009) Random graphs with clustering. *Physical Review Letters*, **103** (5), 05870.

Nier, E., Yang, J., Yorulmazer, T., and Alentorn, A. (2007) Network models and financial stability. *Journal of Economic Dynamics and Control*, **31** (6), 2033–2060.

Peavy, J.W. and Hempel, G.H. (1988) The Penn Square bank failure: effect on commercial bank security returns - a note. *Journal of Banking & Finance*, **12**, 141–150.

Peltonen, T., Scheicher, M., and Vuillemey, G. (2013) *The network structure of CDS market and its determinants*. ECB Working Paper Series, No 1583.

Pelizzon, L. and Schaefer, S. (2005) *Pillar 1 vs. Pillar 2 under risk management*. NBER Working Papers 11666, National Bureau of Economic Research, Inc., October 2005.

Polson, N.G. and Scott, J.G. (2011) *Explosive volatility: a model of financial contagion*. Working paper.

Rochet, J.-C. and Tirole, J. (1996) Interbank lending and systemic risk. *Journal of Money, Credit and Banking*, **28** (4), 733–762.

Slovin, M., Sushka, M.E., and Polonchek, J. (1999) An analysis of contagion and competitive effects at commercial banks. *Journal of Financial Economics*, **54** (2), 197–225.

Smirlock, M. and Kaufold, H. (1987) Bank foreign lending, mandatory disclosure rules, and the reaction of bank stock prices to the Mexican debt crisis. *Journal of Business*, **60**, 347–364.

Soramaki, K., Bech, M.L., Arnold, J., Glass, R.J., and Beyeler, W.E. (2007) The topology of interbank payment flows. *Physica A*, **379**, 317–333.

Tighe, B.P., van Eerd, A.R.T., and Vlugt, T.J.H. (2008) Entropy maximization in the force network ensemble for granular solids. *Physical Review Letters*, **100**, 238001.

Tsuda, K. and Noble, W.S. (2004) Learning kernels from biological networks by maximizing entropy. *Bioinformatics*, **20**(Suppl. 1), i326–i333.

Upper, C. and Worms, A. (2004) Estimating bilateral exposures in the German interbank market: is there a danger of contagion?. *European Economic Review*, **48** (4), 827–849.

Wall, L.D. and Peterson, D.R. (1990) The effect of Continental Illinois' failure on the financial performance of other banks. *Journal of Monetary Economics*, **26** (1), 77–99.

White, H., Kim, T.-H., and Manganelli, S. (2010) *VAR for VaR: measuring systemic risk using multivariate regression quantiles*. Working paper.

5
Detection, Localization, and Tracking of a Single and Multiple Targets with Wireless Sensor Networks

Natallia Katenka

5.1
Introduction and Overview

In this chapter, we will focus on a relatively new technology designed for monitoring of various natural phenomena in space and time, namely wireless sensor networks (WSNs). Applications of WSN include the monitoring of active volcano, alpine environment, distribution of radioactive sources, amphibian population growth, animal behavior, as well as in-home monitoring of elderly patients and intruder detection. Two specific applications related to monitoring of people movement and behavior of zebras in their natural habitat will be further discussed later in this chapter.

Many WSN applications are related to one or more computational network theory tasks. Here, we will focus on three important tasks including detection, localization, and tracking of spatial phenomena. Specifically, we will investigate the problem of target detection by a WSN in Section 5.3; encounter different approaches to a problem of a single target identification (i.e., localization and diagnostic) in Section 5.4; and then extend our findings to the identification and tracking of multiple targets in Sections 5.5 and 5.6, respectively. Collaboratively, detection, localization, and tracking methods proposed in this chapter form a complete framework that can serve as a tool for monitoring the applications in practice, as well as a theoretical basis for computational network theory where dependencies between measurements are induced by the network structure and the nature of observed phenomena.

In order to better understand the nature of the analytical and computational framework proposed in this chapter, it is useful to have some initial sense of the WSN structure, technological constraints, sensor measurements, and data fusion mechanisms (will be discussed in Section 5.2). While the structure and the design of WSN is in itself an important topic of research in statistical and signal processing methods, we present here only a quick overview, meant to provide somewhat a big picture story to a reader.

Typical WSN is built from a large number of devices (sensors) designed to collect the information from surrounding environment, perform some limited

Computational Network Theory: Theoretical Foundations and Applications, First Edition.
Edited by Matthias Dehmer, Frank Emmert-Streib, and Stefan Pickl.

sensor communication and data processing, and route the results to a central node also known as the fusion center. Sensing, processing, communication, and storage capabilities of sensors are usually constrained by their power supply capabilities and corresponding cost.

Technological constraints of sensors together with the application area under consideration determine to a large extent the deployment strategy for the WSN. In industrial applications, for instance, the sensors may be deployed at specific locations of interest; in some environmental applications, the sensors can be deployed on a fairly regular grid. However, in many situations, deterministic deployment is neither feasible nor practical; for example, when the region monitored by the WSN is hardly accessible. In such cases, deployment mechanisms are often equivalent to a random positioning of the sensors. Note that in order to accomplish its task, a well-designed sensor network should satisfy two critical constraints: Coverage (all or most of the region of interest is within the sensing range of at least one sensor) and connectivity (each sensor can communicate with any other sensor either directly or by relaying information through its neighbors).

Additional sensor network design/structure issues may include the estimation of the sensor locations, lossless communication protocols, synchronized transmissions to other sensors and the center node, network size scalability, network and sensors reliability, and so on. Problems related to these issues are out of the scope of this chapter. In what follows in the chapter, we will assume that all the communication and networking issues have been settled in advance, so we will focus only on the collaborative signal processing tasks.

5.2
Data Collection and Fusion by WSN

Suppose that N sensors have been deployed at locations s_i, $i = 1, \dots, N$, over a two-dimensional monitoring region R, which without loss of generality corresponds to the unit square. A target at location $v \in R$ emits a signal captured by the sensors. Denote the energy measured by the ith sensor $E_i = S_i + \epsilon_i$, where $S_i \equiv S_i(v)$ is the signal captured from the target at sensor location i, and ϵ_i, $i = 1 \dots N$ are independent and identically distributed (i.i.d.) random noise observations.[1] It is usually assumed that the signal strength decays monotonically as the function of increasing distance between the sensor and the target increases. For example, Figure 5.1a shows the signal strength of a target located in the center of R exhibiting exponential decay, while the middle panel (Figure 5.1b) shows the same signal corrupted by Gaussian noise.

Based on the observed energy levels E_i, each sensor makes a decision $Y_i \in \{0, 1\}$ regarding the presence ($Y_i = 1$) or absence ($Y_i = 0$) of the target in the monitoring area. Note that each individual sensor decision depends on whether the energy level E_i exceeds a pre-specified threshold τ_i, which determines the *individual sensor's false alarm probability* and can be different for different sensors. The individual sensor's false alarm probability, denoted by γ_i, is the conditional probability that sensor i detects the target given that there is *no* target in the

1) This section © 2008 IEEE. Reprinted, with permission, from Katenka *et al.*, 2008a.

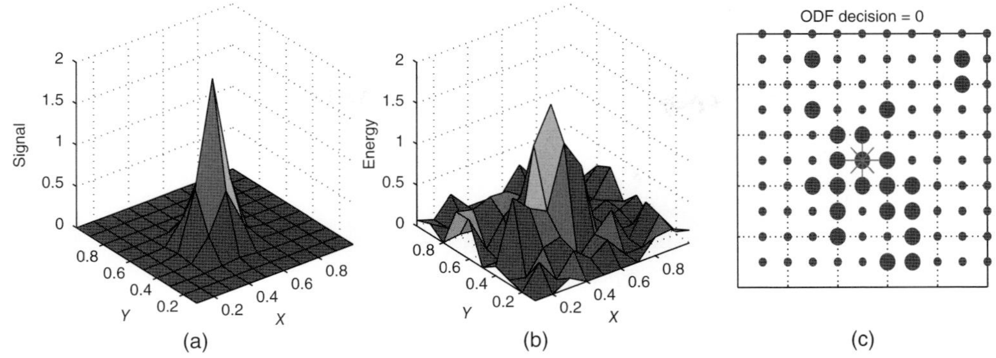

Figure 5.1 (a) Target signal generated by the model $S_i(v) = S_0 \exp(-||s_i - v||^2/\eta^2)$ for a target at location $v = (0.5, 0.5)$ with $S_0 = 2$, $\eta = 0.1$. (b) Target energy contaminated by Gaussian noise of variance $\sigma^2 = 0.16$ (signal-to-noise ratio $S_0/\sigma = 5$). (c) Sensor decisions based on individual false alarm probability (the same for all sensors) $\gamma = 0.2$. Fig. (c) © 2008 IEEE. Reprinted, with permission, from Katenka *et al.*, 2008a. (See inset for color representation of this figure.)

monitored region. Further, we assume that all sensors are identical, and that they all use the same threshold $\tau_i = \tau$ and the same individual false alarm probability $\gamma = P(Y_i = 1|\text{no target in R})$. Figure 5.1c illustrates sensor decisions based on the same γ for all sensors.

To reach a final joint decision, sensors can transmit to the fusion center either the energy readings $\{E_i\}$ or the binary decisions $\{Y_i\}$. Clearly, transmission of binary decisions (or *decision fusion*) offers significant savings in communications costs, since only *positive* one-bit decisions need to be transmitted; while the fusion of energy reading offers significantly better performance in terms of the target detection and identification, especially for low noise levels (see Clouqueur *et al.* (2001) and the rest of the chapter for more details).

In what follows, we will refer to the fusion algorithm based on energy readings $\{E_i\}$ as *value fusion* and the fusion algorithm based on simple initial decisions $\{Y_i\}$ as *ordinary decision fusion* (*ODF*), respectively.

Value Fusion

1) Each sensor i measures an energy value $E_i = S_i + \epsilon_i$;
2) All sensors transmit energy measurements to the fusion center;
3) The fusion center makes final situational assessment (e.g., detection, localization) based on $\{E_i\}$.

Ordinary Decision Fusion

1) Each sensor i measures an energy value $E_i = S_i + \epsilon_i$;
2) Each sensor i makes its own decision $Y_i \in \{0, 1\}$ w.r.t. the sensor threshold τ, $Y_i = I(E_i \geq \tau)$[2];
3) Sensors transmit only positive decisions to the fusion center;
4) The fusion center makes final situational assessment based on set of positive decisions $\{Y_i\}$.

2) $I(\cdot)$ is the indicator function.

There are a number of alternative algorithms that have been proposed in the literature to improve performance of ODF including: Distance weighted voting (Duarte and Hu, 2004) which weighs individual sensor decisions by the inverse of their distance to the target; confidence weighted voting (Sun *et al.*, 2005) which weighs sensor decisions by a measure of confidence based on the neighborhood agreement; and decision fusion algorithm based on majority voting derived only for a 3-sensor system in Klein (1993)[3]. Unfortunately, none of these algorithms provide any analytical solution to guarantee the detection performance of the system. Moreover, it is not clear how proposed techniques can be applied to target localization and/or tracking.

In the face of such concern, we will consider a more rigorously studied alternative, the ODF mechanism called local vote decision fusion (LVDF). This modification has been recently proposed by Katenka, Levina, and Levina (2008a) and offers a full framework applicable toward detection, localization, and tracking of single and multiple targets.

The key idea of LVDF is that each sensor adjusts its initial decision locally by taking a majority vote in its neighborhood, say $U(i)$, that can be defined as either all sensors within a fixed distance r from the sensor i (e.g., communications range), or as a fixed number of its nearest neighbors.

Local Vote Decision Fusion (LVDF)

1) Each sensor i measures an energy value $E_i = S_i + \epsilon_i$;
2) Sensor i makes an *initial* decision Y_i independent of all other sensors and communicates it to all other sensors j in its neighborhood $U(i)$,
3) Subsequently, given a set of decisions $\{Y_j : j \in U(i)\}$, sensor i adjusts its initial decision according to a majority vote; that is, $Z_i = I(\sum_{j\in U(i)} Y_j > M_i/2)$, where $M_i = |U(i)|$ denotes the size of the neighborhood.
4) The positive updated decisions Z_i are communicated to the fusion center, which makes the final situational assessment[3].

Note that in practice, sensors only need to communicate positive decisions in step 1; an absence of communication according to some pre-specified protocol implies that $Y_i = 0$. Also, note that by definition, $i \in U(i)$, so the sensor's own decision is always taken into account.

Figure 5.2 illustrates the advantage of LVDF over ODF for both random and fixed grid deployments. Due to the noise, under ODF more wrong decisions (false positives) are obtained from sensors located far away from the target. These decisions can significantly impact the final decision at the fusion center. On the other hand, under LVDF mechanism many false positives (as well as false negatives) are fixed and that, in turn, can help the fusion center to reach the correct conclusion. Moreover, LVDF can also reduce the overall sensor's communication costs if communications to neighbors are more power-efficient than those to the fusion

3) This section © 2008 IEEE. Reprinted, with permission, from Katenka *et al.*, 2008a.

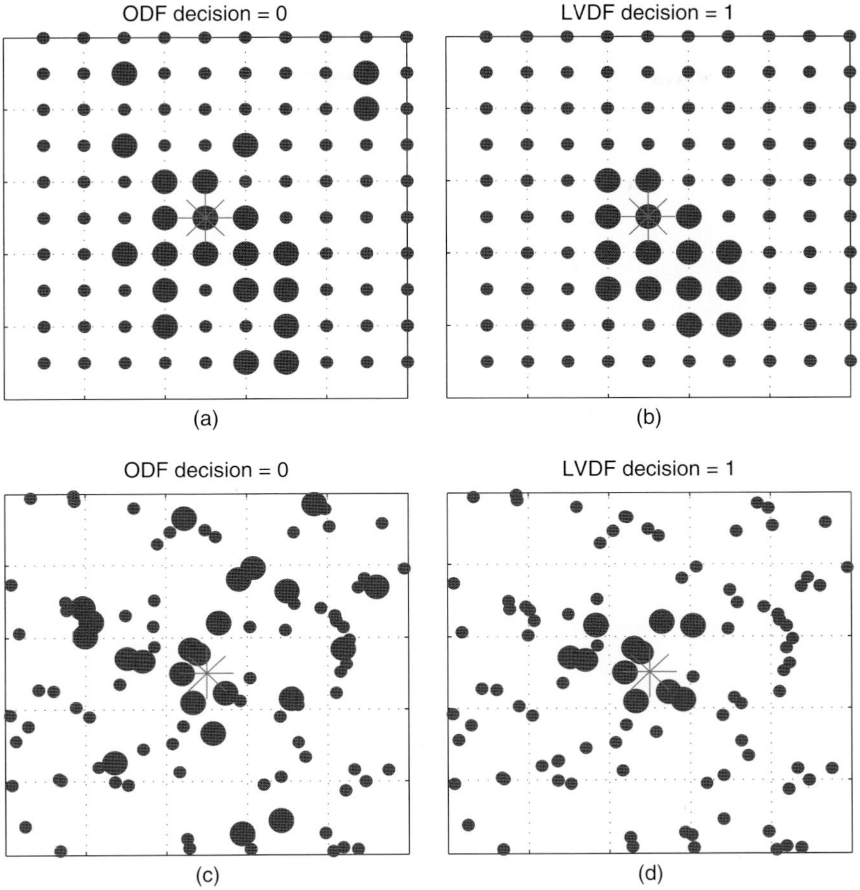

Figure 5.2 Ordinary versus. local vote decision fusion under a square grid design (a,b) and random deployment (c,d). The network is comprised of 100 sensors, with individual sensor false alarm probability $\gamma = 0.2$, system-wide false alarm probability $F = 0.1$ and a target located at the center of the monitored region R. The signal is generated by the model $S_i = S_0 \exp(-||s_i - v||^2/\eta^2)$, with $S_0 = 2$, $\eta = 0.1$, and the measured energy is corrupted by Gaussian noise with $\sigma = 0.4$. © 2008 IEEE. Reprinted, with permission, from Katenka *et al.*, 2008a. (See inset for color representation of this figure.)

center. More detailed comparison of outlined data fusion mechanisms for target detection, localization, and tracking will be presented next.

5.3 Target Detection

In the previous section, we explored three main data fusion algorithms by a WSN, namely value fusion, ODF, and LVDF. In this section, we present a core set of target detection techniques that are derived for each of the fusion algorithms, respectively.

There are two measures of accuracy of the target detection by WSN: The false alarm probability and the detection probability. The *false alarm probability* is the conditional probability that the system (i.e., sensor network) detects the target given that there is no target in the monitored region. The *detection probability* is the conditional probability that the fusion center correctly reports the presence of the target.

The classical approach to this problem goes back to the 1980s and radar-based applications (see Viswanathan and Varshney (1997) for a comprehensive review). In such applications, the target was a remote object, so it was reasonable to assume a specific signal model and also the same false alarm and detection probabilities for all devices (e.g., radars). In such settings, one could derive an optimal decision rule based on classical Bayesian decision theory. Unfortunately, for a WSN with a relatively small target in the middle of a large region, the assumptions of the same detection probabilities for all sensors and/or a known target signal model are unrealistic since the detection probability for each sensor depends on its distance from the target and the signal model can be different for different targets in the monitoring area. For a comprehensive review of more recent target detection algorithms, see Zhu *et al.* (2010) and Ermis and Saligrama (2010).

In this section, we will make no assumptions about the signal model and frame our test of hypothesis as follows:

- H_0: no target is present, or $S_i = 0$ for all i,
- H_1: 'something' is present, or $S_i > 0$ for some i.

This formulation follows the classical (frequentest) approach of treating the S_i's as unknown non-random parameters. Then, the energy readings E_i and the corresponding decisions Y_i are independent, since the only randomness comes from the i.i.d. noise ϵ_i. Note that the advantage of our approach is in its applicability even when no prior knowledge about the target's signal characteristics is available and also when there are multiple targets under surveillance.

Recall that for the convenience of presentation, we have assumed all sensors are identical and use the same threshold $\tau_i = \tau$ for individual sensor decisions $\{Y_i\}$. Then, in terms of our null hypothesis H_0, all energy readings E_i's are i.i.d. and all sensors have the same false alarm probability $\gamma = P_{H_0}(Y_i = 1)$.

Given a target in R, the objective of the sensor network is to maximize its probability of detection D, while controlling the corresponding system-wide false alarm probability F with focus on both individual sensors and the global decision.

5.3.1
Target Detection from Value Fusion (Energies)

We start our exposition from the value fusion algorithm for target detection, where the raw energy measurements are added at the fusion center. According to the value fusion algorithm, sensors make no individual decisions, so in order to guarantee the overall network performance that utilizes the transmission of the energies, one must be able to obtain the threshold T_e for the whole network, given the distribution of the system's false alarm probability F and the level of the noise.

Let G denote the distribution function of the noise levels, that is, $\epsilon_i \sim G$ are i.i.d. with mean 0 and variance σ^2. The system-wide decision about target(s) presence is $I(\sum_{i=1}^{N} E_i \geq T_e)$, where N is the size of the network. When the network size is relatively large and target is absent, that is, $E_i \equiv \epsilon_i$, the probability of a positive decision (the false alarm probability of the system) can be approximated by normal distribution

$$F = \mathbb{P}\left(\sum_{i=1}^{N} E_i \geq T_e | S_i = 0\right) = \mathbb{P}\left(\sum_{i=1}^{N} \epsilon_i \geq T_e\right) \approx 1 - \Phi\left(\frac{T_e}{\sqrt{N\sigma^2}}\right), \quad (5.1)$$

where $\Phi(\cdot)$ denotes the standard normal cumulative distribution function.

5.3.2
Target Detection from Ordinary Decision Fusion

We will continue with the target detection based on the ODF algorithm, where only the positive sensors' decisions are simply added at the fusion center. In order to guarantee the overall system performance of decision fusion for target detection, one must be able to obtain the threshold T for the whole network, given an individual sensor's and the system's false alarm probabilities γ and F, respectively. Here, we assume that γ is determined either by hardware specifications or from the information about background noise levels, whereas F can be controlled by the network's specialist.

In the absence of a target ($S_i \equiv 0$), the system-wide false alarm probability is given by the right tail of the binomial distribution,

$$F = \sum_{i=T}^{N} \binom{N}{i} \gamma^i (1-\gamma)^{N-i}, \quad (5.2)$$

since sensors make individual decisions independently, with $\gamma = G(\tau)$.

In the network size N is large, the above tail probability can be fairly accurately approximated by the normal distribution given by

$$F \approx 1 - \Phi\left(\frac{T - N\gamma}{\sqrt{N\gamma(1-\gamma)}}\right). \quad (5.3)$$

Thus, for individual sensor and system-wide false alarm probabilities γ and F, one can compute the corresponding decision threshold T which roughly translates to the minimum number of sensors required to detect the target for positive system-wide decision. In our problem formulation, however, it is not possible to optimize both thresholds T and F simultaneously. Note also that for ODF, knowledge of the distribution G is not required, as long as γ is known.

The probability of detection depends on the target's and the sensor's locations, signal parameters, and the noise distribution, and is given next

$$D = \sum_{i=T}^{N} \sum_{\pi \in \Gamma} \prod_{j=1}^{i} \left(1 - G(\tau - E_{\pi(j)})\right) \prod_{j=i+1}^{N} G(\tau - E_{\pi(j)}), \quad (5.4)$$

where Γ denotes the set of all permutations of $\{1, \ldots, N\}$. The first product term corresponds to the probability that sensors $\pi(1), \ldots, \pi(i)$ make positive decisions, while the second product term corresponds to the probability that sensors $\pi(i+1), \ldots, \pi(N)$ make negative decisions.

5.3.3 Target Detection from Local Vote Decision Fusion

Next, we will take a few steps to derive the system-wide threshold value T_ℓ for LVDF that guarantees a false alarm probability F. The strategy is to use a normal approximation for large sensor networks where the updated decisions are no longer independent.

First, we will calculate the expected value and variance of the updated decision Z_i under H_0:

$$\mu_i = P(Z_i = 1) = \sum_{j=[M_i/2]+1}^{M_i} \binom{M_i}{j} \gamma^j (1-\gamma)^{M_i - j}, \quad (5.5)$$

where $[x]$ denotes the largest integer smaller than or equal to x. The variance is given by $\sigma_i^2 = \text{Var}(Z_i) = \mu_i(1 - \mu_i)$.

Second, we will compute the covariance between Z_i and Z_j. Note that the dependence between Z_i and Z_j, $j \neq i$ comes from the intersection of their respective neighborhoods $U(i)$ and $U(j)$, as shown on the Figure 5.3. Let n_{ij} denote the number of sensors in the intersection $U(i) \cap U(j)$.

To calculate the covariance between Z_i and Z_j, we first compute $E(Z_iZ_j) = P(Z_i = Z_j = 1)$. Let A be the number of positive decisions in $U(i) \cap U(j)$, B the number of positive decisions in $U(i)$, but not in $U(j)$, and C the number of positive

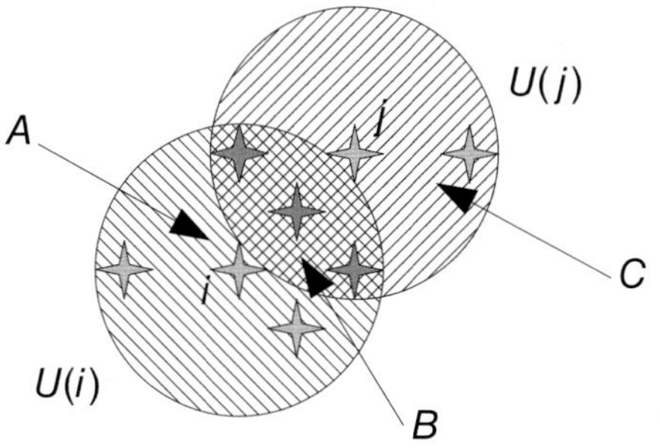

Figure 5.3 Example of sensor neighborhoods with $M_i = 6$, $M_j = 5$, and $n_{ij} = 3$. © 2008 IEEE. Reprinted, with permission, from Katenka *et al.*, 2008a.

decisions in $U(j)$ but not in $U(i)$, and note that A, B, and C are independent. Then, we can write (letting $\binom{a}{b} \equiv 0$ if $b < 0$)

$$E(Z_iZ_j) = \sum_{k=0}^{n_{ij}} P(A = k)\, P(B > \frac{M_i}{2} - k)\, P(C > \frac{M_j}{2} - k), \quad \text{where} \tag{5.6}$$

$$P(A = k) = \binom{n_{ij}}{k} \gamma^k (1-\gamma)^{n_{ij}-k},$$

$$P(B > \frac{M_i}{2} - k) = \sum_{q=[\frac{M_i}{2}]-k+1}^{M_i - n_{ij}} \binom{M_i - n_{ij}}{q} \gamma^q (1-\gamma)^{M_i - n_{ij} - q},$$

$$P(C > \frac{M_j}{2} - k) = \sum_{q=[\frac{M_j}{2}]-k+1}^{M_j - n_{ij}} \binom{M_j - n_{ij}}{q} \gamma^q (1-\gamma)^{M_j - n_{ij} - q}. \tag{5.7}$$

Note that the term A is the probability that enough positive decisions for both sensors i and j to make decisions $Z_i = Z_j = 1$ are present in the intersection of their neighborhoods $U(i) \cap U(j)$. The term B_k is the probability that there are exactly k positive decisions in $U(i) \cap U(j)$ (but not enough to make both Z_i and Z_j positive automatically); and the terms C_k and D_k are the probabilities that there are enough positive decisions outside of the intersection to make $Z_i = 1$ and $Z_j = 1$, respectively.

Thus, the covariance is then given by

$$\mathrm{Cov}(Z_i, Z_j) = [E(Z_iZ_j) - \mu_i\mu_j] I(n_{ij} \neq 0). \tag{5.8}$$

Finally, under the assumption that the target is absent, the system's false alarm probability is given by

$$F = P\left(\sum_{i=1}^{N} Z_i \geq T_\ell\right), \tag{5.9}$$

where T_ℓ denotes the LVDF threshold. The updated decisions $\{Z_i;\ i = 1, \ldots, N\}$ form a dependent random field. The central limit theorem applies to the $\sum Z_i$, both for sensors deployed on a regular grid or at random (see Katenka, Levina, and Levina (2008a) for more details). The following approximation then holds:

$$F \approx 1 - \Phi\left(\frac{T_l - \sum_{i=1}^{N} \mu_i}{\sqrt{\sum_{i=1}^{N} \sigma_i^2 + \sum_{i \neq j, n_{ij} \neq 0} \mathrm{Cov}(Z_i, Z_j)}}\right). \tag{5.10}$$

5.3.3.1 Remark 1: LVDF Fixed Neighbourhood Size[4)]

In some settings (e.g., dense deployments or regular grids), the number of neighbors may be fixed to a pre-specified number with $|U(i)| = M$ for all i. In this case, we have $Z_i = I\left\{\sum_{j \in U(i)} Y_j > \frac{M}{2}\right\}$, which shows that the Z_i's are dependent but now *identically* distributed. Hence, the mean $E(Z_i) = \mu$ and the

4) This section © 2008 IEEE. Reprinted, with permission, from Katenka *et al.*, 2008a.

variance $\mathrm{Var}(Z_i) = \sigma^2$ can be calculated using Eq. (5.5). Then, $E(Z_iZ_j)$ can be calculated from Eq. (5.7) with $M_i = M_j = M$ and the resulting covariance is given by $\mathrm{Cov}(Z_i, Z_j) = [E(Z_iZ_j) - \mu^2]I(n_{ij} \neq 0)$. The normal approximation simplifies to

$$F \approx 1 - \Phi\left(\frac{T_\ell - N\mu}{\sqrt{N\sigma^2 + \sum_{i\neq j, n_{ij}\neq 0} \mathrm{Cov}(Z_i, Z_j)}}\right). \tag{5.11}$$

5.3.3.2 Remark 2: LVDF Regular Grids

In some applications, it may be possible to deploy the sensors along a regular grid. In this case, the false alarm approximation (Eq. (5.10)) further simplifies under the assumption that each sensor has exactly M neighbours to consult including itself (ignoring edge effects). In practice, this can be achieved by ignoring corrected decisions of sensors on the edges, effectively reducing the grid size. On a regular grid, the one-hop neighborhood contains either $M = 5$ (diamond-shaped neighborhood) or 9 neighbors (square neighborhood), depending on whether diagonally located nearest neighbors are included or not, and $M = 7$ (hexagonal neighbourhood), the three most common designs considered in classical random fields theory (Doukhan, 1994) (Figure 5.4)[5].

(i) *Square Neighborhood.* The number of hops (layers) away from the sensor at the center determines the size of the neighborhood. Let m denote the number of layers considered. Then, the size of the square neighborhood $U(i)$ is given by $M = (2m+1)^2$. Let $t = (t_1, t_2)$ be a location shift and $U(i+t)$ the neighborhood of the sensor located at $s_i + t$. Then, the number of common sensors in $U(i)$ and $U(i+t)$ is given by $n_{i,i+t} = (2m+1-|t_1|)(2m+1-|t_2|)$, with $0 \leq |t_1|, |t_2| \leq 2m$, and $n_{i,i+t} = 0$ otherwise. The covariance is given by

$$\mathrm{Cov}(Z_i, Z_{i+t}) = \mathrm{Cov(t)} = [E(Z_iZ_{i+t}) - \mu^2]I(0 \leq |t_1|, |t_2| \leq 2m) \tag{5.12}$$

and the normal approximation of F can be obtained as before.

The previous formula does not reflect the presence of edge effects that are taken into account in the following formula:

$$n_{i,i+t} = \left|U(i) \bigcap U(i+t)\right| = (\max(m_i^L, m_{i+t}^L) - |t_1|)(\max(m_i^W, m_{i+t}^W) - |t_2|), \tag{5.13}$$

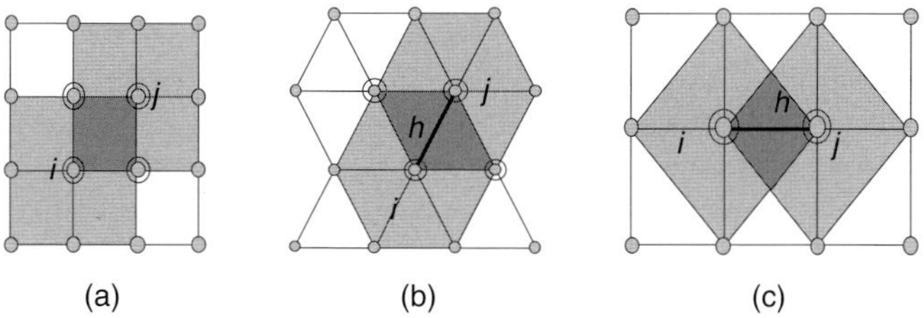

Figure 5.4 Square (a), hexagonal (b), and diamond-shaped (c) neighborhoods on a regular grid.

5) This section © 2008 IEEE. Reprinted, with permission, from Katenka *et al.*, 2008a.

where m_i^L and m_i^W are the length and the width of the neighborhood of sensor i, that is, $M_i = m_i^L \cdot m_i^W$. For each sensor i, the length m_i^L and the width m_i^W can be calculated as follows:

$$\begin{aligned} m_i^L &= \min(t_{i,1}^{max} - t_{i,1}^{min}, 2m + 1), \\ m_i^W &= \min(t_{i,2}^{max} - t_{i,2}^{min}, 2m + 1), \end{aligned} \tag{5.14}$$

where

$$t_{i,1}^{min} = \max(-2m, -(i_1 - 2m + 1)),\ t_{i,1}^{max} = \min(2m, n - i_1), \tag{5.15}$$

$$t_{i,2}^{min} = \max(-2m, -(i_2 - 2m + 1)),\ t_{i,2}^{max} = \min(2m, n - i_2), \tag{5.16}$$

and the sensor location index i is a pair of indexes $i = (i_1, i_2)$ along horizontal and vertical dimensions. The formula for the covariance can be written as:

$$\mathrm{Cov}(Z_i Z_{i+t}) = \begin{cases} E(Z_i Z_{i+t}) - \mu_i \mu_{i+t}^2,\ 0 < t_{i,j}^{min} \le \left|t_j\right| \le t_{i,j}^{max} < 2m,\ j \in 1, 2 \\ \sigma_i^2,\ \left|t_1\right| = \left|t_2\right| = 0 \\ 0,\ \left|t_j\right| > t_{i,j}^{max},\ j \in 1, 2 \end{cases} \tag{5.17}$$

and the normal approximation of F can be obtained as before. However, for large networks, the edge effect is negligible. Simulation results show that there is no significant difference in quality approximation for network size $n \le 10$.

(ii) *Diamond-Shaped Neighborhood.* We only consider the single-layer neighborhood with $M = 5$. The possible values for the size of non-empty intersections of $U(i)$ and $U(j)$ are

$$n_{ij} = |U(i) \cap U(j)| = \begin{cases} 1,\ \|s_i - s_j\| = 2h \\ 2,\ h \le \|s_i - s_j\| \le h\sqrt{2} \end{cases} \tag{5.18}$$

where $2h$ is the size of the diamond's diagonal. The approximation for F can then be straightforwardly obtained.

(iii) *Hexagonal Neighbourhood.* For a hexagonal grid design, let h denote the side of the hexagon. Here, we only consider the single-layer neighborhood with $M = 7$. The only possible values for the size of non-empty intersections of $U(i)$ and $U(j)$ are

$$n_{ij} = |U(i) \cap U(j)| = \begin{cases} 1,\ \|s_i - s_j\| = 2h \\ 2,\ \|s_i - s_j\| = h\sqrt{3} \\ 4,\ \|s_i - s_j\| = h \\ 0,\ \|s_i - s_j\| > 2h \end{cases} \tag{5.19}$$

and the corresponding approximation formula is,

$$F \approx 1 - \Phi\left(\frac{T_\ell - N\mu}{\sqrt{N\left(\sigma^2 + 6\sum_{n_{ij}\in\{1,2,4\}} \mathrm{Cov}(n_{ij})\right)}}\right), \tag{5.20}$$

where the factor of 6 comes again from the symmetry of the grid.

We refer a reader to the original paper (Katenka, Levina, and Levina, 2008a) where the LVDF for target detection was introduced and the extensive performance evaluation study was performed. Here, we only will summarize the main results related to the quality of the false alarm approximation and some related network aspects.

5.3.3.3 Remark 3: Quality of Approximation

The approximation for the value fusion falls into the classical application of the central limit theorem for a sample of independent identically distributed observations, and this is the most accurate. Similarly, the approximation for the ODF is the well-known normal approximation to the binomial distribution, which is also very accurate.

For LVDF, as M increases, the dependencies among the decisions become stronger, and the quality of the approximation deteriorates. On the other hand, as the size n of the grid increases, the approximation improves. Nevertheless, the quality of the approximation remains very good even for moderate network sizes.

Note that the general approximation of LVDF (Eq. (5.10)) depends on sensor locations assumed to be known. However, the examination of the approximation (Eq. (5.10)) shows that it depends on sensor locations only through the *distribution* of neighborhood sizes M_i and their intersections n_{ij}; and, while the actual locations may change, for example, from one random deployment to another, the distribution of neighborhood sizes does not change much. Therefore, the exact sensor locations are not as important as the actual sensor density in the monitored region.

5.3.3.4 Remark 4: Detection Performance

Apparently, the simple local vote-based decision fusion algorithm substantially outperforms ODF in terms of target detection probability and exhibits a competitive performance while compared to the value fusion. LVDF also achieves good results even at a significantly lower signal-to-noise ratio. Therefore, the local vote schemes are particularly beneficial for small targets or signal with low attenuation. However, when designing a WSN employing the LVDF mechanism, one should choose the size of the neighborhood comparable to the size of the smallest target one is interested in detecting, since large targets will be easy to spot.

5.3.3.5 Concluding Remarks

Before closing our discussion of target detection, it cannot hurt to again point out that outlined detection framework provides a rigorously derived analytical approximation for the system-wide decision threshold level T as a function of the system-wide false alarm probability F, for three types of data that can be obtained from the sensor system: Energy measurements, initial sensor decisions, and adjusted sensor decisions. This ensures that one can design a network with a guaranteed false alarm rate using our algorithm. This framework can be directly applied to the monitoring and detection of multiple targets with different signal characteristics. Finally, this framework can be extended to temporal fusion as proposed in Katenka, Levina, and Levina (2008a). We will consider the temporal data for target tracking further in Section 5.6, while next we will turn our attention to Section 5.4 where we will focus on the task of a single target localization and diagnostic.

5.4 Single Target Localization and Diagnostic

Given the energy measurements $E_i = S_i + \epsilon_i$, $i = 1, 2, \ldots, N$, binary initial Y_i or corrected Z_i decisions, the main tasks are: (i) Detect the presence of a target (results described in Section 5.3), (ii) identify target location $v = (v_x, v_y)$ (localization problem), (iii) estimate the strength of the signal that may characterize the type of target present (the problem of target diagnostic), and (iv) with information available over time, track target(s) trajectory through the monitoring region R (tracking problem that will be discussed in Section 5.6).

In this section, we will focus on the tasks of a single target localization and the diagnostic. The roots on these problems go back to the target localization methods applied to radar systems (Abdel-Samad and Tewfik, 1999). The modern localization algorithms for WSNs can be divided into two general classes: Those based on energy sensor readings E_i (Li *et al.*, 2002; Sheng and Hu, 2003; Blatt and Hero, 2006) and those based on binary decisions Y_i (Niu and Varshney, 2004; Noel, Joshi, and Jannett, 2006; Ermis and Saligrama, 2006). Methods that are based on energy readings range from less accurate, non-linear least squares (Li *et al.*, 2002) to a more computation intensive maximum likelihood (ML) estimation coupled with expectation-maximization (EM) algorithm (Sheng and Hu, 2003). The main advantage of proposed methods compared to classical techniques that depended on a direction of arrival (DOA) and/or time delay of arrival (TDOA) (Kaplan, Le, and Molnar, 2001), is that energy-based methods do not require a very accurate synchronization among the sensors. However, these methods proved to be constrained in other ways: For example, methods in Sheng and Hu (2003) require transmission of the mean and variance of the background noise, which often are unknown; methods in Li *et al.* (2002) require transmission of real-value data from all the sensors, which may not be practical, methods in Blatt and Hero (2006) require network structure and capability to collect the data from cluster heads. While a number of studies have been dedicated to the problem of target localization from initial binary decisions, majority of the proposed methods were based on some version of the optimization of the ML function (see Niu and Varshney (2004) and Noel, Joshi, and Jannett (2006)). Later in this section, we will see that the ML function for initial binary decision is a complex non-linear function which is difficult to optimize and that may suffer from existence of local minima, low convergence, and high computational complexity. This problem becomes even more severe in the low signal-to-noise ratio regimes due to a large number of false decisions.

In this section, we develop target localization techniques for the energy sensor readings E_i, the initial decisions Y_i, and more accurate corrected decisions Z_i. Specifically, for the corrected decisions Z_i, that are correlated, we adopt a pseudo-likelihood approach, and develop a localization and signal estimation procedure for LVDF that exhibits the same robustness properties as the LVDF detection algorithm outlined in Section 5.3. We also derive an EM algorithm for ML estimation from binary decisions, for both the original decisions Y_i and Z_i. In addition, we discuss properties of the estimators and provide a bootstrap procedure for uncertainty assessment. And finally, we discuss computational issues including the issue of the starting values of parameter estimates and the computational cost of the different algorithms.

5.4.1 Localization and Diagnostic from Value Fusion (Energies)

Recall that we have assumed that the sensor locations s_i are known or can be estimated prior to localization. Here, we also assume that the attenuation of the target's signal is a known function which is monotonically decreasing in the distance from the target $\delta_i(v) = ||s_i - v||$, and also depends on an attenuation parameter η. That is, the signal at location s_i is given by

$$S_i(v) = S_0 C_\eta(\delta_i(v)), \tag{5.21}$$

with $C_\eta(0) = 1$ and $S_0 \in [0, \infty)$ denoting the signal strength at the target's location v. So, the primary parameters of interest are the target's location v and the signal strength S_0; obviously, the noise variance σ^2 and the attenuation parameter η affect the estimation problem. The noise is assumed to be Gaussian with mean zero and variance σ^2.

A natural starting point is to describe a localization algorithm for a single target based on energy measurements E_i. In the presence of Gaussian background noise with mean zero and variance σ^2, the full log-likelihood of energies, up to an additive constant, is given by:

$$\ell_E(\theta) = -\frac{n}{2}\log\sigma^2 - \frac{1}{2\sigma^2}\sum_{i=1}^{N}\left[E_i - S_0 C_\eta(\delta_i(v))\right]^2. \tag{5.22}$$

Maximizing this over S_0 and σ^2 can be done in closed form. This gives the

$$\hat{S}_0 = \frac{\sum_{i=1}^{N} E_i C_\eta(\delta_i(v))}{\sum_{i=1}^{N} C_\eta^2(\delta_i(v))}, \tag{5.23}$$

$$\hat{\sigma}^2 = \frac{1}{N}\sum_{i=1}^{N}(E_i - \hat{S}_0 C_\eta(\delta_i(v)))^2. \tag{5.24}$$

The other parameters (v and η) can be found by the numerical optimization of Eq. (5.22) with Eqs. (5.23) and (5.24) plugged in.

5.4.2
Localization and Diagnostic from Ordinary Decision Fusion

To localize a target from initial binary decisions, we, first, define the vector of unknown parameters as

$$\theta = (v_x, v_y, S_0, \sigma, \eta). \tag{5.25}$$

Then, decisions $\{Y_i\}$ are independent Bernoulli random variables with probability of success given by

$$\mathbb{P}(Y_i = 1) \equiv \alpha_i(\theta) = 1 - F(A_i(\theta)), \tag{5.26}$$

where $F(\cdot)$ denotes the cumulative distribution function (not necessarily Gaussian) and $A_i(\theta)$ is the standardized excess energy level given by

$$A_i(\theta) = \frac{\tau - S_0 C_\eta(\delta_i(v))}{\sigma}. \tag{5.27}$$

The log-likelihood function of $\{Y_i\}$ is given by:

$$\ell_Y(\theta) = \sum_{i=1}^{N} \left[Y_i \log \alpha_i(\theta) + (1 - Y_i) \log(1 - \alpha_i(\theta)) \right]. \tag{5.28}$$

There are two options for obtaining estimates of the unknown parameters: direct numerical maximization of the log-likelihood function (Eq. (5.28)) (no closed form solution exists) or the EM algorithm (Dempster, Laird, and Rubin, 1997). We treat initial binary decisions $Y_i = I(E_i > \tau)$ as incomplete data of the true energy readings E_i. So, EM consists here of an expectation step (E-step), where expected likelihood of the full data E_i conditional on the available data Y_i is obtained, and a maximization step (M-step) where the parameters are estimated by maximizing the likelihood from the E-step[6]. In fact, the likelihood (Eq. (5.22)) serves as a basis for E-step and M-step of the algorithm. Moreover, since the likelihood (Eq. (5.22)) is a curved exponential family in θ, the M-step shows that there are just two quantities that need to be computed in the E-step: $\hat{E}_i = \mathbb{E}[E_i | Y] = \mathbb{E}[E_i | Y_i]$ and $\hat{E}_i^2 = \mathbb{E}[E_i^2 | Y] = \mathbb{E}[E_i^2 | Y_i]$. Note that each E_i only depends on Y_i rather than all Y because S_i is not random, and ϵ_i's are independent. Hence, the expectations can be computed as follows:

$$\mathbb{E}[E_i | Y_i = 0] = \frac{\int_{-\infty}^{\tau} x p_{E_i}(x)\, dx}{\int_{-\infty}^{\tau} p_{E_i}(x)\, dx} = S_0 C_\eta(\delta_i(v)) - \frac{\sigma \exp\left(-\frac{A_i(\theta)^2}{2}\right)}{\sqrt{2\pi}\Phi\left(-\frac{A_i(\theta)^2}{2}\right)} \tag{5.29}$$

Combining the similar computations for $\mathbb{E}[E_i | Y_i = 1]$ and $\mathbb{E}[E_i^2 | Y_i]$ gives the *E-step.*

$$\hat{E}_i = S_0 C_\eta(\delta_i(v)) + \frac{\sigma \exp\left(-\frac{A_i(\theta)^2}{2}\right)}{\sqrt{2\pi}} B_i(\theta, Y) \tag{5.30}$$

$$\hat{E}_i^2 = S_0 C_\eta(\delta_i(v))(\tau - \hat{E}_i) - \hat{E}_i \tau, \tag{5.31}$$

6) This section © Taylor & Francis Ltd, http://www.tandfonline.com. Reprinted, with permission, from Katenka *et al.*, 2008b.

where

$$B_i(\theta, Y) = \frac{Y_i - 1}{\Phi(-\frac{A_i(\theta)^2}{2})} + \frac{Y_i}{1 - \Phi(-\frac{A_i(\theta)^2}{2})} . \tag{5.32}$$

Note that by nature, the EM is an iterative algorithm which, as a rule, is computationally more expensive than direct numerical optimization of the likelihood; however; it produces much more accurate results (see Katenka, Levina, and Michailidis 2008b). Note also that both the EM and the optimization of the likelihood require good initial values for parameter estimates. We will briefly discuss this issue later in the section.

5.4.3 Localization and Diagnostic from Local Vote Decision Fusion[7)]

Since adjusted decisions Z_i produced by the LVDF algorithm are correlated, the direct form of the likelihood function presented in Eq. (5.28) is invalid for Z_i. Alternatively, we will adopt a pseudo-likelihood formulation (Besag, 1986), by assuming that all adjusted decisions Z_i are *independent*. Further, we make a simplifying assumption for neighbors $j \in U(i)$, $\mathbb{P}(Y_j = 1) \approx \mathbb{P}(Y_i = 1)$.
Letting $\beta_i(\theta) = \mathbb{P}(Z_i = 1)$, this gives

$$\beta_i(\theta) = \mathbb{P}(\sum_{j \in U(i)} Y_j \geq \frac{M}{2}) \approx \sum_{k=[M/2]}^{M} \binom{M}{k} \alpha_i^k (1 - \alpha_i)^{M-k}. \tag{5.33}$$

The pseudo-loglikelihood function for the adjusted decisions Z_i is given by:

$$\ell_Z(\theta) = \sum_{i=1}^{N} \left[Z_i \log \beta_i(\theta) + (1 - Z_i) \log(1 - \beta_i(\theta))\right] . \tag{5.34}$$

ML estimates based on Eq. (5.34) can again be obtained through direct maximization. For the EM algorithm, the M-step is the same as before. The E-step requires calculating the first and second conditional moments $\mathbb{E}[E_i|Z]$ and $\mathbb{E}[E_i^2|Z]$. We first compute the conditional distribution of E_i given all the decisions Z. Write

$$\mathbb{P}[E_i|Z] = \frac{1}{\mathbb{P}(Z)} \sum_{k=0,1} \mathbb{P}(E_i, Z|Y_i = k)\mathbb{P}(Y_i = k) \tag{5.35}$$

$$= \frac{1}{\mathbb{P}(Z)} \sum_{k=0,1} \mathbb{P}(E_i|Y_i = k)\mathbb{P}(Z|Y_i = k)\mathbb{P}(Y_i = k) \tag{5.36}$$

where the last equality follows because conditional on the value of Y_i the energy reading E_i is independent of the vector of corrected decisions Z (recall again that all randomness comes from the noise ϵ_i, not the signal). Integrating Eq. (5.36) gives

$$\mathbb{E}[E_i|Z] = \sum_{k=0,1} \mathbb{E}(E_i|Y_i = k)\mathbb{P}(Y_i = k|Z) \tag{5.37}$$

$$\mathbb{E}[E_i^2|Z] = \sum_{k=0,1} \mathbb{E}(E_i^2|Y_i = k)\mathbb{P}(Y_i = k|Z) \tag{5.38}$$

7) This section © Taylor & Francis Ltd, http://www.tandfonline.com. Reprinted, with permission, from Katenka *et al.*, 2008b.

Since we have already obtained $\mathbb{E}[E_i|Y_i]$ and $\mathbb{E}[E_i^2|Y_i]$ in the E-step for ODF, all that remains to be calculated is,

$$\mathbb{P}(Y_i = 1|Z) = \frac{\mathbb{P}(Y_i = 1)\mathbb{P}(Z|Y_i = 1)}{\mathbb{P}(Z)} \tag{5.39}$$

$$\approx \alpha_i \prod_{j=1}^{N} \frac{\mathbb{P}(Z_j|Y_i = 1)}{\mathbb{P}(Z_j)} \tag{5.40}$$

$$= \alpha_i \prod_{j:i \in U(j)} \frac{\mathbb{P}(Z_j|Y_i = 1)}{\mathbb{P}(Z_j)}, \tag{5.41}$$

where Eq. (5.39) is the Bayes rule, Eq. (5.40) is the pseudo-likelihood approximation, and Eq. (5.41) follows because only corrected decisions that come from a neighborhood containing sensor i depend on Y_i. Once again, using the assumption $\alpha_j \approx \alpha_i$ for $j \in U(i)$, we get

$$\tilde{\beta}_{ji} = \mathbb{P}(Z_j = 1|Y_i = 1) = \mathbb{P}\left(\sum_{k \in U(j), k \neq i} Y_k \geq \frac{M}{2} - 1 \right)$$

$$\approx \sum_{q=[M/2-1]}^{M-1} \binom{M-1}{q} \alpha_j^q (1-\alpha_j)^{M-1-q} \tag{5.42}$$

and finally

$$\mathbb{P}(Y_i = 1|Z) = \alpha_i \prod_{j:i \in U(j)} \left(\frac{\tilde{\beta}_{ji}}{\beta_j} \right)^{Z_j} \left(\frac{1-\tilde{\beta}_{ji}}{1-\beta_j} \right)^{1-Z_j} \tag{5.43}$$

substituting Eq. (5.62) into Eqs. (5.37) and (5.38) completes the E-step for the LVDF decisions.

5.4.4 Hybrid Maximum Likelihood Estimates

Hybrid maximum–likelihood (HML) estimation has been motivated by the situations when a mixture of energy readings and binary decisions may be transmitted to the fusion center, thereby combining benefits of saving sensor power consumption and collecting most informative energies. The main idea of hybrid methods is to use energy information from the sensors with positive decisions and model energies for the rest of the network. By using energy readings from sensors with positive initial or updated decisions, we both reduce significantly communication cost compared to transmission of the full energy measurements from all sensor nodes and improve the decision-based localization.

Hybrid expectation maximization (*HEM*) algorithm is an extension of the original EM algorithm. Since each E_i only depends on Y_i, for ODF, hybrid EM formulas are given as:

If $Y_i = 1$,

$$\hat{E}_i = E_i, \tag{5.44}$$

$$\hat{E}_i^2 = E_i^2. \tag{5.45}$$

Otherwise ($Y_i = 0$),

$$\hat{E}_i = S_0 C_\eta(\delta_i(v)) - \frac{\sigma \exp\left(-\frac{A_i(\tau)^2}{2}\right)}{\sqrt{2\pi}\Phi(-\frac{A_i(\tau)^2}{2})} \tag{5.46}$$

$$\hat{E}_i^2 = S_0^2 C_\eta^2(\delta_i(v))(\tau - \hat{E}_i) - \hat{E}_i \tau. \tag{5.47}$$

Analogously, for LVDF hybrid EM version, we model only the energies that correspond to $Z_i = 0$ using Eqs. (5.37), (5.38), (5.62) and use the available energies for $Z_i = 1$. Although HEM proves to be competitive in terms of the accuracy of localization and less computationally expensive than the original EM algorithm, it sometimes fails to converge. Another option is to replace energies corresponding to zero decisions by the threshold τ and maximize the energy-based likelihood (Eq. (5.22)), which avoids iterative computations, but suffers in the accuracy of location and signal estimates. We refer to this method as hybrid maximum-likelihood estimation (HML).

5.4.5 Properties of Maximum-Likelihood Estimates

In what follows, we briefly describe the properties of the ML and EM estimates for the ODF and the LVDF mechanisms. While the properties of the energy-based methods fall into the classical statistical theory and can be easily derived, the properties of the estimate based on binary decisions need more attention.

1) Under the following assumptions on the log-likelihood function: (i) the log-likelihood function is distinct when $\theta_1 \neq \theta_2$, (ii) the true parameter θ_0 is in the interior of the parameter space, and (iii) the log-likelihood function is differentiable in θ, the estimate $\hat{\theta}$ is *consistent*.
2) Under additional assumptions on the log-likelihood function and its derivatives, asymptotic normality of the estimates can also be established, which can be used to provide a measure of uncertainty for the estimates. Whether these assumptions hold will depend on the exact form of the signal decay function C. The EM algorithm will converge to a local maximum of the energy likelihood (Eq. (5.22)); additional properties can be established depending on the function C.
3) Assuming the conditions on the likelihood function, that guarantee asymptotic normality hold, one can obtain confidence regions for the parameters of interest. We show next how to construct a two-dimensional confidence region for the main parameter of interest, target location v. Let $\hat{v} = (\hat{v}_x, \hat{v}_y)$ be the coordinates of the estimate of the true target location, with $\hat{v} \sim \mathcal{N}(v, \Sigma_v)$, with $\Sigma_v = \text{Var}(\hat{v})$. A two-dimensional confidence region Q satisfies $\mathbb{P}(v \in Q) = 1 - \zeta$, with $1 - \zeta$ denoting the confidence level. Standardizing the location estimate yields

$$\tilde{v} = \Sigma_v^{-1/2}(\hat{v} - v) \sim \mathcal{N}(0, I_2), \tag{5.48}$$

which in turn implies that the desired confidence region $\tilde{Q}$ for $\tilde{v}$ is a circle of radius r that satisfies $\mathbb{P}(\|\tilde{v}\|^2 \leq r^2) = 1 - \zeta$. The appropriate value of r is given by the $(1 - \zeta)$-quantile of the χ^2 distribution with two degrees of freedom. The region $\tilde{Q}$ can then be inverted to obtain Q using Eq. (5.48).

This procedure requires an estimate of the covariance matrix $\Sigma = \mathrm{Var}(\hat{\theta})$. The estimation can be done (i) asymptotically using Cramer–Rao bound, which may be inaccurate for smaller samples, particularly for the pseudo-likelihood; or (ii) numerically through a parametric bootstrap procedure (Efron and Tibshirani, 1994), as follows.

1) Energies are simulated from the posited model with parameters set to the ML estimates: simulate M samples from the assumed signal attenuation model to obtain
$$E^*_{i,m} = \hat{S}_0 C_{\hat{\eta}}(\delta_i(\hat{v})) + \epsilon^*_{i,m}, \tag{5.49}$$
where $\epsilon^*_{i,m} \sim \mathcal{N}(0, \hat{\sigma}^2)$ are i.i.d. noise, $i = 1, \ldots, N$, $m = 1, \ldots, M$.
2) The simulated energies are used to obtain bootstrap estimates of the parameters of interest $\hat{\theta}_m$, $m = 1, \ldots, M$.
3) The empirical covariance of the estimates $\hat{\theta}_m$ across the M samples gives an estimate for Σ[8].

5.4.5.1 Remark 1: Accuracy of Target Localization[8]

The accuracy of the various algorithms for target localization was evaluated in Katenka, Levina, and Michailidis (2008b). As expected, the LVDF localization algorithms clearly outperform their ODF counterparts. Surprisingly, the results also show that in the low SNR regime, LVDF algorithms clearly outperform the "gold standard" ML(E) and for the medium and high SNR regimes, they exhibit a competitive performance. The HEM algorithms tend to be the most accurate, followed by EM, while both ML and HML tend to be less accurate. All algorithms do somewhat better on the signal model where the signal decay is slower and allows more sensors to pick up the target. It is also worth noting that for the ODF-based algorithms, the EM version significantly outperforms the one based on numerical optimization. The poor performance of ML using Y_i, particularly at low SNR, is primarily due to the sensitivity of the numerical solver to the selection of starting values, which in the case of the adjusted decisions is not an issue due to the de-noising nature of LVDF. As expected, for larger values of SNR, the accuracy of all the algorithms improves, and for random deployments the pattern remains the same but all methods are somewhat less accurate.

5.4.5.2 Remark 2: Starting Values for Localization

All the decision based algorithms are iterative in nature and require good starting values for the parameters of interest otherwise a poor choice of starting values

8) This section © Taylor & Francis Ltd, http://www.tandfonline.com. Reprinted, with permission, from Katenka *et al.*, 2008b.

can slow down convergence and/or lead to poor quality estimates. Notice that the starting values have to be a function of the information available for the method, and a good initial guess for the target's location is the centroid of the positive decisions, given by

$$v_0(Y) = \frac{\sum_i s_i I(Y_i = 1)}{\sum_i I(Y_i = 1)}$$

for ODF and

$$v_0(Z) = \frac{\sum_i s_i I(Z_i = 1)}{\sum_i I(Z_i = 1)}$$

for LVDF. Because LVDF eliminates many distant false positives, $v_0(Z)$ tends to be significantly more accurate than $v_0(Y)$.

For the benchmark ML(E), where all energies are available, a natural choice of starting value is the location $v_0(E)$ of the maximum energy reading $\max_i E_i$. All methods improve at higher SNR, but the starting value for Y is, on average, much further from the truth than the starting value for Z; for energies, the starting value based on maximum energy works well at higher SNRs, but not a very low SNR.

Note if better starting values are available from some prior information or external knowledge, performance of all methods will improve.

5.4.5.3 Remark 3: Robustness to Model Misspecification

The performance of all algorithms may change when the true signal model and/or the noise distribution are misspecified. Via a sequence of simulations in Katenka, Levina, and Michailidis (2008b), we explored sensitivity in performance of the different energy and decision-based algorithms.

In the experiment, where we misspecified the signal model, the performance of the algorithms relative to each other remains exactly the same (results not shown). The performance of both ML(E) and LVDF is very robust, whereas ODF performs somewhat worse, though the differences are small. These differences may be larger for more drastically different models.

In the simulation, where the true noise comes from a t-distribution with three degrees of freedom, while Gaussian distribution is assumed by the algorithms, the LVDF algorithms performed well and proved to be robust; the ODF errors were also similar; the energy-based ML(E), however, proved to be the most sensitive to distribution misspecification.

5.4.5.4 Remark 4: Computational Cost[9)]

On an average, the LVDF algorithms converge faster than their ODF counterparts; however, it takes the optimization about 1/10 of the iterations to converge on an average, compared to the EM versions (recall that the M-step requires a numerical optimization; the number of iterations shown for EM is the sum of the optimization iterations at each M-step and the EM iterations). Given the significantly higher accuracy of the EM algorithms, this represents the usual trade-off between computational complexity and accuracy. The hybrid EM algorithms converge faster than their EM counterparts as one would expect.

9) This section © Taylor & Francis Ltd, http://www.tandfonline.com. Reprinted, with permission, from Katenka *et al.*, 2008b.

5.4.5.5 Concluding Remarks

Although the localization of a stationary single target by WSN may prove to be a naive and overly simplified task, a number of the techniques described in this section serve as useful building blocks for a complex computational framework that allows tracking of unknown number of moving targets that can be different in size and signal characteristics. Specifically, in what comes next, we first will use analogous ML and EM-based approach to extend the localization of a single target to a multiple target localization in Section 5.5, and then, we will describe how to build an effective tracking system for WSN with specific constraints and targets under consideration in Section 5.6.

5.5 Multiple Target Localization and Diagnostic

In this section, we will extend the algorithms that estimate target locations at a given point in time based on energy/binary data developed by Katenka, Levina, and Michailidis (2008b) and described in the previous section from the case of one target to the case of multiple targets. First, we discuss localization with the number of targets assumed known, and present a method for selecting the number of targets.

Consider a WSN comprised of N identical sensors deployed at locations s_i, $i = 1, 2, \dots, N$ over a two-dimensional monitoring region R. Suppose that p targets are present in R at locations x_j, $j = 1, \dots, p$. Each target j emits a signal of strength $S_0^{(j)}$ at the target location. The signal attenuation is a decreasing function $C(\cdot)$ of the distance from the target to sensor located at s_i. Therefore, the sensor located at s_i receives signal from target j given by

$$S_i^{(j)}(t) = S_0^{(j)} C_{\eta_j}(\delta_i(x_j)), \tag{5.50}$$

where $\delta_i(x_j(t)) = \|s_i - x_j\|$ is the distance from the target to sensor i, and η_j is a scaling parameter representing the effective target size[10)].

At some fixed point in time, each sensor obtains an *energy* reading comprised of all p individual signals and corrupted by random noise:

$$E_i = \sum_{j=1}^{p} S_i^{(j)} + \epsilon_i, \quad i = 1, \dots, N, \tag{5.51}$$

where errors ϵ_i are assumed to be independent and identically distributed with mean zero and variance σ^2.

The collected energy readings E_i are either directly transmitted to the fusion center or converted to binary decisions $Y_i = I(E_i \geq \tau_i)$, using a pre-specified threshold τ_i, which is related to the individual sensor's false alarm probability; or the corrected decision $Z_i = I(\sum_{j \in U(i)} Y_i \geq M_i/2)$.

10) This section © Taylor & Francis Ltd, http://www.tandfonline.com. Reprinted, with permission, from Katenka *et al.*, 2013.

In the localization of multiple targets, the ultimate goal is to estimate the following parameter vector,

$$\theta = (p, x_1, ..., x_p, S_0^{(1)}, ..., S_0^{(p)}, \eta_1, ..., \eta_p, \sigma), \tag{5.52}$$

based either on energy readings (E_i) or binary decisions (Y_i or corrected Z_i).

5.5.1 Multiple Target Localization from Energies[11)]

In the presence of Gaussian mean zero, variance σ^2 background noise, the log-likelihood of the energies at a fixed point in time is given by:

$$\ell_E(\theta) = -\frac{N}{2}\log 2\pi\sigma^2 - \frac{1}{2\sigma^2}\sum_{i=1}^{N}\left[E_i - \sum_{j=1}^{p} S_0^{(j)} C_{\eta_j}\left(\delta_i\left(x_j\right)\right)\right]^2. \tag{5.53}$$

With other parameters fixed, maximizing over $S_0^{(j)}$ is equivalent to solving the following system of linear equations:

$$\sum_{l=1}^{p} S_0^{(l)} \sum_{i=1}^{N} C_{\eta_l}\left(\delta_i\left(x_l\right)\right) C_{\eta_j}\left(\delta_i\left(x_j\right)\right) = \sum_{i=1}^{N} E_i C_{\eta_j}\left(\delta_i\left(x_j\right)\right), \; j = 1 \ldots p. \tag{5.54}$$

Maximizing with respect to the noise variance σ^2 can be done in closed form, if all other parameters are fixed, as

$$\sigma^2 = \frac{1}{N}\sum_{i=1}^{N}\left[E_i - \sum_{j=1}^{p} S_0^{(j)} C_{\eta_j}\left(\delta_i(x_j)\right)\right]^2. \tag{5.55}$$

5.5.2 Multiple Target Localization from Binary Decisions

Again, we notice that the initial decisions $Y_i \in \{0, 1\}$ are independent Bernoulli random variables with probability of success given by $\mathbb{P}(Y_i = 1) \equiv \alpha_i(\theta) = 1 - F(A_i(\theta))$, where $F(\cdot)$ is the cumulative distribution function of ϵ_i/σ (not necessarily Gaussian) and $A_i(\theta)$ represents the standardized excess energy level given by

$$A_i(\theta) = \frac{\tau - \sum_{j=1}^{p} S_0^{(j)} C_{\eta_j}(\delta_i(x_j))}{\sigma}. \tag{5.56}$$

The log-likelihood function of $\{Y_i\}$ can be computed analogously to a single-target scenario:

$$\ell_Y(\theta) = \sum_{i=1}^{N}\left[Y_i \log \alpha_i(\theta) + (1 - Y_i)\log(1 - \alpha_i(\theta))\right]. \tag{5.57}$$

Since there is no closed form solution for any of the parameters, we again need to solve numerically maximization of the log-likelihood function or use an EM algorithm, but for significantly more parameters even when only a few targets are under consideration.

11) This section © Taylor & Francis Ltd, http://www.tandfonline.com. Reprinted, with permission, from Katenka *et al.*, 2013.

The EM algorithm proceeds as follows: Under the Gaussian assumption on the noise, the M-step is defined by Eqs. (5.53)–(5.55). The E-step equations are given by

$$\mathbb{E}[E_i|\vec{Y}] = \hat{E}_i = \sum_{j=1}^{p} S_0^{(j)} C_{\eta_j} \left(\delta_i(x_j)\right) + \frac{\sigma}{\sqrt{2\pi}} \exp\left(-\frac{A_i(\theta)^2}{2}\right) B_i(\theta, Y_i) \,, \quad (5.58)$$

$$\mathbb{E}[E_i^2|\vec{Y}] = \hat{E}_i^2 = \sum_{j=1}^{p} S_0^{(j)} C_{\eta_j} (\delta_i(x_j))(\tau - \hat{E}_i) - \hat{E}_i \tau \,, \quad (5.59)$$

where

$$B_i(\theta, Y_i) = \frac{Y_i - 1}{\Phi(-\frac{A_i(\theta)^2}{2})} + \frac{Y_i}{1 - \Phi(-\frac{A_i(\theta)^2}{2})} \,. \quad (5.60)$$

As usual, the E-step and the M-step are alternated until convergence.

5.5.3 Multiple Target Localization from Corrected Decisions[12)]

Since corrected decisions Z_i are not independent, we employ a pseudo-likelihood estimation approach (Besag, 1986), and to simplify calculations, we will further assume that the success probabilities of initial decisions are approximately the same within the neighbourhood, that is, for $j \in U(i)$, $\mathbb{P}(Y_j = 1) = \mathbb{P}(Y_i = 1)$. Note these are the same assumptions that we have made for a single-target scenario.

The pseudo-loglikelihood function at a fixed time point for the corrected decisions Z_i is given by,

$$\ell_Z(\theta) = \sum_{i=1}^{N} \left[Z_i \log \beta_i(\theta) + (1 - Z_i) \log(1 - \beta_i(\theta)) \right] \,, \quad (5.61)$$

where $\beta_i(\theta) = \mathbb{P}(Z_i = 1)$ is approximated by

$$\beta_i(\theta) \approx \sum_{k=[M_i/2]}^{M_i} \binom{M_i}{k} \alpha_i(\theta)^k (1 - \alpha_i(\theta))^{M_i - k} \,,$$

and M_i is the number of sensors in the neighborhood of sensor i, which is typically the same for all i. Again, we can either maximize the likelihood directly or apply the EM algorithm. The M-step is the same as that used for the initial decisions Y_i. The E-step requires calculating the first and second conditional moments $\mathbb{E}[E_i|\vec{Z}]$, $\mathbb{E}[E_i^2|\vec{Z}]$, which can be written as $\mathbb{E}[E_i|\vec{Z}] = \sum_{k=0,1} \mathbb{E}(E_i|Y_i = k)\mathbb{P}(Y_i = k|Z)$, and an analogous formula holds for the second moment. The moments conditional on Y were calculated in Eqs. (5.58) and (5.59), and $\mathbb{P}(Y_i = k|\vec{Z})$ can be approximately computed using the Bayes rule. For example, for $k = 1$ we have

$$\mathbb{P}(Y_i = 1|\vec{Z}) = \alpha_i \prod_{j:i \in U(j)} \left(\frac{\tilde{\beta}_{ji}}{\beta_j}\right)^{Z_j} \left(\frac{1 - \tilde{\beta}_{ji}}{1 - \beta_j}\right)^{1 - Z_j} \,, \quad (5.62)$$

12) This section © Taylor & Francis Ltd, http://www.tandfonline.com. Reprinted, with permission, from Katenka *et al.*, 2013.

where

$$\tilde{\beta}_{ji} = \mathbb{P}\left(\sum_{k\in U(j),k\neq i} Y_k \geq \frac{M_i}{2} - 1 \right) \approx \sum_{q=[M_i/2-1]}^{M_i-1} \binom{M_i - 1}{q} \alpha_j^q (1-\alpha_j)^{M_i-1-q}. \quad (5.63)$$

5.5.3.1 Remark 1: Hybrid Estimation

Note that given the extended framework for the localization of multiple targets from energies, initial and corrected decisions earlier, one can also use the hybrid methods where only the most informative energy reading is transmitted from the sensors with positive initial or corrected decisions.

5.5.3.2 Remark 2: Starting Values

Since many of the employed algorithms are iterative in nature, they prove sensitive to starting values, as discussed in the previous section. A comprehensive investigation in Katenka, Levina, and Michailidis (2013) suggests that using the centroids of positive decisions after clustering them using, for example, the K-means method performs well. Some clustering methods, however, can fail to provide a good starting values; for example, hierarchical clustering methods.

5.5.3.3 Estimating the Number of Targets

So far, we have assumed that the number of multiple targets under consideration is known. Next, we will focus on the problem of estimating the number of targets in situations when it is unknown. To estimate the number of targets present in the monitoring region, we will use the Bayes information criterion (BIC), a common way to select a "model complexity" parameter (e.g., the number of targets) within a likelihood framework.

In practice, to pick the best p, we have to maximize the following criterion,

$$\text{BIC} = -2\ell_p(\hat{\theta}) + 2(4p+1)\log N, \quad (5.64)$$

where $4p+1$ gives the total number of parameters to be estimated in the full model with p targets (signal amplitude, two plane coordinates, and the attenuation parameter for each target, and the noise variance σ^2), and $\hat{\theta}$ is the ML estimate of the parameters assuming p targets are present. Note that if some parameters are assumed known or equal for all targets, the factor $4p+1$ is adjusted accordingly. In the next section, we will discuss how BIC can be used adaptively while tracking multiple targets over time.

5.5.3.4 Concluding Remarks

The results of an extensive simulation study performed in Katenka, Levina, and Michailidis (2013) for the multiple target localization are fairly consistent with the results obtained for the single target localization in Katenka, Levina, and Michailidis (2008b). Specifically, the LVDF methods clearly outperform the methods based on the original decisions. While the ML(E) and the HEM(Z) estimates are the most accurate, ML(E) performs slightly better for higher signal-to-noise ratios and far apart targets, and HEM(Z)performs better for close targets and low SNR. In fact, for a very low SNR of 3, when the targets are close together, the LVDF methods outperform ML(E) with the HEM(Z) exhibiting the best performance. In general, the localization of two or more targets that are very close to each other (i.e., have intersecting signal attenuation) is a difficult task, especially when only binary decisions are available. That is why tracking becomes extremely challenging when there are multiple targets with intersecting trajectories, particularly when the number of targets is unknown.

5.6 Multiple Target Tracking

Our last section in this chapter is dedicated to multiple-target tracking which, unlike target detection and localization, is a significantly harder problem, because a good tracking framework should be applicable to situations where multiple targets moving in complicated, possibly intersecting patterns, may appear and disappear in the monitored area over time. Moreover, some applications (e.g., surveillance systems) require tracking results reported in real time.

In general, the problem of target tracking by WSNs has received a lot of attention among researchers in different disciplines due to its complexity and importance. Some of the proposed methods require restrictive assumptions, for example, one target can generate at most one sensor measurement during a time period, or that a sensor can receive a signal from one target only; and many of the proposed methods require a very complex computations. We refer a reader to Tinati and Rezaii (2009), Teng, Snoussi, and Richard (2010), and Katenka, Levina, and Michailidis (2013) for a comprehensive review of the most recent results in tracking multiple targets.

The goal of this section is to address multi-target tracking and to create an efficient tracking system by building up on algorithms developed for target detection and localization. Specifically, in this section, we extend our likelihood-based framework for the localization of a single target described in Section 5.4 and multiple targets presented in Section 5.5 and to tracking over time, and incorporate information about the number of targets, targets' speed and acceleration. Note that the framework that we will describe in this section appropriately handles multiple targets that appear or disappear over time, and does not assume any prior information on their number.

Again, we consider a WSN comprised of N identical sensors deployed at locations s_i, $i = 1, 2, \dots, N$ over a two-dimensional monitoring region R. Now, we suppose that p targets *move* in R over time and have positions $x_j(t)$, $j = 1, \dots, p$ at times $t = t_1, t_2, \dots$. Each target j emits a signal (e.g., infrared, temperature, acoustical) of strength $S_0^{(j)}(t)$ at the target location. The signal attenuates with distance from the target according to a decreasing function C, and thus a sensor located at s_i receives signal from target j given by

$$S_i^{(j)}(t) = S_0^{(j)}(t)C_{\eta_j}(\delta_i(x_j(t))) , \tag{5.65}$$

where $\delta_i(x_j(t)) = \|s_i - x_j(t)\|$ is the distance from the target to sensor i, and η_j is a potentially time varying scaling parameter which represents the effective target size.

At time t, each sensor obtains an *energy* reading comprised of all p individual signals and corrupted by random noise:

$$E_i(t) = \sum_{j=1}^{p} S_i^{(j)}(t) + \epsilon_i(t), \quad i = 1, \dots , N, \tag{5.66}$$

where errors $\epsilon_i(t)$ are assumed to be independent in time and space with mean zero and variance $\sigma^2(t)$. The collected energy readings $E_i(t)$ are either directly transmitted to the fusion center or converted to binary decisions $Y_i(t) = I(E_i(t) \geq \tau_i)$, using a pre-specified threshold τ_i, which is related to the individual sensor's false alarm probability. Here, we assume that all sensors are identical and $\tau_i \equiv \tau$.

In tracking applications, our ultimate goal is to estimate the following parameter vector over time:

$$\theta(t) = (p(t), x_1(t), ..., x_p(t), S_0^{(1)}(t), ..., S_0^{(p)}(t), \eta_1(t), ..., \eta_p(t), \sigma(t)), \quad t = 1, 2, \dots \tag{5.67}$$

based either on energy readings $(E_i(t))$, initial decisions $Y_i(t)$, or corrected decision $Z_i(t)$ obtained at the fusion center on a synchronized schedule, at discrete points in time, which is usually the case in practice.

To build a complete tracking framework, one should consider all requirements of the actual application at hand. Specifically, one needs to know in advance if the application requires offline tracking (i.e., collecting and analyzing all the sensor measurements at all the time point $t_1, \dots , t_{N_T}$) or online tracking (i.e., estimating parameters at time t as soon as the data for that time point becomes available). Another important modeling consideration is whether to model a dependency structure in the noise over time. In our context, the noise is primarily receiver noise, so there is no reason to assume dependence.

If no dependence is assumed for the energies $E_i(t_n)$ and the decisions $Y_i(t_n)$ (or $Z_i(t_n)$), the respective log-likelihood functions are given by:

$$\ell(\theta) = \sum_{n=1}^{N_T} \ell(\theta(t_n)). \tag{5.68}$$

The number of parameters to be estimated is $(4p + 1) \times$ (number of time periods), which may become computationally challenging if either the number of targets or the number of time periods increases substantially.

It is important to note that incorporating the application context in this way may reduce the number of parameters to estimate which, in turn, may lead to both more accurate estimation and reduced computational cost. Particularly, it is important to make reasonable assumptions about which parameters are likely to change over time and which are likely to remain constant. For example, consider the following three different cases for signal amplitude:

1) Signal amplitudes $S_0^{(j)}(t)$ are changing over time and are different for each target;
2) Signal amplitudes $S_0^{(j)}(t) \equiv S_0^{(j)}$ are constant over time, but different for each target;
3) Signal amplitudes $S_0^{(j)}(t) \equiv S_0$ are the same for all targets and constant over time[13].

Whether the signal changes over time depends a lot on the type of sensor used. For instance, if WSN uses infrared sensors to detect humans, one can assume in practice that the signal amplitudes are the same for all targets and constant over time. Similarly, we assume that the variance $\sigma^2(t) \equiv \sigma^2$ is constant and the signal attenuation parameter η does not depend on time and is the same for all targets[13].

In this case, the formulas described in Section 5.5 are simplified, and current estimates of global parameters can be updated at every time step by incorporating new data; the common signal amplitude, for example, can be updated at each time step t_n as follows:

$$\hat{S}_0(t_n) = \frac{\sum_{i=1}^{N} \sum_{k=1}^{n} E_i(t_k) \sum_{j=1}^{p} C_{\hat{\eta}_j} \left(\delta_i(\hat{x}_j(t_k))\right)}{\sum_{i=1}^{N} \sum_{k=1}^{n} \sum_{j=1}^{p} C_{\hat{\eta}_j}^2 \left(\delta_i(\hat{x}_j(t_k))\right)}. \tag{5.69}$$

In many applications, it is reasonable to assume fairly regular target trajectories, which suggests that some kind of trajectory smoothing could be beneficial in estimating it. To guarantee some degree of smoothness in the trajectory, we will use a penalized-likelihood approach, which in general can be represented as

$$\begin{aligned} \ell(\{\theta(t),\, t \in [0,\, t_{n+1}]\}) &- \sum_{j=1}^{p} \lambda_j \int_0^{t_{n+1}} [\ddot{x}_j(t)]^2 dt \\ &- \sum_{j=1}^{p} \rho_j \int_0^{t_{n+1}} [\ddot{S}_0^{(j)}(t)]^2 dt, \end{aligned} \tag{5.70}$$

where $\ddot{x}_j(t)$ denotes the acceleration of the target and $\ddot{S}_0^{(j)}(t)$ the second derivative of the amplitude of the signal. The second term vanishes if we assume the signal remains constant over time.

Assuming independent errors over time, the log-likelihood of the observations can be decomposed into a sum. Thus, for online tracking, we approximate $\ell(\{\theta(t),\, t \in [0,\, t_{n+1}]\})$ with $\sum_{q=1}^{n} \ell(\hat{\theta}(t_q)) + \ell(\theta(t_{n+1}))$, and only use the last term. Similarly, if we approximate the integral in Eq. (5.70) by second-order differences, only the two previous time points affect the penalty terms that involve $\theta(t_{n+1})$. Thus, given the parameter estimates for times up to t_n, and assuming constant signal, we estimate parameters at time point t_{n+1} by maximizing

13) This section © Taylor & Francis Ltd, http://www.tandfonline.com. Reprinted, with permission, from Katenka *et al.*, 2013.

$$\ell(\tilde{\theta}(t_{n+1})) - \sum_{j=1}^{p} \lambda_j \left[\left(\frac{x_j^{(1)}(t_{n+1}) - \hat{x}_j^{(1)}(t_n)}{t_{n+1} - t_n} - \frac{\hat{x}_j^{(1)}(t_n) - \hat{x}_j^{(1)}(t_{n-1})}{t_n - t_{n-1}} \right)^2 \right.$$
$$\left. + \left(\frac{x_j^{(2)}(t_{n+1}) - \hat{x}_j^{(2)}(t_n)}{t_{n+1} - t_n} - \frac{\hat{x}_j^{(2)}(t_n) - \hat{x}_j^{(2)}(t_{n-1})}{t_n - t_{n-1}} \right)^2 \right] \frac{(t_{n+1} - t_n)}{(t_n - t_{n-1})^2}, \tag{5.71}$$

where we write $x = (x^{(1)}, x^{(2)})$ for the two planar coordinates of x, and, with our assumptions, $\tilde{\theta}(t_{n+1}) = (x_1(t), \ldots, x_p(t), S_0, \eta, \sigma^2)$. The estimation procedure iterates over different parameters; we use estimates that average over time for the global parameters (such as Eq. (5.69) for S_0), and estimate the coordinates $x_j(t)$ for the current time point only with the global estimates plugged in.

Penalized ML provides the user with flexibility to enforce smoothness on trajectory estimates as needed and to incorporate prior knowledge about the expected trajectories to improve estimation; however, it requires some prior knowledge about target trajectories while choosing the smoothing parameters λ_j. This parameter can be specified for each target separately, or assumed to be the same for all targets. As a rule of thumb, one should apply a larger degree of smoothing to the targets following regular trajectories with relatively constant speed. Unfortunately, an appropriate criteria for automatic choice of λ_j is beyond the scope of this chapter.

Finally, to account for situations when targets may appear or disappear over time, we propose the following algorithm. At each time point, the number of targets present is estimated by using the BIC as described in Section 5.5 in conjunction with prior information about the number of targets and target locations.

The number of targets $\hat{p}(t_{n+1})$ is chosen by optimizing the BIC over the values $\hat{p}(t_n) \pm k$, where k is a small number reasonable for the application (we used $k = 1$ in simulations and $k = 2$ in the real data examples)[14].

Once the targets are localized, the locations of the identified targets at t_{n+1} are matched to those estimated at the previous time slot using the Hungarian algorithm, to connect the trajectories. If $\hat{p}(t_{n+1}) < \hat{p}(t_n)$, tracking of unmatched targets is discontinued; if $\hat{p}(t_{n+1}) > \hat{p}(t_n)$, new targets start to be tracked. This is the most flexible version of the algorithm that allows for targets appearing and disappearing at arbitrary times; when additional information is available, one can adjust this procedure accordingly[14].

To illustrate the difficulties encountered in real applications, we will consider two different scenarios. Figure 5.5a illustrates the scenario when the information about one of the three targets following parallel trajectories becomes lost due to sensor failure. At time slots $t = 7, 8$ (out of a total of 11) the information about the second target is lost, and the target "reappears" at $t = 9$. Another difficulty is illustrated in Figure 5.5b and c, where two targets travel very close to each other for a period of time, and the third target is also briefly lost. These issues make it challenging to estimate the number and the location of the targets correctly[14].

14) This section © Taylor & Francis Ltd, http://www.tandfonline.com. Reprinted, with permission, from Katenka *et al.*, 2013.

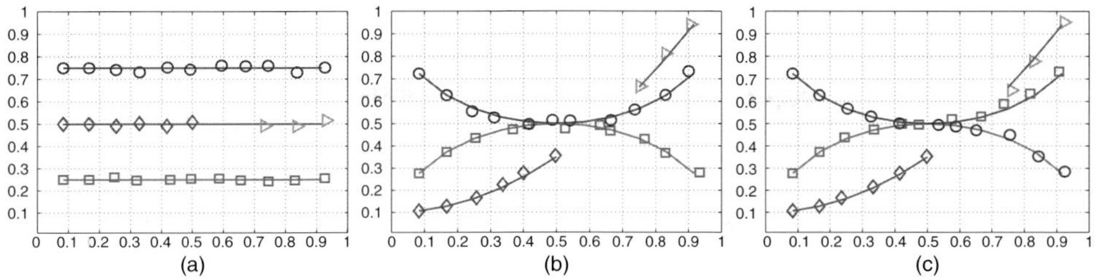

Figure 5.5 True trajectories (solid lines) and positions estimated by ML(Z) at each time point for three targets with SNR = 5. (a) The signal from the second target is briefly lost; (b) Two targets come close together and the third target briefly loses signal; (c) Another noise realization/solution for (b). © Taylor & Francis Ltd, http://www.tandfonline.com. Reprinted, with permission, from Katenka *et al.*, 2013. (See inset for color representation of this figure.)

As we expected, in both scenarios, our described framework behaves as follows: When the signal is lost, it stops tracking the target in question and then starts tracking it as a new target once the signal is recovered (Figure 5.5a). When two targets come close together and possibly meet, the two noise realizations show that the target labels are assigned arbitrarily once the targets separate. Since in scenarios (b) and (c) all targets have the same size and signal amplitudes, they are indistinguishable; when targets have different signal characteristics they are labeled correctly (results not shown). More extensive simulation study has been performed in Katenka, Levina, and Michailidis (2013).

5.7 Applications and Case Studies

In this section, we will focus on two case studies that motivated work in this chapter. The first case study is based on the data that were collected as part of the Network Embedded Systems Technology (NEST) project[15)] (Chen *et al.*, 2006; Oh *et al.*, 2006), the purpose of which was to create a wireless sensor system for tracking targets (e.g., people) traversing a monitored region. The second study is based on the data from ZebraNet project[16)], the purpose of which was to track a zebra herd in its natural habitat at Sweetwaters Game Reserve near Nanyuki, Kenya (Liu *et al.*, 2004; Wang *et al.*, 2005).

Although both projects are somewhat similar in nature and purpose, the implementation is very different. The implementation of the ZebraNet project was an improved version of the existing animal monitoring tactic which requires catching and tagging animals in captivity and then monitoring labeled subjects in their natural habitat. In the deployment of ZebraNet, sensors with GPS capabilities were attached to the zebras and then the GPS readings (i.e., time and locations)

15) NEST project is developed at the University of California at Berkeley.

16) ZebraNet is an inter-disciplinary joint animal-monitoring study of the Departments of Electrical Engineering, Ecology and Evolutionary Biology, and Computer Science at Princeton University (*http://www.princeton.edu/~ mrm/zebranet.html*).

were recorded for each animal carrying the device. As part of the NEST project, a prototype system comprised of wireless (infrared) sensors placed in the field was deployed and tested on three experiments tracking one, two, or three people crossing the monitored area. Unlike the ZebraNet, this implementation did not require any additional devices placed on the participants; and even though it did require, in general, estimating the number of targets was less invasive and was not restricted to a particular number of targets. In general, the ZebraNet project found that placing sensing collars on zebras was not at all effective. There were multiple instances when animals managed to remove or lose the collars, not to mention other frequent hardware failures. Thus, it seems reasonable to consider a stationary sensor network of the NEST type for environmental monitoring instead.

In what follows next, we first describe and analyze the data collected from the three experiments from the NEST project, and then we will describe a simulation experiment that would illustrate the performance of our tracking framework if a similar to the NEST wireless sensor platform could be implemented for animal monitoring.

5.7.1 The NEST Project

NEST prototype sensor system was comprised of 144 sensors placed at regular 12×12 grid and spaced about 5 m apart. Each sensor had an 8-m sensing radius and a 10% false alarm probability. In a testing experiment whose goal was to track one, two or three people crossing the monitored area, the sensors reported their decisions on the presence or absence of target(s) to the fusion center that made the final decision. All positive decisions from a trial involving a single person traversing the monitored area are shown in Figure 5.6a. Note that different positive decisions were made at different times. The goal is to infer the path of the target(s) through the monitored area, which requires detecting the presence and estimating the number of targets, locating them at a particular point in time, and tracking their positions over time. Note that target(s) detection was declared if at

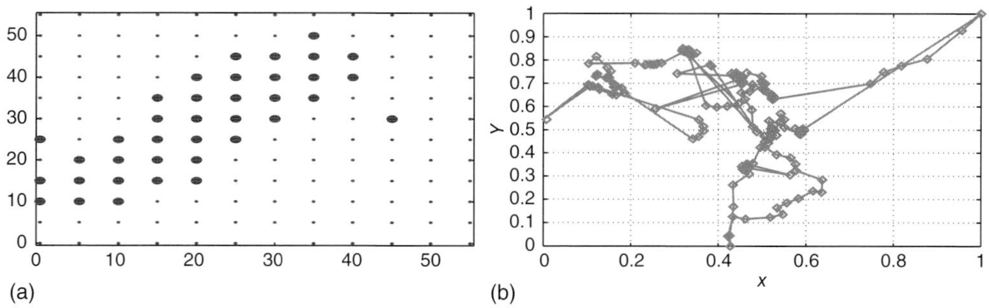

Figure 5.6 (a) The activation pattern of NEST sensors by a person traversing the monitored area. (b) The trajectory of a single zebra in the monitored area. © Taylor & Francis Ltd, http://www.tandfonline.com. Reprinted, with permission, from Katenka *et al.*, 2013. (See inset for color representation of this figure.)

least three sensors recorded positive decisions. Note also that the "true" trajectories in each experiment were recovered from video recordings.

Due to the nature of the sensors (infrared) in the NEST project, we will assume the same signal model for all targets in all experiments:

$$S_i(t) = S_0/(1 + (\delta_i(x_j(t))/\eta))^3. \tag{5.72}$$

Due to the nature of the available data in the form of positive decisions for each sensor, we choose the ML(Z)-based tracking algorithm as the most accurate and suitable for binary data. Taking into account square grid deployment of the sensors, the algorithm was applied with a neighborhood of $M = 9$ sensors.

The results of estimation and the comparison of the different techniques are provided in Figure 5.7 and Table 5.1, respectively. In fact, the results summarized in Table 5.1 show that the adaptive version of ML(Z) outperforms the one with the number of targets fixed a priori, with the most apparent improvement in accuracy for the scenario with three people; and the effect of smoothing ($\lambda = 0$ vs $\lambda = 1$)

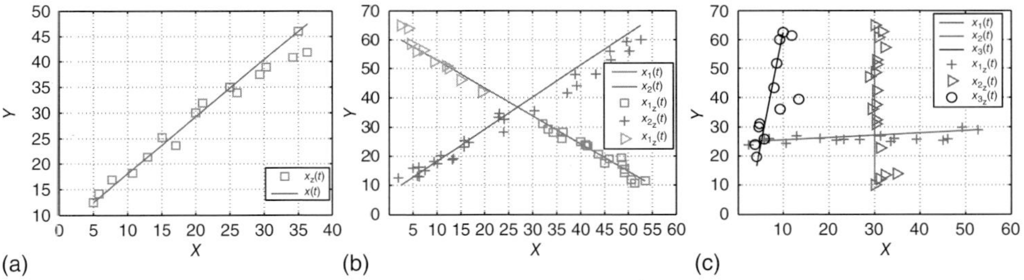

Figure 5.7 Estimated and true trajectories for (a) one, (b) two, and (c) three NEST targets. © Taylor & Francis Ltd, http://www.tandfonline.com. Reprinted, with permission, from Katenka *et al.*, 2013. (See inset for color representation of this figure.)

Table 5.1 Average distances from the true trajectories and estimated SNR.

		Fixed *p*, $\lambda = 0$			**Adaptive, $\lambda = 0$**	
Target	Sc.1	Sc.2	Sc.3	Sc.1	Sc.2	Sc.3
1	1.6553	5.9369	14.8523	1.6551	5.8832	2.9071
2	n/a	9.4583	11.0036	n/a	7.0816	1.8923
3	n/a	n/a	18.2545	n/a	n/a	4.0922
SNR	2.0961	1.7785	1.3643	2.0968	1.9652	2.0634
		Fixed *p*, $\lambda = 1$			**Adaptive, $\lambda = 1$**	
Target	Sc.1	Sc.2	Sc.3	Sc.1	Sc.2	Sc.3
1	1.5505	5.7325	17.5941	1.5505	5.2864	1.5156
2	n/a	8.5750	10.0829	n/a	6.8767	1.9553
3	n/a	n/a	19.3467	n/a	n/a	2.3720
SNR	2.0909	1.7663	1.3514	2.0909	2.0541	2.0374

is not as significant since the estimates obtained from corrected decisions are already quite accurate without smoothing.

5.7.2 The ZebraNet Project[17]

Unlike the highly controlled trajectories of people crossing the field in the NEST project, the trajectories of zebras in their natural habitat are highly irregular. The trajectory of one zebra over the course of about a day is shown in Figure 5.6b. This example illustrates that in order to perform well in realistic environments, the tracking algorithms need to be able to handle complicated trajectories with abrupt changes in direction. In what follows, we consider a more complicated experiment that involves tracking of four zebras with highly irregular, simultaneously crossing trajectories.

The GPS data (time and locations) were collected from GPS devices that were fitted as collars on four zebras, selected for their varying behavioral patterns. Specifically, we have data on a bachelor male (id.6), actively searching for a mate, a female leader of the herd (id.10), a passive female with a characteristic of a very small home range (id.14), and another female zebra (id.8). Note that the zebras' locations and a time stamp were recorded every 8 min for approximately 10 days, but due to hardware problems there are many missing values in the data, so we only use the time frame when the movements of all four zebras were recorded, which is just over 24 h long.

In order to test the proposed algorithms in this application, the following simulated sensor experiment was designed. It would likely be impossible to place sensors on a regular grid in this environment, and the habitat area also contains some inaccessible areas, such as large water puddles. Thus, we simulated a random deployment with 400 sensors, uniformly distributed in the unit square, and mapped the true locations of the zebras available from the ZebraNet data to this monitored region. The original monitored region is roughly 5×5 km, so in the simulated network, the average distance from a sensor to its closest neighbor is approximately 150 m. One arbitrarily selected random deployment was fixed in what follows; additional simulations show that averaged results over many random deployments are very similar. Further, in order to approximate the real-world inaccessible areas, two regions were selected in the unit squares and sensors located inside them (12 out of 400) were deleted from the network. The resulting network is shown in Figure 5.8a.

The emitted signals were generated according to the following model:

$$S_i^{(j)}(t) = S_0 \exp(-\delta_i(x_j(t), \eta_1, \eta_2)^2) \tag{5.73}$$

$$= S_0 \exp\left(-\left\{ \frac{(s_{i,1} - x_{j,1}(t))^2}{\eta_1^2} + \frac{(s_{i,2} + x_{j,2}(t))^2}{\eta_2^2} \right\} \right). \tag{5.74}$$

17) This section © Taylor & Francis Ltd, http://www.tandfonline.com. Reprinted, with permission, from Katenka *et al.*, 2013.

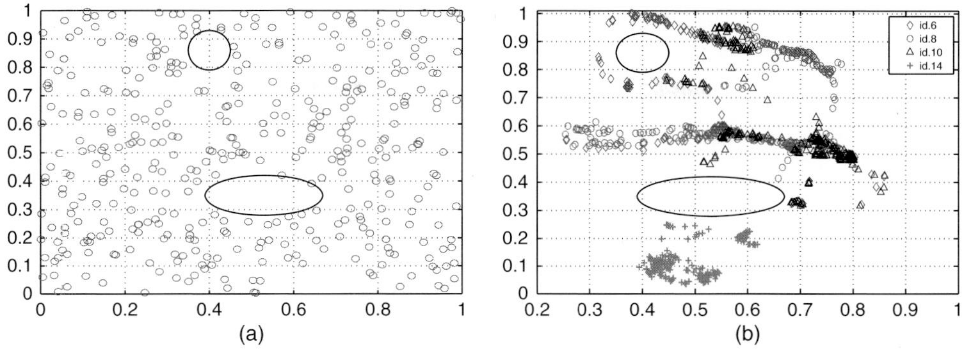

Figure 5.8 (a) Random sensor deployment. (b) The recorded locations of the four zebras scaled and plotted on the unit square. © Taylor & Francis Ltd, http://www.tandfonline.com. Reprinted, with permission, from Katenka *et al.*, 2013. (See inset for color representation of this figure.)

Allowing for different attenuation parameters, η_1 and η_2 represents more realistic scenarios where the signal attenuation can depend on direction (e.g., with wind). In the simulation experiments, we set SNR to the most challenging of the previously considered settings (SNR = 3), and $S_0 = 2$. For the signal attenuation parameters, we generated data from two different cases: $\eta_1 = \eta_2 = 0.1$, and $\eta_1 = 0.14, \eta_2 = 2$. These particular values were chosen to keep the size of the area, where the signal is detectable, approximately the same.

Figure 5.8b shows the profile of the true animal trajectories. It can be seen that the trajectories of the zebras with ids 6, 8, and 10 intersect at some points in time, while zebra 14 remains isolated. The tracking results using the HEM(Z) algorithm indicate that all the zebras are fairly well tracked at almost all points in time. Nevertheless, the more active zebras (id 6 and 10) prove the hardest to track.

Table 5.2 compares the performance of the three main algorithms when the number of targets is fixed in advance (at the true $p = 4$) to estimating it adaptively, with and without smoothing. For smoothing, an optimal λ was picked from the set $\{0, 100, 500, 1000, 5000\}$ for each zebra. As in our previous results, adaptively estimating the number of targets yields significantly improved tracking. Further, smoothing the trajectories yields some improvement; however, the complex nature of the underlying trajectories, coupled with the sparse sampling over time, limits the gains from smoothing.

More detailed information on the estimated number of targets is provided in Table 5.3, which shows the percentage of time points where the number of targets was estimated as $2, \ldots, 6$. Recall that the SNR is set to 3, and these results are consistent with earlier simulation results: When the two zebras are very close together, the number of targets is likely to be estimated as 3 rather than 4, and particularly so by ML(Z) which only has binary information available; extra targets are also sometimes picked up due to high noise levels, but they tend to be quickly

Table 5.2 Average distance from the true zebra trajectories (one unit of distance is approximately 5 km), for the case of isotropic signal attenuation ($\eta_1 = \eta_2 = 0.1$).

	Fixed $p = 4$, $\lambda = 0$			Adaptive, $\lambda = 0$		
id	ML(E)	ML(Z)	HEM(Z)	ML(E)	ML(Z)	HEM(Z)
6	0.0903	0.1278	0.0981	0.0226	0.0414	0.0295
8	0.0832	0.1124	0.0939	0.0276	0.0389	0.0457
10	0.1113	0.1333	0.1111	0.0296	0.0390	0.0281
14	0.0552	0.0631	0.0577	0.0246	0.0453	0.0320
	Fixed $p = 4$, optimal λ			**Adaptive, optimal λ**		
id	ML(E)	ML(Z)	HEM(Z)	ML(E)	ML(Z)	HEM(Z)
6	0.0840	0.1135	0.0753	0.0226	0.0332	0.0252
8	0.0786	0.0953	0.0885	0.0240	0.0339	0.0310
10	0.0913	0.1094	0.0903	0.0236	0.0277	0.0253
14	0.0435	0.0593	0.0529	0.0230	0.0394	0.0265

Table 5.3 The distribution of the estimated number of targets for zebra tracking (%), for the case of isotropic signal attenuation.

	2	3	4	5	6
ML(E)	1.48	10.65	58.89	22.48	6.50
ML(Z)	17.22	51.57	24.07	6.25	0.89
HEM(Z)	0.59	15.08	46.17	30.47	7.69

dropped. Overall, it seems that these algorithms would be appropriate for tracking animals in natural environments, and their tracking performance can be further improved if additional discriminating information about the targets is available.

5.8 Final Remarks

In this chapter, we reviewed a set of tools for a single and multiple-target localization and tracking with information obtained from wireless sensors. Based on the results, we learned that methods based on corrected binary decisions significantly outperform methods based on uncorrected decisions, and in noisy environments outperform even the ML estimation based on full energies. Moreover, the methods proposed for corrected decisions proved to be robust to the misspecification of the signal model and also to the misspecification of the noise distribution.

In practice, physical constraints often force irregular deployment patterns with obstacles, and our ZebraNet results together with simulations involving randomly deployed sensors strongly indicate that the main findings continue to hold. Obviously, the density of sensors around the target affects the quality of its estimated location. Hence, designing sensor network deployments that provide adequate coverage for tracking applications is a topic of current research. Another interesting aspect of tracking is the sampling resolution over time.

For more details, the reader is suggested to the use resources mentioned in this chapter and references therein. A list of references on a section level follows.

References

Abdel-Samad, A.A. and Tewfik, A.H. (1999) Search strategies for radar target localization. *Proceedings of International Conference on Image Processing*, vol. 3, p. 862.

Besag, J. (1986) On the statistical analysis of dirty pictures. *Journal of the Royal Statistical Society. Series B*, **48**, 259.

Blatt, D. and Hero, A.O. III (2006) Detection, energy-based sensor network source localization via projection onto convex sets. *IEEE Transactions on Signal Processing*, **54**, 3614.

Chen, P., Oh, S., Manzo, M., Sinopoli, B., Sharp, C., Whitehouse, K., Tolle, G., Jeong, J., Dutta, P., Hui, J., Shaffert, S., Kim, S., Taneja, J., Zhu, B., Roosta, T., Howard, M., Culler, D., and Sastry, S. (2006) Experiments in instrumenting wireless sensor networks for real-time surveillance. *Proceedings of IEEE International Conference on Robotics and Automation*, p. 3128.

Clouqueur, T., Ramanathan, P., Saluja, K.K., and Wang, K.C. (2001) Value-fusion versus decision-fusion for fault-tolerance in collaborative target detection in sensor networks. *Proceedings of 4th Annual Conference on Information Fusion.*

Dempster, A.P., Laird, N.M., and Rubin, D.B. (1997) Maximum likelihood from incomplete data via the EM algorithm. *Journal of the Royal Statistical Society. Series B (Methodological)*, **39**, 1.

Doukhan, P. (1994) *Mixing: Properties and Examples*, Springer-Verlag.

Duarte, M. and Hu, Y.H. (2004) Distance-based decision fusion in a distributed wireless sensor network. *Telecommunication Systems*, **26**, 339.

Efron, B. and Tibshirani, R.J. (1994) *An Introduction to the Bootstrap*, CRC Press.

Ermis, E.B. and Saligrama, V. (2006) Detection and localization in sensor networks using distributed FDR. *Proceedings of Conference on Information Sciences and Systems.*

Ermis, E.B. and Saligrama, V. (2010) Distributed detection in sensor networks with limited range multimodal sensors, *IEEE Transactions on Signal Processing*, vol. 9, IEEE Press, p. 843.

Kaplan, L.M., Le, Q., and Molnar, P. (2001) Maximum-likelihood methods for bearings-only target localization. *Proceedings of IEEE International Conference on Acoustics, Speech, and Signal Processing*, vol. 5, p. 3001.

Katenka, N., Levina, E., and Michailidis, G. (2008a) Local vote decision fusion for target detection in wireless sensor networks. *IEEE Transactions on Signal Processing*, **56**, 329.

Katenka, N., Levina, E., and Michailidis, G. (2008b) Robust target localization from binary decisions in wireless sensor networks. *Technometrics*, **50**, 448.

Katenka, N., Levina, E., and Michailidis, G. (2013) Tracking multiple targets using binary decisions from wireless sensor networks. *Journal of the American Statistical Association (JASA) Applications and Case Studies*, **108**, 398.

Klein, L. (1993) A Boolean Algebra Approach to Multiple Sensor Voting Fusion. *IEEE Transactions on Aerospace and Electronic Systems*, **29**, 317.

Li, D., Wong, K.D., Hu, Y.H., and Sayeed, A.M. (2002) Detection, classification, and tracking of targets. *IEEE Signal Processing Magazine*, **19**, 17.

Liu, T., Sadler, C., Zhang, P., and Martonosi, M. (2004) Implementing software on resource-constrained mobile sensors: experiences with impala and ZebraNet. *Proceedings of the 2nd International Conference on Mobile Systems, Applications, and Services*, p. 256.

Niu, R. and Varshney, P.K. (2004) Target location estimation in wireless sensor networks using binary data. *Conference on Information Sciences and Systems.*

Noel, M., Joshi, P., and Jannett, T. (2006) Improved maximum likelihood estimation of target position in wireless sensor networks using particle swarm optimization. *Proceedings of the 3rd International Conference on Information Technology: New Generations.*

Oh, S., Chen, P., Manzo, M., and Sastry, S. (2006) Instrumenting wireless sensor networks for real-time surveillance. *Proceedings of IEEE International Conference on Robotics and Automation.*

Sheng, X. and Hu, Y.H. (2003) Energy based acoustic source localization. *Proceedings of the 3rd International Workshop on Information Processing in Sensor Networks*, vol. 2634, p. 286.

Sun, T., Chen, L.J., Han, C.C., and Gerla, M. (2005) Reliable sensor networks for planet exploration. *Proceedings of IEEE International Conference on Networking, Sensing and Control.*

Teng, J., Snoussi, H., and Richard, C. (2010) Decentralized variational filtering for target tracking in binary sensor networks. *IEEE Transactions on Mobile Computing*, vol. 9, IEEE Educational Activities Department, p. 1465.

Tinati, M.A. and Rezaii, T.Y. (2009) Multi-target tracking in wireless sensor networks using distributed joint probabilistic data association and average consensus filter. *Proceedings of the International Conference on Advanced Computer Control*, vol. 5, IEEE Computer Society, p. 51.

Viswanathan, R. and Varshney, P.K. (1997) Distributed detection with multiple sensors: part I-Fundamentals. *Proceedings of the IEEE*, **85** (1), 54–63.

Wang, Y., Jain, S., Martonosi, M., and Fall, K. (2005) Erasure coding based routing for opportunistic networks. *Proceedings of the 2nd International Conference on Mobile Systems, Applications, and Services.*

Zhu, M., Ding, S., Wu, Q., Brooks, R., Rao, N.S.V., and Iyengar, S.S. (2010) Fusion of threshold rules for target detection in wireless sensor networks. *ACM Transactions on Sensor Networks*, vol. 6, ACM, p. 18:1.

6
Computing in Dynamic Networks*

Othon Michail, Ioannis Chatzigiannakis, and Paul G. Spirakis

6.1
Introduction

6.1.1
Motivation-State of the Art

Distributed computing systems are more and more becoming dynamic. The static and relatively stable models of computation can no longer represent the plethora of recently established and rapidly emerging information and communication technologies. In recent years, we have seen a tremendous increase in the number of new mobile computing devices. Most of these devices are equipped with some sort of communication, sensing, and mobility capabilities. Even the Internet has become mobile. The design is now focused on complex collections of heterogeneous devices that should be robust, adaptive, and self-organizing, possibly moving around and serving requests that vary with time. Delay-tolerant networks are highly dynamic, infrastructure-less networks whose essential characteristic is a possible absence of end-to-end communication routes at any instant. Mobility may be *active*, when the devices control and plan their mobility pattern (e.g., mobile robots), or *passive*, in opportunistic-mobility networks, where mobility stems from the mobility of the carriers of the devices (e.g., humans carrying cell phones) or a combination of both (e.g., the devices have partial control over the mobility pattern like, for example, when GPS (global positioning system) devices provide route instructions to their carriers). Thus, it can vary from being completely predictable to being completely unpredictable. Gossip-based communication mechanisms, e-mail exchanges, peer-to-peer networks, and many other contemporary communication networks all assume or induce some sort of highly-dynamic communication network.

* Supported in part by the project "Foundations of Dynamic Distributed Computing Systems" (FOCUS) which is implemented under the "ARISTEIA" Action of the Operational Programme "Education and Lifelong Learning" and is cofunded by the European Union (European Social Fund) and Greek National Resources.

Computational Network Theory: Theoretical Foundations and Applications, First Edition.
Edited by Matthias Dehmer, Frank Emmert-Streib, and Stefan Pickl.

The formal study of dynamic communication networks is hardly a new area of research. There is a huge amount of work in distributed computing that deals with causes of dynamicity such as failures and changes in the topology that are rather slow and usually eventually stabilize (like, e.g., in self-stabilizing systems (Dolev, 2000)). However, the low rate of topological changes that is usually assumed is unsuitable for reasoning about truly dynamic networks. Even graph-theoretic techniques need to be revisited: The suitable graph model is now that of a *dynamic graph* (a.k.a., *temporal graph* or *time-varying graph*) (see e.g., Mertzios *et al.* (2013); Kempe, Kleinberg, and Kumar (2000); Casteigts *et al.* (2012); Holme and Saramäki (2012)), in which each edge has an associated set of time-labels indicating availability times. Though static graphs have been extensively studied, for their temporal generalization, we are still far from having a concrete set of structural and algorithmic principles. Additionally, it is not yet clear how the complexity of combinatorial optimization problems is affected by introducing to them a notion of time. In an early but serious attempt to answer this question, Orlin (1981) observed that many dynamic languages derived from **NP**-complete languages can be shown to be **PSPACE**-complete. Among the other few things that we do know is that the max-flow min-cut theorem holds with unit capacities for time-respecting paths (Berman, 1996). However, there are fundamental properties of classical graphs that do not easily carry over to their temporal counterparts. For example, Kempe, Kleinberg, and Kumar (2000) found out that there is no analog of Menger's theorem[1] for arbitrary temporal networks with one label on every edge, which additionally renders the computation of the number of node-disjoint s–t paths **NP**-complete. In a very recent work Mertzios *et al.* (2013), the authors achieved a reformulation of Menger's theorem which is valid for all temporal graphs and introduced several interesting cost minimization parameters for optimal temporal network design. One is the *temporality* of a graph G, in which the goal is to create a temporal version of G minimizing the maximum number of labels of an edge, and the other is the *temporal cost* of G, in which the goal is to minimize the total number of labels used. Optimization of these parameters is performed subject to some *connectivity constraint*. They proved several upper and lower bounds for the temporality of some very basic graph families such as rings, directed acyclic graphs, and trees, as well as a trade-off between the temporality and the maximum label of rings. Furthermore, they gave a *generic method* for computing a lower bound of the temporality of an arbitrary graph, with respect to (w.r.t.) the constraint of preserving a time-respecting analog of every simple path of G. Finally, they proved that computing the temporal cost, the constraint of preserving at least one time-respecting path from u to v, whenever v is reachable from u in G, is **APX**-hard. Even the standard network diameter metric is no more suitable and has to be replaced by a dynamic/temporal version. In a dynamic star graph in which all leaf nodes but one

1) *Menger's theorem*, which is the analog of the max-flow min-cut theorem for udirected graphs, which states that the maximum number of node-disjoint s–t paths is equal to the minimum number of nodes needed to separate s from t (see e.g., Bollobás 1998).

go to the center, one after the other in a modular way, any message from the node that enters last the center to the node that never enters the center needs $n-1$ steps to be delivered, where n is the size (number of nodes) of the network; that is, the *dynamic diameter* is $n-1$, while on the other hand, the classical diameter is just 2 (Avin, Koucký, and Lotker, 2008) (see also Kuhn and Oshman (2011)).

Distributed systems with worst-case dynamicity were first studied in O'Dell and Wattenhofer (2005). Their outstanding novelty was to assume a communication network that may change arbitrarily from time to time subject to the condition that each instance of the network is connected. They studied asynchronous communication and considered nodes that can detect local neighborhood changes; these changes cannot happen faster than it takes for a message to transmit. They studied *flooding* (in which one node wants to disseminate one piece of information to all nodes) and *routing* (in which the information need only reach a particular destination node t) in this setting. They described a uniform protocol for flooding that terminates in $O(Tn^2)$ rounds using $O(\log n)$ bit storage and message overhead, where T is the maximum time it takes to transmit a message. They conjectured that without identifiers (ids) flooding is impossible to solve within the above resources. Finally, a uniform routing algorithm was provided that delivers to the destination in $O(Tn)$ rounds using $O(\log n)$ bit storage and message overhead.

Computation under worst-case dynamicity was further studied in a series of works by Kuhn *et al.* in the synchronous case. In Kuhn, Lynch, and Oshman (2010), the network was assumed to be T*-interval connected* meaning that any time-window of length T has a static connected spanning subgraph (persisting throughout the window). Among others, *counting* (in which nodes must determine the size of the network) and *all-to-all token dissemination* (in which n different pieces of information, called tokens, are handed out to the n nodes of the network, each node being assigned one token, and all nodes must collect all n tokens) were solved in $O(n^2/T)$ rounds using $O(\log n)$ bits per message, almost-linear-time randomized approximate counting was established for $T = 1$, and two lower bounds on token dissemination were given. Dutta *et al.* (2013) improved one of these lower bounds and presented offline centralized algorithms for the token dissemination problem. Several variants of *coordinated consensus* in 1-interval connected networks were studied in Kuhn, Moses, and Oshman (2011). Two interesting findings were that in the absence of a good initial upper bound on n, eventual consensus is as hard as computing deterministic functions of the input and that *simultaneous consensus* can never be achieved in less than $n-1$ rounds in any execution. Haeupler (2011) is a recent work that presents information spreading algorithms in worst-case dynamic networks based on *network coding*. An *open* setting (modeled as high churn) in which nodes constantly join and leave has very recently been considered in Augustine *et al.* (2012). For an excellent introduction to distributed computation under worst-case dynamicity, see Kuhn and Oshman (2011). Some very thorough surveys on dynamic networks are Scheideler (2002), Casteigts *et al.* (2012), and Holme and Saramäki (2012).

Here, we are mostly concerned with: (i) (Section 6.4) (Michail, Chatzigiannakis, and Spirakis, 2012, 2013) that studied the fundamental naming and counting problems (and some variations) in dynamic networks that are anonymous, unknown, and possibly dynamic. In *counting*, nodes must determine the size of the network n and in *naming*, they must end up with unique identities. Networks are *anonymous* because all nodes begin from identical states possibly in addition to a unique leader node and *unknown* because nodes have no a priori knowledge of the network (in addition to some minimal knowledge when necessary) including ignorance of n. The network dynamicity model in these papers was the one of Kuhn, Lynch, and Oshman (2010). (ii) (Section 6.5) The worst-case distributed model of Michail, Chatzigiannakis, and Spirakis (2014), in which the requirement for continuous connectivity was first dropped. That work proposed a set of metrics for capturing the speed of information spreading in a dynamic network that may be disconnected at every instant and efficient algorithms were developed.

Another notable model for dynamic distributed computing systems is the *population protocol* (*PP*) model (Angluin *et al.*, 2006). In that model, the computational agents are passively mobile, interact in ordered pairs, and the connectivity assumption is a *strong global fairness condition* according to which all events that may always occur, occur infinitely often. These assumptions give rise to some sort of structureless interacting automata model. The usually assumed *anonymity* and *uniformity* (i.e., n is not known) of protocols only allow for commutative computations that eventually stabilize to a desired configuration. Several computability issues in this area have already been established. Constant-state nodes on a complete interaction network (and several variations) compute the *semilinear predicates* (Angluin *et al.*, 2007). Semilinearity persists up to $o(\log\log n)$ local space but not more than this (Chatzigiannakis *et al.*, 2011). If constant-state nodes can additionally leave and update fixed-length pair-wise marks then, the computational power dramatically increases to the commutative subclass of $\mathbf{NSPACE}(n^2)$ (Michail, Chatzigiannakis, and Spirakis, 2011a). Interestingly, when operating under a uniform random scheduler, PPs are formally equivalent to *chemical reaction networks* (*CRNs*), which model chemistry in a *well-mixed solution* and are widely used to describe information processing occurring in natural cellular regulatory networks (Doty, 2014). However, CRNs and PPs can only capture the dynamics of molecular counts and not of structure formation. Then, Michail and Spirakis (2014) studied the fundamental problem of *network construction* by a distributed computing system. They initiated this study by proposing and studying a very *simple*, yet sufficiently generic model for distributed network construction. To this end, the authors assumed (as in Angluin *et al.* (2006) and Michail, Chatzigiannakis, and Spirakis (2011a)) the computationally weakest type of processes, that is, finite automata, and also a very minimal adversarial communication model. The model of (Michail and Spirakis, 2014) may be viewed as an extension of PPs and CRNs aiming to capture the stable structures that may occur in a well-mixed solution. They gave protocols (optimal in some cases) and lower bounds for several basic network construction problems such as *spanning line*, *spanning ring*, *spanning star*, and *regular network*, and they proved several *universality* results by presenting generic protocols that are capable of simulating a turing machine (TM) and exploiting it in order to construct a large class of networks. For introductory texts to this area, see Aspnes and Ruppert (2007) and Michail, Chatzigiannakis, and Spirakis (2011b).

6.1.2 Structure of the Chapter

In this chapter, our focus is on computational network analysis from a theoretical point of view. In particular, we study the *propagation of influence and computation in dynamic distributed computing systems*. We focus on a *synchronous message passing* communication model with bidirectional links. Our network dynamicity assumption is a *worst-case dynamicity* controlled by an adversary scheduler, which has received much attention recently. Section 6.2 formally defines the dynamic network models under consideration and the problems studied throughout. Section 6.3 discusses the central notion of causal influence and the 1-interval connectivity model. In Section 6.4, we study the fundamental *naming* and *counting* problems (and some variations) in networks that are *anonymous, unknown,* and possibly dynamic. Network dynamicity is modeled here by the *1-interval connectivity model* (Kuhn, Lynch, and Oshman, 2010), in which communication is synchronous and a (worst-case) adversary chooses the edges of every round subject to the condition that each instance is connected. Then, in Section 6.5, we replace the assumption that the network is connected at every instant by minimal *temporal connectivity* conditions. These conditions only require that *another causal influence occurs within every time-window of some given length.* Based on this basic idea, we define several novel metrics for capturing the speed of information spreading in a dynamic network. We present several results that correlate these metrics. Moreover, we investigate *termination criteria* in networks in which an upper bound on any of these metrics is known. We exploit these termination criteria to provide efficient (and optimal in some cases) protocols that solve the fundamental *counting* and *all-to-all token dissemination* (or *gossip*) problems. In Section 6.6, we propose another model of worst-case temporal connectivity, called *local communication windows,* that assumes a fixed underlying communication network and restricts the adversary to allow communication between local neighborhoods in every time-window of some fixed length. We prove some basic properties and provide a protocol for counting in this model. Finally, in Section 6.7, we conclude and discuss some interesting future research directions.

6.2 Preliminaries

6.2.1 The Dynamic Network Model

A *dynamic network* is modeled by a dynamic graph $G = (V, E)$, where V is a set of n nodes (or processors) and $E : \mathbb{N} \rightarrow \mathcal{P}(E')$ (wherever we use $\mathbb{N}$ we mean $\mathbb{N}_{\geq 1}$)

is a function mapping a round number $r \in \mathbb{N}$ to a set $E(r)$ of bidirectional links drawn from $E' = \{\{u, v\} : u, v \in V\}$.[2] Intuitively, a dynamic graph G is an infinite sequence $G(1), G(2), \ldots$ of *instantaneous graphs*, whose edge sets are subsets of E' chosen by a *worst-case adversary*. A *static network* is just a special case of a dynamic network in which $E(i+1) = E(i)$ for all $i \in \mathbb{N}$. The set V is assumed throughout this section to be *static*, that is it remains the same throughout the execution.

A dynamic graph/network $G = (V, E)$ is said to be *1-interval connected*, if, for all $r \in \mathbb{N}$, the static graph $G(r)$ is connected (Kuhn, Lynch, and Oshman, 2010). Note that this allows the network to change arbitrarily from round to round always subject to the condition that it remains connected. In Section 6.4, we focus on 1-interval connected dynamic networks which also implies that we deal with connected networks in the static-network case.

In Section 6.4, we assume that nodes in V are *anonymous*, by which we mean they do not initially have any ids and also, we assume that they do not know the topology or the size of the network, in addition to some minimal knowledge when necessary (i.e., we say that *the network is unknown*). In several cases, and in order to break symmetry, we may assume a unique *leader node* (or *source*) l. If this is the case, then we assume that l starts from a unique initial state l_0 (e.g., 0) while all the other nodes start from the same initial state q_0 (e.g., $\perp$). All nodes, but the leader, execute identical programs. In Section 6.5, we assume that nodes in V have unique identities (ids) drawn from some namespace $\mathcal{U}$ (we assume that ids are represented using $O(\log n)$ bits) and again that they do not know the topology or the size of the network, in addition to some minimal necessary knowledge to allow for terminating computations (usually an upper bound on the time it takes for the information to make some sort of progress). Any such assumed knowledge will be clearly stated. In all cases, nodes have unlimited local storage (though they usually use a reasonable portion of it).

Communication is *synchronous message passing* (Lynch, 1996; Attiya and Welch, 2004), meaning that it is executed in discrete steps controlled by a global clock that is available to the nodes and that nodes communicate by sending and receiving messages (usually of length that is some reasonable function of n, like e.g., $\log n$). Thus, all nodes have access to the current round number via a local variable that we usually denote by r. We use the terms *round, time,* and *step* interchangeably to refer to the discrete steps of the system. Naturally, real rounds begin to count from 1 (e.g., first round) and we reserve time 0 to refer to the initial state of the system. We consider two different models of message transmission. One is *anonymous broadcast*, in which, in every round r, each node u generates a single message $m_u(r)$ to be delivered to all its current neighbors in $N_u(r) = \{v : \{u, v\} \in E(r)\}$. The other is *one-to-each* in which a different message $m_{(u,i)}(r)$, $1 \leq i \leq d_u(r)$, where $d_u(r) := |N_u(r)|$ is the degree of u in round r, may be generated for each neighbor v_i.

2) By $\mathcal{P}(S)$ we denote the *powerset* of the set S, that is the set of all subsets of S.

In every round, the adversary first chooses the edges for the round; for this choice, it can see the internal states of the nodes at the beginning of the round. In the one-to-each message transmission model, we additionally assume that the adversary also reveals to each node u a set of locally unique edge-labels $1, 2, \ldots, d_u(r)$, one for each of the edges currently incident to it. Note that these labels can be reselected arbitrarily in each round so that a node cannot infer what the internal state of a neighbor is, based solely on the corresponding local edge-name. Then, each node transitions to a new state based on its internal state (containing the messages received in the previous round) and generates its messages for the current round: In anonymous broadcast, a single message is generated and in one-to-each a different message is generated for each neighbor of a node. Note that, in both models, a node does not have any information about the internal state of its neighbors when generating its messages. Deterministic algorithms are only based on the current internal state to generate messages. This implies that the adversary can infer the messages that will be generated in the current round before choosing the edges. The messages are then delivered to the corresponding neighbors. In one-to-each, we assume that each message m_i received by some node u is accompanied with u's local label i of the corresponding edge, so that a node can associate a message sent through edge i with a message received from edge i. These messages will be processed by the nodes in the subsequent round so we typically begin rounds with a "receive" command referring to the messages received in the previous round. Then, the next round begins.

6.2.2
Problem Definitions

We investigate the computability of the following fundamental problems for distributed computing in the context of dynamic networks.

k-Labeling. An algorithm is said to solve the k-labeling problem if whenever it is executed on a network comprising n nodes, each node u eventually terminates and outputs a *label* (or *name* or *id*) id_u so that $|\{id_u : u \in V\}| \geq k$.

Naming. The naming problem is a special case of the k-labeling problem in which it must additionally hold that $k = n$. This, in turn, implies that $id_u \neq id_v$ for all distinct $u, v \in V$ (so, unique labels are required for the nodes).

Minimal (Consecutive) Naming. It is a special case of naming in which it must additionally hold that the n nodes output the labels $\{0, 1, \ldots, n-1\}$.

Counting Upper Bound. Nodes must determine an upper bound k on the network size n.

Counting. A special case of counting upper bound in which it must hold that $k = n$.

All-to-All Token Dissemination (or Gossip). There is a token assignment function $I : V \to \mathcal{T}$ that assigns to each node $u \in V$ a single token $I(u)$ from some domain $\mathcal{T}$ s.t. $I(u) \neq I(v)$ for all $u \neq v$. An algorithm solves all-to-all token dissemination if for all instances (V, I), when the algorithm is executed in any dynamic graph $G = (V, E)$, all nodes eventually terminate and output $\bigcup_{u \in V} I(u)$. We assume that each token in the nodes' input is represented using $O(\log n)$ bits. The nodes know that each node starts with a unique token but they do not know n.

6.3 Spread of Influence in Dynamic Graphs (Causal Influence)

Probably, the most important notion associated with a dynamic network/graph is the *causal influence,* which formalizes the notion of one node "influencing" another through a chain of messages originating at the former node and ending at the latter (possibly going through other nodes in between). Recall that we denote by (u, t) the state of node u at time t and usually call it the *t-state of u*. The pair (u, t) is also called a *time-node.* We again use $(u, r) \rightsquigarrow (v, r')$ to denote the fact that node u's state in round r influences node v's state in round r'. Formally:

Definition 6.1 (Lamport 1978) *Given a dynamic graph $G = (V, E)$, we define an order $\to \subseteq (V \times \mathbb{N}_{\geq 0})^2$, where $(u, r) \to (v, r + 1)$ iff $u = v$ or $\{u, v\} \in E(r + 1)$. The* causal order $\rightsquigarrow \subseteq (V \times \mathbb{N}_{\geq 0})^2$ *is defined to be the reflexive and transitive closure of* $\to$.

Obviously, for a dynamic distributed system to operate as a whole, there must exist some upper bound on the time needed for the information to spread through the network. This is a very weak guarantee as without it, global computation is in principle impossible. An abstract way to talk about information spreading is via the notion of the *dynamic diameter*. The *dynamic diameter* (also called *flooding time,* e.g., in Clementi *et al.* (2008) and Baumann, Crescenzi, and Fraigniaud (2009)) of a dynamic graph, is an upper bound on the time required for each node to causally influence (or, equivalently, to be causally influenced by) every other node; formally, the dynamic diameter is the minimum $D \in \mathbb{N}$ s.t. for all times $t \geq 0$ and all $u, v \in V$ it holds that $(u, t) \rightsquigarrow (v, t + D)$. A small dynamic diameter allows for fast dissemination of information. Throughout, we do not allow nodes to know the dynamic diameter of the network. We only allow some minimal knowledge (that will be explained every time) based on which nodes may infer bounds on the dynamic diameter.

A class of dynamic graphs with small dynamic diameter is that of *T-interval connected* graphs. As already stated, T-interval connectivity was proposed in Kuhn, Lynch, and Oshman (2010) as an elegant way to capture a special class of dynamic networks, namely *those that are connected at every instant.* Intuitively, the parameter T represents the rate of connectivity changes. Formally, a dynamic graph

$G = (V, E)$ is said to be *T-interval connected*, for $T \geq 1$, if, for all $r \in \mathbb{N}$, the static graph $G_{r,T} := (V, \bigcap_{i=r}^{r+T-1} E(r))$ is connected (Kuhn, Lynch, and Oshman, 2010); that is, in every time-window of length T, a connected spanning subgraph is preserved. In one extreme, if $T = 1$, then the underlying connected spanning subgraph may change arbitrarily from round to round and in the other extreme, if T is ∞, then a connected spanning subgraph must be preserved forever. Recall that T-interval connected networks have the very nice feature to allow for constant propagation of information. For example, 1-interval connectivity guarantees that the state of a node causally influences the state of another uninfluenced node in every round (if one exists). To get an intuitive feeling of this fact, consider a partitioning of the set of nodes V to a subset V_1 of nodes that know the r-state of some node u and to a subset $V_2 = V \backslash V_1$ of nodes that do not know it. Connectivity asserts that there is always an edge in the cut between V_1 and V_2; consequently, if nodes that know the r-state of u broadcast in every round, then in every round, at least one node moves from V_2 to V_1. This is formally captured by the following Lemma from Kuhn, Lynch, and Oshman (2010):

Lemma 6.1 (Kuhn, Lynch, and Oshman 2010) *For any node $u \in V$ and time $r \geq 0$, in a 1-interval connected network, we have*

1) $|\{v \in V : (u, 0) \rightsquigarrow (v, r)\}| \geq \min\{r + 1, n\}$,
2) $|\{v \in V : (v, 0) \rightsquigarrow (u, r)\}| \geq \min\{r + 1, n\}$.
Note that $v \in \mathrm{past}_{(u,t')}(t)$ *iff* $u \in \mathrm{future}_{(v,t)}(t')$.

Proof. If $t = 0$, we have $\mathrm{future}_{(u,0)}(0) = \mathrm{past}_{(u,0)}(0) = \{u\} \Rightarrow |\mathrm{future}_{(u,0)}(0)| = |\mathrm{past}_{(u,0)}(0)| = 1 \geq \min\{0 + 1, n\} = 1$ and both statements hold in the base case.

1) Assume that $|\mathrm{future}_{(u,0)}(i)| \geq \min\{i + 1, n\}$ for some $i > 0$. If $\min\{i + 1, n\} = n$ then clearly $\min\{i + 2, n\} = n$ and the statement also holds for time $i + 1$. If $\min\{i + 1, n\} < n$ then the set $T = V \backslash \mathrm{future}_{(u,0)}(i)$ is non-empty. Connectivity in round $i + 1$ implies that there is some edge $\{v, w\}$ in the cut between $\mathrm{future}_{(u,0)}(i)$ and T (such an edge joins the set of nodes whose state at time i has been influenced by the initial state of u and those that have not). This, in turn, implies that w is influenced during round $i + 1$ so that $|\mathrm{future}_{(u,0)}(i + 1)| \geq= |\mathrm{future}_{(u,0)}(i)| + 1$ (increases by at least 1). So, the statement remains true for time $i + 1$. Informally, the set of nodes that have been influenced by the initial state of u increases by at least 1 in each round (while this set is smaller than V) due to connectivity and clearly in $n - 1$ rounds, all nodes must have been influenced by the initial state of any other node.
2) This one is a little more subtle. The reason is that here $|\mathrm{past}_{(u,r)}(0)|$ does not necessarily increase as r increases (may remain the same in some rounds). For an example, consider a first round in which a node u is connected to $n - 2$ nodes $\{v_1, v_2, \ldots, v_{n-2}\}$ and a node w is connected only to v_{n-2}. At time 1, we have $|\mathrm{past}_{(u,1)}(0)| = n - 1$. From now on, the graph maintains the following static structure: The graph is the Hamiltonian path $u, v_1, v_2, \ldots, v_{n-2}, w$. Clearly, the initial state of w must travel across the path to influence u, so no new influence occurs at u for $n - 3$ rounds.

Let us now prove this statement. Assume that $|\text{past}_{(u,i)}(0)| \geq \min\{i+1, n\}$ for some $i > 0$. The only interesting case is when $|\text{past}_{(u,i)}(0)| = i + 1 < n$ (if it is $> i + 1$ then the statement trivially holds for the next round). Again, $T = V \setminus \text{past}_{(u,i)}(0)$ is non-empty. Due to case 1, the initial configuration of the set T needs $V \setminus T = i + 1$ rounds to influence all nodes in $\text{past}_{(u,i)}(0)$. Thus, again, the initial state of some node in T influences u during round $i + 1$ and the statement follows. □

Before proving the lemma, let us first define two very useful sets. For all times $0 \leq t \leq t'$, we define by $\text{past}_{(u,t')}(t) := \{v \in V : (v, t) \rightsquigarrow (u, t')\}$ (Kuhn, Moses, and Oshman, 2011) the *past set of a time-node* (u, t') *from time* t and by $\text{future}_{(u,t)}(t') := \{v \in V : (u, t) \rightsquigarrow (v, t')\}$ the *future set of a time-node* (u, t) *at time* t'. In words, $\text{past}_{(u,t')}(t)$ is the set of nodes whose t-state (i.e., their state at time t) has causally influenced the t'-state of u and $\text{future}_{(u,t)}(t')$ is the set of nodes whose t'-state has been causally influenced by the t-state of u. If $v \in \text{future}_{(u,t)}(t')$ we say that at time t' node v *has heard of/from* the t-state of node u. If it happens that $t = 0$, we say simply that v has heard of u.

For a distributed system to be able to perform global computation, nodes need to be able to determine for all times $0 \leq t \leq t'$ whether $\text{past}_{(u,t')}(t) = V$. If nodes know n, then a node can easily determine at time t' whether $\text{past}_{(u,t')}(t) = V$ by counting all different t-states that it has heard of so far (provided that every node broadcasts at every round all information it knows). If it has heard the t-states of all nodes, then the equality is satisfied. If n is not known, then various techniques may be applied (which is the subject of this section). By *termination criterion*, we mean any locally verifiable property that can be used to determine whether $\text{past}_{(u,t')}(t) = V$.

Remark 6.1 *Note that any protocol that allows nodes to determine whether* $\text{past}_{(u,t')}(t) = V$ *can be used to solve the counting and all-to-all token dissemination problems. The reason is that if a node knows at round* r *that it has been causally influenced by the initial states of all other nodes, then it can solve counting by writing* $|\text{past}_{(u,r)}(0)|$ *on its output and all-to-all dissemination by writing* $\text{past}_{(u,r)}(0)$ *(provided that all nodes send their initial states and all nodes constantly broadcast all initial states that they have heard of so far).*

6.4
Naming and Counting in Anonymous Unknown Dynamic Networks

In this section, we study the fundamental naming and counting problems (and some variations) in networks that are anonymous, unknown, and possibly dynamic. In *counting*, nodes must determine the size of the network n and in *naming*, they must end up with unique identities. By *anonymous*, we mean that all nodes begin from identical states possibly in addition to a unique leader node and by *unknown*, that nodes have no a priori knowledge of the network (in addition to some minimal knowledge when necessary) including ignorance of n. Network dynamicity is modeled by the 1-interval connectivity model (Kuhn, Lynch, and Oshman, 2010), in which communication is synchronous and a worst-case adversary chooses the edges of every round subject to the condition that each instance is connected. We first focus on static networks with broadcast where we show that a unique leader suffices in order to solve counting in linear time. Then, we focus on dynamic networks with broadcast. We conjecture that dynamicity renders nontrivial computation impossible. In view of this, we let the nodes know an upper bound on the maximum degree that will ever appear and show that in this case the nodes can obtain an upper bound on n. Finally, we replace broadcast with *one-to-each*, in which a node may send a different message to each of its neighbors. Interestingly, this natural variation gives us the ability to state a correct naming protocol for this kind of dynamic distributed systems.

6.4.1
Further Related Work

The question concerning which problems can be solved by a distributed system when all processors use the same algorithm and start from the same state has a long story with its roots dating back to the seminal work of Angluin (1980), who investigated the problem of establishing a "center." Further investigation led to the classification of computable functions (Yamashita and Kameda, 1996; Attiya, Snir, and Warmuth, 1988). Boldi and Vigna (1999) removed the, until then, standard assumption of knowing the network size n and provided characterizations of the relations that can be computed with arbitrary knowledge. Other well-known studies on unknown networks have dealt with the problems of robot-exploration and map-drawing of an unknown graph (Deng and Papadimitriou, 1990; Albers and Henzinger, 2000) and on information dissemination (Awerbuch *et al.*, 1990). Fraigniaud *et al.* (2000) assumed a unique leader in order to break symmetry and assign short labels as fast as possible. To circumvent the further symmetry introduced by broadcast message transmission, they also studied other natural message transmission models as sending only one message to a single neighbor. Recently, and independent of our work, Chalopin, Métivier, and Morsellino (2012) have studied the problem of naming anonymous networks in the context of snapshot computation. Finally, Aspnes, Fich, and Ruppert (2006) studied the relative powers of reliable anonymous distributed systems with different communication mechanisms: anonymous broadcast, read-write registers, or read-write registers plus additional shared-memory objects.

6.4.2
Static Networks with Broadcast

We, here, assume that the network is described by a static graph $G = (V, E)$, where $E \subseteq \{\{u, v\} : u, v \in V\}$. Moreover, the message transmission model is broadcast, that is, in every round, each node u generates a single message to be delivered to all its neighbors. Note that any impossibility result established for static networks is also valid for dynamic networks as a static network is a special case of a dynamic network.

First of all, note that if all nodes start from the same initial state then, if we restrict ourselves to deterministic algorithms, naming is impossible to solve in general static networks, even if nodes know n. The reason is that in the worst-case, they may be arranged in a ring (in which each node has precisely two neighbors) and it is a well-known fact (Angluin, 1980; Lynch, 1996; Attiya and Welch, 2004) that, in this case, in every round r, all nodes are in identical states.

We show now that impossibility persists even if we allow a unique leader and even if nodes have complete knowledge of the network.

Theorem 6.1 (Michail, Chatzigiannakis, and Spirakis 2013) *Naming is impossible to solve by deterministic algorithms in general anonymous (static) networks with broadcast even in the presence of a leader and even if nodes have complete knowledge of the network.*

Proof. Imagine a star graph in which the leader has $n - 1$ neighbors (it is the *center*) and every other node has only the leader as its unique neighbor (they are the *leaves*). All leaf nodes are in the same initial state and receive the same first message m_1 from the center. So, they all transition to the same new state and generate the same outgoing message. It is straightforward to verify, by induction on the number of rounds, that in every round r all leaf nodes are in identical states. In fact, in any network in which some node is connected to at least two terminal nodes, that is, nodes with no further neighbors, those terminal nodes will forever be in identical states. □

An obvious generalization is that, under the same assumptions as in the statement of the theorem, it is impossible to solve k-labeling for any $k \geq 3$.

We now turn our attention to the simpler counting problem. First, we establish the necessity of assuming a unique leader.

Theorem 6.2 (Michail, Chatzigiannakis, and Spirakis 2013) *Without a leader, counting is impossible to solve by deterministic algorithms in general anonymous networks with broadcast.*

Proof. For the sake of contradiction, assume that an algorithm A solves it. Then, it solves it on a static ring R_1 of size n with the first node terminating in $k \geq n$ rounds. Now, consider a ring R_2 of size $k + 1$. All nodes in both rings are initially in the same identical initial state $\perp$. Thus, any node in R_2 has the same k-neighborhood (states of nodes in distance at most k) as any node in R_1 which implies that after k rounds, these two nodes will be in the same state (see e.g., Lemma 3.1 in Attiya, Snir, and Warmuth (1988)). Thus, a node in R_2 terminates after k rounds and outputs n, which is a contradiction. □

In view of Theorem 6.2, we assume again a unique leader in order to solve counting. Recall that the *eccentricity* of a node u is defined as the greatest geodesic distance between u and v, over all $v \in V \backslash \{u\}$, where "distance" is equivalent to "shortest path." We first describe a protocol *Leader_Eccentricity* (inspired by the

Wake&Label set of algorithms of Fraigniaud *et al.* (2000)) that assigns to every node a label equal to its distance from the leader and then we exploit this to solve counting. We assume that all nodes have access to the current round number via a variable r.

Protocol Leader_Eccentricity. The leader begins with $label \leftarrow 0$ and $max\_asgned \leftarrow 0$ and all other nodes with $label \leftarrow \perp$. In the first round, the leader broadcasts an *assign* (1) message. Upon reception of an *assign* (i) message, a node that has $label = \perp$ sets $label \leftarrow i$ and broadcasts to its neighbors an *assign* ($i + 1$) message and an *ack* (i) message. Upon reception of an *ack* (i) message, a node with $label \neq \perp$ and $label < i$ broadcasts it. Upon reception of an *ack* (i) message, the leader sets $max\_asgned \leftarrow i$ and if $r > 2 \cdot (max\_asgned + 1)$ then it broadcasts a *halt* message, outputs its label, and halts. Upon reception of a *halt* message, a node broadcasts *halt*, outputs its label, and halts.

Theorem 6.3 (Michail, Chatzigiannakis, and Spirakis 2013) *In Leader_Eccentricity nodes output $\epsilon + 1$ distinct labels where ϵ is the eccentricity of the leader. In particular, every node outputs its distance from the leader.*

Proof. At time 2, nodes at distance 1 from the leader receive *assign* (1) and set their label to 1. By induction on distance, nodes at distance i get label i at round $i + 1$. In the same round, they send an ack that must arrive at the leader at round $2i + 1$. If not then, there is no node at distance i. □

We now use *Leader_Eccentricity* to solve counting in anonymous unknown static networks with a leader. We additionally assume that at the end of the *Leader_Eccentricity* process each node u knows the number of neighbors $up(u) = |\{\{v, u\} \in E : label(v) = label(u) - 1\}|$ it has to its upper level (it can store this during the *Leader_Eccentricity* process by counting the number of *assign* messages that arrived at it from its upper level neighbors). Moreover, we assume that all nodes know the leader's eccentricity ϵ (just have the leader include *max_asgned* in its *halt* message). Finally, let, for simplicity, the first round just after the completion of the above process be round $r = 1$. For this, we just need all nodes to end concurrently the *Leader_Eccentricity* process. This is done by having node with label i that receives or creates (this is true for the leader) a *halt* message in round r halt in round $(r + max\_asgned - i)$. Then the nodes just reset their round counters.

Protocol Anonymous_Counting. Nodes first execute the modified *Leader_Eccentricity*. When $\epsilon - r + 1 = label(u)$, a non-leader node u receives a possibly empty (in case of no lower-level neighbors) set of $partial\_count_i$ ($rval_i$) messages and broadcasts a *partial_count* $((1 + \sum_i rval_i)/up(u))$ message. When $r = \epsilon + 1$, the leader receives a set of $partial\_count_i$ ($rval_i$) messages, sets $count \leftarrow 1 + \sum_i rval_i$, broadcasts a *halt* (*count*) message, outputs *count*, and halts. When a non-leader u receives a *halt* (*count*) message, it outputs *count* and halts.

For a given round r denote by $rval_i(u)$ the ith message received by node u.

Theorem 6.4 (Michail, Chatzigiannakis, and Spirakis 2013)
Anonymous_Counting solves the counting problem in anonymous static networks with broadcast under the assumption of a unique leader. All nodes terminate in $O(n)$ *rounds and use messages of size* $O(\log n)$.

Proof. By induction on the round number r, in the beginning of round $r \geq 2$, it holds that $\sum_{u:label(u)=\epsilon-r+1}\left(1+\sum_i rval_i(u)\right) = |\{u : label(u) \geq \epsilon - r + 1\}|$. Clearly, in round $\epsilon + 1$ it holds that $count = 1 + \sum_i rval_i(leader) = |\{u : label(u) \geq 0\}| = n$. □

6.4.3 Dynamic Networks with Broadcast

We now turn our attention to the more general case of 1-interval connected dynamic networks with broadcast. We begin with a conjecture stating that dynamicity renders nontrivial computation impossible (see also O'Dell and Wattenhofer (2005) for a similar conjecture in a quite different setting). Then we naturally strengthen the model to allow some computation.

Conjecture 1 (Michail, Chatzigiannakis, and Spirakis 2013) It is impossible to compute (even with a leader) the predicate $N_a \geq 1$, that is "exists an a in the input," in general anonymous unknown dynamic networks with broadcast.

The conjecture is essentially based on the following fact. Even in a dynamic network, it can be the case that two nodes that are initially in the same state a can for any number of rounds T have the same T-neighborhood, which means that the whole history of received messages is the same in both nodes and thus they always transition to identical states. This is, for example, true in a symmetric tree rooted at the leader (e.g., a tree with k identical lines leaving the root) in which the two nodes are in each round in equal distance from the root (even if this distance changes from round to round by moving the two nodes back and forth). In dynamic networks, it is also the case that for a node u to causally influence the leader with its t-state, all nodes that receive the t-state of u should continuously broadcast it at least until the leader receives it (then they could probably stop by receiving an ack or by using some known upper bound on the delivery time). Potentially, $O(n)$ nodes can receive the t-state of u before it is delivered to the leader. It seems that if the leader could at some point decide that the received messages originate from two distinct nodes that are forever in identical states then it would also decide the same on a dynamic network containing only one of these nodes, as in both cases, the whole network could be full of messages of the same kind. So, it seems impossible for the leader to determine whether the network contains at least two as and such a process is necessary for the leader to count the size of the network. To determine whether there are no as at all, in the absence of as, the leader should somehow determine that it has been causally influenced by the whole network, which in turn requires counting.

In view of Theorem 6.1, which establishes that we cannot name the nodes of a static, and thus also of a dynamic network if broadcast communication is assumed, and of the above conjecture, implying that in dynamic networks, we cannot count even with a leader,[3] we start strengthening our initial model.

Let us now assume that there is a unique leader l that knows an upper bound d on maximum degree ever to appear in the dynamic network, that is $d \geq \max_{u \in V, r \in \mathbb{N}}\{d_u(r)\}$. We keep the broadcast message transmission.

Note first that impossibility of naming persists. However, we show that obtaining an upper bound on the size of the network now becomes possible, though exponential in the worst-case.

Protocol Degree_Counting. The leader stores in d the maximum degree that will ever appear and begins with $label \leftarrow 0$, $count \leftarrow 1$, $latest\_event \leftarrow 0$, $max\_label \leftarrow 0$, and $r \leftarrow 0$ while all other nodes begin with $label \leftarrow \perp$, $count \leftarrow 0$, and $r \leftarrow 0$. In the beginning of each round, each node increments by one its round counter r. The leader in each round r broadcasts *assign* (r). Upon reception of an *assign* (r_label) message, a node with $label = \perp$ sets $label \leftarrow r\_label$ and from now in each round r broadcasts *assign* (r) and *my_label* ($label$). A node with $label = \perp$ that did not receive an *assign* message sends an *unassigned* (r) message. All nodes continuously broadcast the maximum *my_label* and *unassigned* messages that they have received so far. Upon reception of an *unassigned* (i) message, the leader, if $i > latest\_event$, it sets $count \leftarrow 1$ and, for $k = 1, \ldots, i$, $count \leftarrow count + d \cdot count$, $max\_label \leftarrow i$, and $latest\_event \leftarrow r$ and upon reception of a *my_label* (j) message, if $j > max\_label$, it sets $count \leftarrow 1$ and, for $k = 1, \ldots, j$, $count \leftarrow count + d \cdot count$, $latest\_event \leftarrow r$, and $max\_label \leftarrow j$ (if receives both i, j it does it for $\max\{i, j\}$). When it holds that $r > count + latest\_event - 1$ (which must eventually occur) then the leader broadcasts a *halt* ($count$) message for $count$ rounds and then outputs $count$ and halts. Each node that receives a *halt* (r_count) message, sets $count \leftarrow r\_count$, broadcasts a *halt* ($count$) message for $count$ rounds and then outputs $count$ and halts.

Theorem 6.5 (Michail, Chatzigiannakis, and Spirakis 2013) *Degree_Counting solves the counting upper bound problem in anonymous dynamic networks with broadcast under the assumption of a unique leader. The obtained upper bound is $O(d^n)$ (in the worst-case).*

Proof. In the first round, the leader assigns the label 1 to its neighbors and obtains an *unassigned* (1) message from each one of them. So, it sets $count \leftarrow (d + 1)$ (in fact, note that in the first step it can simply set $count \leftarrow d_u(1) + 1$, but this is minor), $latest\_event \leftarrow 1$, and $max\_label \leftarrow 1$. Now, if there are further nodes, at most by round $count + latest\_event - 1$ it must have received an *unassigned* (i) message with $i > latest\_event$ or a *my_label* (j) with $j > max\_label$. Note that the

3) This is implied because if we could count, we could have a node wait at most $n - 1$ rounds until it hears of an a (provided that all nodes that have heard of an a forward it) and if no reject.

reception of an *unassigned* (i) message implies that at least $i + 1$ distinct labels have been assigned because as long as there are unlabeled nodes, one new label is assigned in each round to at least one node (this is implied by Lemma 6.1 and the fact that all nodes with labels constantly assign new labels). Initially, one node (the leader) assigned to at most d nodes label 1. Then, the $d + 1$ labeled nodes assigned to at most $(d + 1)d$ unlabeled nodes the label 2, totalling $(d + 1) + (d + 1)d$, and so on.

In the worst-case, each label in $\{0, 1, \ldots, n - 1\}$ is assigned to precisely one node (e.g., consider a static line with the leader in the one endpoint). In this case, the nodes count $O(d^n)$. □

We point out that if nodes have access to more drastic initial knowledge such as an upper bound e on the *maximum expansion*, defined as $\max_{u,r,r'}\{|\mathrm{future}_{u,r}(r' + 1)| - |\mathrm{future}_{u,r}(r')|\}$ (maximum number of concurrent new influences ever occurring), where $\mathrm{future}_{(u,r)}(r') := \{v \in V : (u, r) \rightsquigarrow (v, r')\}$, for $r \leq r'$, then essentially the same protocol as above provides an $O(n \cdot e)$ upper bound.

6.4.4 Dynamic Networks with One-to-Each

The result of Theorem 6.1, in the light of (i) Conjecture 1, and (ii) the assumption of a broadcast message transmission model, indicates that nontrivial computations in anonymous unknown dynamic networks are impossible even under the assumption of a unique leader. We now relax our assumptions so that we can state a correct naming protocol. We start by relaxing the assumption of a broadcast message transmission medium by offering to nodes access to a *one-to-each message transmission mechanism*. We also assume a unique leader, as without it, even under a one-to-each model, naming is impossible to solve.

First Version – Protocol *Fair* We first present a protocol, that we call *Fair*, in which the unique leader assigns distinct labels to each node of the network. The labels assigned are tuples (r, h, i), where r is the round during which the label was assigned, h is the label of the leader node, and i is a unique number assigned by the leader. The labels can be uniquely ordered first by r, then by h and finally by i (in ascending order).

Each node maintains the following local variables: *clock*, for counting the rounds of execution of the protocol (implemented due to synchronous communication, see Section 6.2.1), *label*, for storing the label assigned by the leader, *state*, for storing the local state that can be set to $\{anonymous, named, leader\}$, and *counter*, for storing the number of labels generated. All nodes are initialized to $clock \leftarrow 0$, $id \leftarrow (0, \bot, \bot)$, $state \leftarrow anonymous$, and $counter \leftarrow 0$ except from the leader that is initialized to $clock \leftarrow 0$, $id \leftarrow (0, 1, 1)$, $state \leftarrow leader$, and $counter \leftarrow 1$.

Each turn, the leader u consults the one-to-each transmission mechanism and identifies a set of locally unique edge-labels $1, 2, \ldots, d(u)$, one for each of the edges incident to it.[4] The leader iterates the edge-label set and transmits to each neighboring node a different message m_i, $1 \leq i \leq d(u)$ that contains the unique label (*clock*, *label*, *counter* + *i*). When the transmission is complete, it increases the variable *counter* by $d(u)$. All the other nodes of the network do not transmit any messages (or transmit a null message if message transmission is compulsory).

All nodes under *state* = *anonymous*, upon receiving a (non-null) message set the local *label* to the contents of the message and change *state* to *named*. All the other nodes of the network simply ignore all the messages received.

At the end of the turn, all nodes do *clock* + + (where "++" is interpreted as "increment by one").

Recall that a naming assignment is correct if *all nodes* are assigned *unique* labels. It is clear that *Fair* is a non-terminating correct protocol, given the following *fairness assumption*: the leader node at some point has become directly connected with each other node of the network (i.e., eventually meets all nodes).

Lemma 6.2 *With one-to-each transmission, under the fairness assumption, and in the presence of a unique leader, protocol Fair eventually computes a unique assignment for all the nodes in any anonymous unknown dynamic network.*

Second Version – Protocol *Delegate* We now proceed by presenting a stronger protocol *Delegate* (based on *Fair*) that is correct even without the fairness assumption. To achieve correctness, the leader node delegates the role of assignment of labels to all the nodes that it encounters. Thus, without loss of generality, even if the leader does not encounter all other nodes of the network, due to the *connectivity property*, all nodes will eventually hear from the leader. Therefore, all nodes will either receive a unique label from the leader or from another labeled node. The uniqueness among the labels generated is guaranteed since each label can be traced back to the node that issued it using the h parameter.

In *Delegate*, the nodes maintain the same variables as in *Fair*. In each turn, the leader performs the same actions as in *Fair*. Also, similarly to *Fair*, each node that is in *state* = *anonymous* does not transmit any message (or transmits a null message if message transmission is compulsory). Each node u that is in *state* = *named* performs similar actions as the leader node and transmits to each edge-label i a message containing the unique label ($clock_u$, $label_u$, $counter_u + i$) and then increases the variable $counter_u$ by $d(u)$. All nodes under *state* = *anonymous*, upon receiving one or more (non-null) messages that contain a label, select the message that contains the lowest label (i.e., the one with the lowest h parameter) and set the local *label* to the contents of the message and change *state* to *named*. At the end of the turn all nodes do *clock* + +.

Lemma 6.3 (Michail, Chatzigiannakis, and Spirakis 2013) *With one-to-each transmission, and in the presence of a unique leader, protocol Delegate correctly*

4) Recall from Section 6.4.1 that these edge-labels can be reselected arbitrarily in each round (even if the neighbors remain the same) by the adversary so that a node cannot infer what the internal state of a neighbor is, based solely on the corresponding local edge-name.

computes a unique assignment for all the nodes in any anonymous unknown dynamic network.

Third Version – Protocol *Dynamic_Naming* (Terminating) Protocol *Fair* computes a correct naming assignment (based on different assumptions) but does not terminate. Essentially, the nodes continue to transmit labels for ever. We now describe a protocol that we call *Dynamic_Naming* that manages to terminate. *Dynamic_Naming* is an $O(n)$-time protocol that assigns unique ids to the nodes and informs them of n. As usual, there is a unique leader l with id 0 while all other nodes have id $\perp$.

The idea here is as follows. All nodes that have obtained an id assign ids and these ids are guaranteed to be unique. Additionally, we have nodes that have obtained an id to acknowledge their id to the leader. Thus, all nodes send their ids and all nodes continuously forward the received ids so that they eventually arrive at the leader (simple flooding mechanism). So, at some round r, the leader knows a set of assigned ids $K(r)$. We describe now the termination criterion. If $|K(r)| \neq |V|$ then in at most $|K(r)|$ additional rounds the leader must hear (be causally influenced) from a node outside $K(r)$ (to see why, see Lemma 6.1). Such a node either has an id that the leader first hears of, or has no id yet. In the first case, the leader updates $K(r)$ and in the second waits until it hears of a new id (which is guaranteed to appear in the future). On the other hand, if $|K(r)| = |V|$ no new info will ever arrive at the leader in the future and the leader may terminate after the $|K(r)|$-round waiting period ellapses. This protocol solves the naming problem in anonymous unknown dynamic networks under the assumptions of one-to-each message transmission and of a unique leader. All nodes terminate in $O(n)$ rounds and use messages of size $\Theta(n^2)$.

Protocol *Dynamic_Naming*. Initially, every node has three variables $count \leftarrow 0$, $acks \leftarrow \emptyset$, and $latest\_unassigned \leftarrow 0$ and the leader additionally has $latest\_new \leftarrow 0$, $time\_bound \leftarrow 1$, and $known\_ids \leftarrow \{0\}$. A node with $id \neq \perp$ for $1 \leq i \leq k$ sends $assign$ $(id, count + i)$ message to its ith neighbor and sets $count \leftarrow count + k$. In the first round, the leader additionally sets $known\_ids \leftarrow \{0, (0, 1), (0, 2), \ldots, (0, k)\}$, $latest\_new \leftarrow 1$, and $time\_bound \leftarrow 1 + |known\_ids|$. Upon receipt of l $assign$ messages (rid_j), a node with $id = \perp$ sets $id \leftarrow \min_j\{rid_j\}$ (in number of bits), $acks \leftarrow acks \cup id$, sends an ack ($acks$) message to all its k current neighbors, for $1 \leq i \leq k$ sends $assign$ $(id, count + i)$ message to its ith neighbor, and sets $count \leftarrow count + k$. Upon receipt of l ack messages ($acks_j$), a nonleader sets $acks \leftarrow acks \cup (\bigcup_j acks_j)$ and sends ack ($acks$). A node with $id = \perp$ sends $unassigned$ ($current\_round$). Upon receipt of $l \geq 0$ $unassigned$ messages (val_j), a node with $id \notin \{0, \perp\}$ sets $latest\_unassigned \leftarrow \max\{latest\_unassigned, \max_j\{val_j\}\}$ and sends $unassigned$ ($latest\_unassigned$). Upon receipt of l ack messages ($acks_j$), the leader if $(\bigcup_j acks_j) \backslash known\_ids \neq \emptyset$ sets $known\_ids \leftarrow known\_ids \cup (\bigcup_j acks_j)$, $latest\_new \leftarrow current\_round$ and $time\_bound \leftarrow current\_round + |known\_ids|$ and upon receipt of l $unassigned$ messages (val_j), it sets $latest\_unassigned \leftarrow$

$\max\{latest\_unassigned, \max_j\{val_j\}\}$. If, at some round r, it holds at the leader that $r > time\_bound$ and $latest\_unassigned < latest\_new$, the leader sends a *halt* ($|known\_ids|$) message for $|known\_ids| - 1$ rounds and then outputs *id* and halts. Any node that receives a *halt* (n) message, sends *halt* (n) for $n - 2$ rounds and then outputs *id* and halts.

Denote by $S(r) = \{v \in V : (l, 0) \rightsquigarrow (v, r)\}$ the set of nodes that have obtained an id at round r and by $K(r)$ those nodes in $S(r)$ whose id is known by the leader at round r, that is $K(r) = \{u \in V : \exists r' \text{ s.t. } u \in S(r') \text{ and } (u, r') \rightsquigarrow (l, r)\}$.

Theorem 6.6 (Michail, Chatzigiannakis, and Spirakis 2013) *Dynamic_Naming solves the naming problem in anonymous unknown dynamic networks under the assumptions of one-to-each message transmission and of a unique leader. All nodes terminate in $O(n)$ rounds and use messages of size $\Theta(n^2)$.*

Proof. Unique names are guaranteed as in *Delegate*. Termination is as follows. Clearly, if $V\backslash K(r) \neq \emptyset$, either $|K(r + |K(r)|)| \geq |K(r)| + 1$ or $(u, r) \rightsquigarrow (l, r + |K(r)|)$ for some $u \in V\backslash S(r)$. The former is recognized by the leader, by the arrival of a new id and the latter by the arrival of an *unassigned* (*timestamp*) message, where $timestamp \geq r$. On the other hand, if $K(r) = V$ then $|K(r + |K(r)|)| = |K(r)|$ and $\nexists u \in V\backslash S(r)$ s.t. $(u, r) \rightsquigarrow (l, r + |K(r)|)$ as $V\backslash S(r) = \emptyset$. Finally, note that connectivity implies that $|S(r + 1)| \geq \min\{|S(r)| + 1, n\}$ which in turn implies $O(n)$ rounds until unique ids are assigned. Then another $O(n)$ rounds are required until nodes terminate. □

Clearly, by executing a simple $O(n)$-time process after *Dynamic_Naming*, we can easily reassign minimal (consecutive) names to the nodes. The leader just floods a list of (*old_id*, *new_id*) pairs, one for each node in the network.

Fourth Version – Protocol *Individual_Conversations* (Logarithmic Messages) Though *Dynamic_Naming* is a correct and time-efficient terminating protocol for the naming problem, it still has an important drawback. The messages sent may be of size $\Omega(n^2)$. We now refine *Dynamic_Naming* to arrive at a more involved construction that reduces the message size to $\Theta(\log n)$ by paying a small increase in termination time. We call this fourth version of our naming protocols *Individual_Conversations*. We only give that main idea here.

Protocol *Individual_Conversations* [Main Idea]. To reduce the size of the messages (i) the assigned names are now of the form $k \cdot d + id$, where *id* is the id of the node, d is the number of *unique consecutive* ids that the leader knows so far, and $k \geq 1$ is a name counter (ii) Any time that the leader wants to communicate to a remote node that has a unique id, it sends a message with the id of that node and a timestamp equal to the current round. The timestamp allows all nodes to prefer this message from previous ones so that the gain is twofold: The message is delivered and no node ever issues a message containing more than one id. The remote node then can reply in the same way. For the assignment formula to work, nodes that obtain ids are not allowed to further assign ids until the leader freezes all named nodes and reassigns to them unique consecutive ids. During freezing, the leader is informed of any new assignments by the named nodes and terminates if all report that no further assignments were performed.

Theorem 6.7 (Michail, Chatzigiannakis, and Spirakis 2013)
Individual_Conversations solves the (minimal) naming problem in $O(n^3)$ rounds using messages of size $\Theta(\log n)$.

Proof. Though *Dynamic_Naming* is a correct and time-efficient terminating protocol for the naming problem it still has an important drawback. The messages sent may be of size $\Omega(n^2)$. There are two reasons for this increased message size. One is the method of assigning ids, in which the id of a node is essentially set to a pair containing the id of its first parent and a counter. By induction on assignments, in which the leader assigns to a single node, that node assigns to another node, the third node to a fourth one, and so on, it is easy to see that ids may become n-tuples and thus have size $O(n)$. The other reason is that, for a node to acknowledge to the leader its assigned id, that node and all nodes that receive it must continuously broadcast it until the leader receives it (otherwise, delivery is not guaranteed by our dynamic network model). As $O(n)$ nodes may want to acknowledge at the same time, it follows that some node may need to continuously broadcast $O(n)$ ids each of size $O(n)$, thus $O(n^2)$. We now refine *Dynamic_Naming* to arrive at a more involved construction that reduces the message size to $\Theta(\log n)$ by paying a small increase in termination time. We call this protocol *Individual_Conversations*. Due to the many low-level details of the protocol, we adopt a high-level but at the same time precise and clear verbal presentation.

One refinement concerns the method of assigning ids. We notice that if some d nodes have the unique consecutive ids $D = \{0, 1, 2, \dots, k-1\}$, then we can have node with id $j \in D$ assign ids $k \cdot d + j$, for all $k \geq 1$. For example, if we have nodes $\{0, 1, 2, 3\}$, then node 0 assigns ids $\{4, 8, 12, \dots\}$, node 1 assigns $\{5, 9, 13, \dots\}$, node 2 assigns $\{6, 10, 14, \dots\}$, and node 3 assigns $\{7, 11, 15, \dots\}$. Clearly, the assignments are unique and in the worst-case $k, d, j = O(n)$, which implies that the maximum assigned id is $O(n^2)$ thus its binary representation is $\Theta(\log n)$. So, if we could keep the assigning nodes to have unique consecutive ids while knowing the maximum existing id (so as to evaluate the id-generation formula), we could get logarithmic ids.

Even if we could implement the above assignment method, if nodes continued to constantly forward all ids that they ever hear of then, we would not do better than message sizes $O(n \log n)$ (a node forwards $O(n)$ ids each of size $O(\log n)$). Clearly, another necessary improvement is to guarantee communication between the leader and some node with unique id j that the leader knows of, that is, a pairwise conversation. It is important that a conversation is initiated by the leader so that we do not get multiple nodes trying to initiate a conversation with the leader, as this would increase the communication complexity. The leader sends a *request*(*rem_id*, *current_round*) message, where *rem_id* is the id of the remote node and *current_round* is a timestamp indicating the time in which the request for conversation was initiated. Upon receipt of a *request*(r_id, *timestamp*) message, all nodes such that $id \neq r\_id$ forward the message if it is the one with the largest timestamp that they have ever heard of. All nodes keep forwarding the message with the largest timestamp. When the remote node receives the message, it replies with *report*(*id*, *current_round*), where *id* is its own id. Now, all nodes will forward the report as it is the one with the largest timestamp and the report will eventually reach the leader who can reply with another request, and so on. Note that a node that participates in a conversation need not know how much time it will take for the other node to reply. It only needs to have a guarantee that the reply will eventually arrive. Then, it can recognize that this is the correct reply by the type, the id-component, and the timestamp of the received message. A nice property of 1-interval connected graphs is that it guarantees any such reply to arrive in $O(n)$ rounds if all nodes that receive it keep broadcasting it (which is the case here, due to the timestamps). So, in order to keep the message sizes low, we must implement the above communication method in such a way that the leader always participates in a single conversation, so that a single message ever floods the whole network (in particular, the most recently created one).

Now, let us further develop our id-assignment method. Clearly, in the first round the leader can keep id 0 for itself and assign the unique consecutive ids $\{1, 2, \dots, d_l(1)\}$ to its $|d_l(1)|$ neighbors in round 1. Clearly, each node with id j in $K(1) = \{0, 1, \dots, |d_l(1)|\}$ can further assign the unique ids $k \cdot |K(1)| + j$, for $k \geq 1$. As before, we can have a node stick to the smallest id that it hears from its neighbors but we additionally need that node to remember those ids that it rejected in a *rejected* list. However, note that, if nodes outside $K(1)$ that obtain a unique id are not allowed to further assign ids, then we do not guarantee that all nodes will eventually obtain an id. The reason is that the adversary can forever hide the set $K(1)$ from the rest of the graph via nodes that have obtained an id and do not further assign ids (i.e., all nodes in $K(1)$ may communicate only to nodes in $K(1)$ and to nodes that have obtained an id but do not assign and all nodes that do not have an id may communicate only to nodes that do not have an id and to nodes that have obtained an id but do not assign, which is some sort of a deadlock). So, we must somehow also allow to nodes that obtain ids to further assign ids. The only way to do this while keeping our assignment formula is to restructure the new assignments so that they are still unique and additionally consecutive. So, for example, if nodes in $K(1)$ have at some point assigned a set of ids T, then the leader should somehow reassign to nodes in T the ids $\{|K(1)|, |K(1)| + 1, \dots, |K(1)| + |T| - 1\}$.

So, at this point, it must be clear that the leader must first allow to the nodes that have unique consecutive ids (including itself) to perform some assignments. Then at some point, it should freeze the assigning nodes and ask them one after the other to report the assignments that they have performed so far. Then, assuming that it has learned all the newly assigned unique ids, it should communicate with those nodes to reassign to them the next available unique consecutive ids and also, it should inform all nodes with id of the maximum consecutive id that has been assigned so far. Now that all nodes with id have unique consecutive ids and know the maximum assigned, they can all safely use the id-assignment formula. In this manner, we have managed to also allow to the new nodes, to safely assign unique ids. Finally, the leader unfreezes the nodes with ids one after the other, allows them to assign some new ids, and at some point, freezes them again to repeat the above process which we may call a cycle.

A very important point that we should make clear at this point is that, in 1-interval connected graphs, a new assignment is only guaranteed if at least for one round all nodes that have ids send assignment messages to all their incident edges. As, if some node with id, selected to issue no-assignment message to some of its edges, then the adversary could make that edge be the only edge that connects nodes with ids to nodes without ids, and it could do the same any time an edge is not used. Fortunately, this is trivially guaranteed in the solution we have developed so far. When the leader unfreezes the last node with id, even if it chooses to start freezing the nodes in the subsequent round, provided that at least for that round it does not freezes itself, then in that round all nodes including itself are not frozen, thus all take an assignment step in that round (sending assignment messages to all their incident edges). This guarantees that for at least one round, all assign at the same time which in turn guarantees at least one new delivery, provided that there are still nodes without ids.

Another point that is still blur is the following. When the leader gets all reports from all nodes that were assigning ids during this cycle, it cannot know which ids have been assigned but only which ids have been possibly assigned. The reason is that when a node u assigns some ids then it is not guaranteed that in the next round it will have the same neighbors. So, it can be the case that some of its neighbors choose to stick to a smaller id sent by some other node and u never notices it. So, we have each node that assigns ids to remember the ids that have possibly been assigned and each node that is assigned an id to remember those ids that it rejected. Note that when a node u tries to assigns an id by sending it via a local edge, then, in the next round when it receives from that local edge, it can tell whether that id was possibly assigned by simply having all nodes send their id in every round. If the received id from that edge was $\perp$, then the corresponding neighbor did not have an id, thus it must have been assigned some id even if that was not the one sent by u. In any case, the id sent by u will either be assigned or stored in the *rejected* list of that node. On the other hand, if the received id was not equal to $\perp$, then the neighbor already had an id, u knows that its assignment was for sure unsuccessful and may reuse this id in future assignments. The problem now is that, if the leader tries to initiate a conversation with an arbitrary id from those that have been possibly assigned, it can very well be the case that this id was not assigned and the leader may have to wait for a reply forever. Fortunately, this can be solved easily by having the unique node that has stored this id in its *rejected* list to reply not only when it gets a *request* message containing its own id but also when it gets a message containing an id that is also in its *rejected* list. Another way is the following. As the leader has first collected all possibly delivered ids, it can order them increasingly and start seeking that smallest id. As the nodes stick to the smallest they hear, the smallest of all possibly assigned was for sure selected by some node. Then, that node may inform the leader of some rejected ids, which the leader will remove from its ordering and then the leader may proceed to seek for the next id that has remained in its ordered list. It is not hard to see that this method guarantees that the leader always seeks for existing ids.

Finally, the termination criterion is more or less the same as in *Dynamic_Naming*. The leader knows that, if it allows all nodes with ids at a common assignment step, then, provided that there are nodes without ids, at least one new assignment must take place. Clearly, if all nodes report that they performed no assignments, then the leader can terminate (and tell others to terminate) knowing that all nodes must have obtained an id. In the termination phase, it can reassign, for a last time, unique consecutive ids and inform all the nodes of n. □

6.4.5 Higher Dynamicity

Given some high-dynamicity assumption (some sort of fairness), naming can be solved under broadcast communication. Intuitively, to break the symmetry that is responsible for the impossibility of Conjecture 1, we require that, given sufficient time, a node has influenced every other node in different rounds. Formally, there must exist k (not necessarily known to the nodes) s.t $(\mathrm{arrival}_{(u,r)}(v), \mathrm{arrival}_{(u,r+1)}(v), \ldots, \mathrm{arrival}_{(u,r+k-1)}(v)) \neq (\mathrm{arrival}_{(u,r)}(w), \mathrm{arrival}_{(u,r+1)}(w), \ldots, \mathrm{arrival}_{(u,r+k-1)}(w))$, $\forall u \in V, r \geq 0, v, w \in V \backslash \{u\}$, where $\mathrm{arrival}_{(u,r)}(v) := \min\{r' > r : (u,r) \rightsquigarrow (v,r')\}$ (first time that v is causally influenced by the r-state of u). We also allow nodes to have time to acknowledge to their neighbors (formally, we may duplicate each instance of the dynamic graph, i.e., make it persist for two rounds).

The idea is to have the leader name its first $d_l(1)$ neighbors, say with id 1. What the leader can exploit is that it knows the number of 1s in the network as it knows its degree in round 1. Now, every node v named 1 counts $\mathrm{arrival}_{(l,i)}(v)$ for all $i \geq 2$. This is achieved by having the leader continuously send an $(l, current\_round)$ pair, unnamed nodes constantly forward it, and having every node named 1 set $\mathrm{arrival}_{(l,i)}(v)$ to the round in which an (l, i) pair was first delivered. It is clear that, due to the above high-dynamicity assumption, the vector $s(v) = (1, \mathrm{arrival}_{(l,2)}(v), \mathrm{arrival}_{(u,3)}(v), \ldots, \mathrm{arrival}_{(u,k+2)}(v))$ (in k rounds) will be a unique *id*. As the named nodes do not know k, we have them continuously send $(s, current\_round)$ pairs, where s is the above vector, and all other nodes continuously forward these pairs. At some point, the leader must hear from $d_l(1)$ different s vectors with equal timestamps and then it knows that the 1s have obtained unique ids. Now, the leader can stop them from further changing their ids. Then it allows them (including itself) to concurrently assign id 2 for at least one step. The assigning nodes count the number of assignments that they perform (in a variable *count* initially 0). This is done by having a node u that was assigned id 2 in round r to respond to its neighbors the number l of nodes that tried to assign 2 to it. Then each of the assigning 1s sets *count* $\leftarrow$ *count* $+ 1/l$. When the leader freezes the 1s, they report their *count* variable and by summing them the leader learns the number, j, of 2s assigned. Then, the leader sends again $(l, current\_round)$ pairs and waits to receive j different s vectors with equal timestamps. The process continues in such cycles until at some point all existing unique ids report that they did not manage to assign the current id being assigned.

6.5
Causality, Influence, and Computation in Possibly Disconnected Synchronous Dynamic Networks

In this section, we study the *propagation of influence and computation in dynamic distributed computing systems that are possibly disconnected at every instance.* We focus on a *synchronous message passing* communication model with and bidirectional links. Our network dynamicity assumption is again a *worst-case dynamicity* controlled by an adversary scheduler. However, we replace the usual (in worst-case dynamic networks) assumption that the network is connected at every instance by minimal *temporal connectivity* conditions. Our conditions only require that *another causal influence occurs within every time-window of some given length.* Based on this basic idea, we define several novel metrics for capturing the speed of information spreading in a dynamic network. Moreover, we investigate *termination criteria* in networks in which an upper bound on any of these metrics is known. We exploit our termination criteria to give protocols that solve the fundamental *counting* and *all-to-all token dissemination* (or *gossip*) problems.

6.5.1
Our Metrics

As already stated, in this section, we aim to deal with dynamic networks that are allowed to have disconnected instances. To this end, we define some novel generic metrics that are particularly suitable for capturing the speed of the information propagation in such networks.

6.5.1.1 The Influence Time

Recall that the guarantee on the propagation of information resulting from instantaneous connectivity ensures that any time-node (u, t) influences another node *in each step* (if an uninfluenced one exists). From this fact, we extract two novel generic influence metrics that capture the maximal time until another influence (outgoing or incoming) of a time-node occurs.

We now formalize our first influence metric.

Definition 6.2 (Outgoing Influence Time) *We define the* outgoing influence time *(*oit*) as the minimum $k \in \mathbb{N}$ s.t. for all $u \in V$ and all times $t, t' \geq 0$ s.t. $t' \geq t$ it* *holds that*

$$|\text{future}_{(u,t)}(t' + k)| \geq \min\{|\text{future}_{(u,t)}(t')| + 1, n\}.$$

Intuitively, the oit is the maximal time until the t-state of a node influences the state of another node (if an uninfluenced one exists) and captures the speed of information spreading.

Our second metric is similarly defined as follows.

Definition 6.3 (Incoming Influence Time) *We define the* incoming influence time *(iit) as the minimum $k \in \mathbb{N}$ s.t. for all $u \in V$ and all times $t, t' \geq 0$ s.t. $t' \geq t$ it holds that*

$$|\text{past}_{(u,t'+k)}(t)| \geq \min\{|\text{past}_{(u,t')}(t)| + 1, n\}.$$

We can now say that the oit of a T-interval connected graph is 1 and that the iit can be up to $n - 2$. However, is it necessary for a dynamic graph to be T-interval connected in order to achieve unit oit? First, let us make a simple but useful observation:

Proposition 6.1 (Michail, Chatzigiannakis, and Spirakis 2014) *If a dynamic graph $G = (V, E)$ has oit (or iit) 1 then every instance has at least $\lceil n/2 \rceil$ edges.*

Proof. $\forall u \in V$ and $\forall t \geq 1$ it must hold that $\{u, v\} \in E(t)$ for some v. In other words, at any time t, each node must have at least one neighbor since otherwise it influences (or is influenced by) no node during round t. A minimal way to achieve this is by a perfect matching in the even-order case and by a matching between $n - 3$ nodes and a linear graph between the remaining 3 nodes in the odd-order case. □

Proposition 6.1 is easily generalized as: if a dynamic graph $G = (V, E)$ has oit (or iit) k then for all times t, it holds that $|\bigcup_{i=t}^{t+k-1} E(i)| \geq \lceil n/2 \rceil$. The reason is that now any node must have a neighbor in any k-window of the dynamic graph (and not necessarily in every round).

Now, inspired by Proposition 6.1, we define a minimal dynamic graph that at the same time satisfies oit 1 and always disconnected instances:

The Alternating Matchings Dynamic Graph. Take a ring of an even number of nodes $n = 2l$, partition the edges into two disjoint perfect matchings A and B (each consisting of l edges) and alternate round after round between the edge sets A and B (Figure 6.1).

Proposition 6.2 (Michail, Chatzigiannakis, and Spirakis 2014) *The alternating matchings dynamic graph has oit 1 and any node needs precisely $n/2$ rounds to influence all other nodes.*

Proof. Take any node u. In the first round, u influences its left or its right neighbor on the ring depending on which of its two adjacent edges become available first.

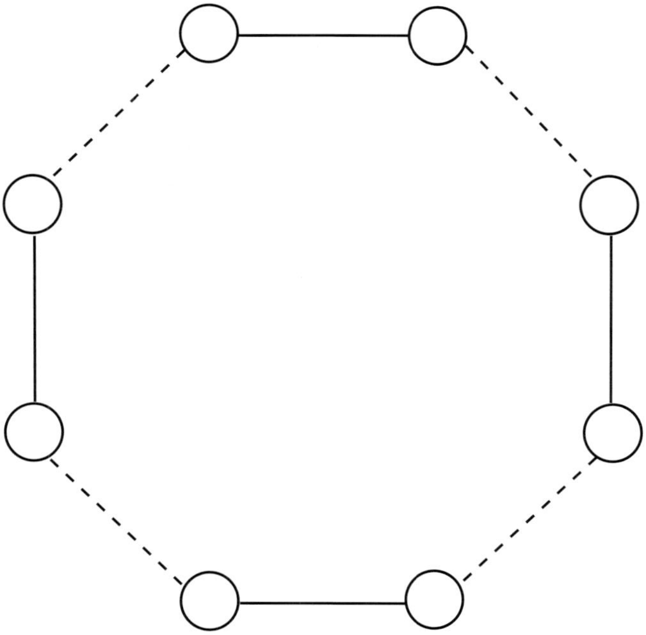

Figure 6.1 The alternating matchings dynamic graph for $n = 8$. The solid lines appear every odd round $(1, 3, 5, \ldots)$ while the dashed lines every even round $(2, 4, 6, \ldots)$.

Thus, including itself, it has influenced two nodes forming a line of length 1. In the next round, the two edges that join the endpoints of the line with the rest of the ring become available and two more nodes become influenced; the one is the neighbor on the left of the line and the other is the neighbor on the right. By induction on the number of rounds, it is not hard to see that the existing line always expands from its endpoints to the two neighboring nodes of the ring (one on the left and the other on the right). Thus, we get exactly two new influences per round, which gives oit 1 and $n/2$ rounds to influence all nodes. □

In the alternating matchings construction, any edge reappears every second step but not faster than this. We now formalize the notion of the *fastest edge reappearance* (fer) of a dynamic graph.

Definition 6.4 *The fastest edge reappearance (fer) of a dynamic graph $G = (V, E)$ is defined as the minimum $p \in \mathbb{N}$ s.t., $\exists e \in \{\{u, v\} : u, v \in V\}$ and $\exists t \in \mathbb{N}$, $e \in E(t) \cap E(t + p)$.*

Clearly, the fer of the alternating matchings dynamic graph described earlier is 2, because no edge ever reappears in one step and some, at some point, (in fact, all and always) reappears in two steps. In Section 6.5.2, by invoking a geometric edge-coloring method, we generalize this basic construction to a more involved dynamic graph with oit 1, always disconnected instances, and fer equal to $n-1$.[5)]

We next give a proposition associating dynamic graphs with oit (or iit) upper bounded by K to dynamic graphs with connected instances.

Proposition 6.3 (Michail, Chatzigiannakis, and Spirakis 2014) *Assume that the oit or the iit of a dynamic graph, $G=(V,E)$, is upper bounded by K. Then, for all times $t \in I\!N$ the graph $(V, \bigcup_{i=t}^{t+K\lfloor n/2\rfloor-1} E(i))$ is connected.*

Proof. It suffices to show that for any partitioning (V_1, V_2) of V, there is an edge in the cut labeled from $\{t,\dots,t+K\lfloor n/2\rfloor-1\}$. W.l.o.g. let V_1 be the smaller one, thus $|V_1| \leq \lfloor n/2\rfloor$. Take any $u \in V_1$. By definition of oit, $|\text{future}_{(u,t)}(t+K\lfloor n/2\rfloor-1)| \geq |\text{future}_{(u,t)}(t+K|V_1|-1)| \geq |V_1|+1$ implying that some edge in the cut has transferred u's t-state out of V_1 at some time in the interval $[t, t+K\lfloor n/2\rfloor-1]$. The proof for the iit is similar. □

6.5.1.2 The Moi (Concurrent Progress)

Consider now the following influence metric:

Definition 6.5 *Define the* maximum outgoing influence *(moi) of a dynamic graph $G=(V,E)$ as the maximum k for which $\exists u \in V$ and $\exists t, t' \in I\!N$, $t' \geq t$, s.t. $|\text{future}_{(u,t)}(t'+1)| - |\text{future}_{(u,t)}(t')| = k$.*

In words, the moi of a dynamic graph is the maximum number of nodes that are ever concurrently influenced by a time-node.

Here, we show that one cannot guarantee at the same time unit oit and at most one outgoing influence per node per step. In fact, we conjecture that unit oit implies that some node disseminates in $\lfloor n/2\rfloor$ steps.

We now prove an interesting theorem stating that if one tries to guarantee unit oit, then he/she must necessarily accept that at some steps, more than one outgoing influences of the same time-node will occur leading to faster dissemination than $n-1$ for this particular node.

Theorem 6.8 (Michail, Chatzigiannakis, and Spirakis 2014) *The moi of any dynamic graph with $n \geq 3$ and unit oit is at least 2.*

Proof. For $n=3$, just notice that unit oit implies that, at any time t, some node has necessarily two neighbors. We, therefore, focus on $n \geq 4$. For the sake of contradiction, assume that the statement is not true. Then, at any time t any node u is connected to exactly one other node v (at least one neighbor is required for oit 1 – see Proposition 6.1 – and at most one is implied by our assumption). Unit

5) It is interesting to note that in dynamic graphs with a static set of nodes (i.e., V does not change), if at least one change happens each time, then every instance $G(t)$ will eventually reappear after at most $\sum_{k=0}^{\binom{n}{2}} \binom{\binom{n}{2}}{k}$ steps. This counts all possible different graphs of n vertices with k edges and sums for all $k \geq 0$. Thus, the fer is bounded from above by a function of n.

oit implies that, at time $t+1$, at least one of u, v must be connected to some $w \in V \backslash \{u, v\}$, let it be v. Proposition 6.1 requires that also u must have an edge labeled $t+1$ incident to it. If that edge arrives at v, then v has two edges labeled $t+1$. If it arrives at w, then w has two edges labeled $t+1$. So it must arrive at some $z \in V \backslash \{u, v, w\}$. Note now that, in this case, the $(t-1)$-state of u first influences both w, z at time $t+1$ which is contradictory, consequently the moi must be at least 2. □

In fact, notice that the above theorem proves something stronger: every second step, at least, half of the nodes influence at least two new nodes each. This, together with the fact that it seems to hold for some basic cases, makes us suspect that the following conjecture might be true:

Conjecture 2 (Michail, Chatzigiannakis, and Spirakis 2014) If the oit of a dynamic graph is 1 then $\forall t \in \mathbb{N}, \exists u \in V$ s.t. $|\text{future}_{(u,t)}(t + \lfloor n/2 \rfloor)| = n$.

That is, if the oit is 1 then, in every $\lfloor n/2 \rfloor$-window, some node influences all other nodes (e.g., influencing two new nodes per step).

6.5.1.3 The Connectivity Time

We now propose another natural and practical metric for capturing the temporal connectivity of a possibly disconnected dynamic network that we call the *connectivity time* (ct).

Definition 6.6 (Connectivity Time) *We define the* connectivity time *(ct) of a dynamic network $G = (V, E)$ as the minimum $k \in \mathbb{N}$ s.t. for all times $t \in \mathbb{N}$ the static graph $(V, \bigcup_{i=t}^{t+k-1} E(i))$ is connected.*

In words, the ct of a dynamic network is the maximal time of keeping the two parts of any cut of the network disconnected. That is to say, in every ct-window of the network an edge appears in every (V_1, V_2)-cut. Note that, in the extreme case in which the ct is 1, every instance of the dynamic graph is connected and we thus obtain a 1-interval connected graph. On the other hand, greater ct allows for different cuts to be connected at different times in the ct-round interval and the resulting dynamic graph can very well have disconnected instances. For an illustrating example, consider again the alternating matchings graph from Section 6.5.1.1. Draw a line that crosses two edges belonging to matching A partitioning the ring into two parts. Clearly, these two parts communicate every second round (as they only communicate when matching A becomes available), thus the ct is 2 and every instance is disconnected. We now provide a result associating the ct of a dynamic graph with its oit.

Proposition 6.4 (Michail, Chatzigiannakis, and Spirakis 2014) *(i) oit $\leq$ ct but (ii) there is a dynamic graph with oit 1 and ct $= \Omega(n)$.*

Proof. (i) We show that for all $u \in V$ and all times $t, t' \in \mathbb{N}$ s.t. $t' \geq t$ it holds that $|\text{future}_{(u,t)}(t' + \text{ct})| \geq \min\{|\text{future}_{(u,t)}(t')| + 1, n\}$. Assume $V \backslash \text{future}_{(u,t)}(t') \neq \emptyset$

(as the other case is trivial). In at most ct rounds at least one edge joins $\text{future}_{(u,t)}(t')$ to $V\backslash\text{future}_{(u,t)}(t')$. Thus, in at most ct rounds $\text{future}_{(u,t)}(t')$ increases by at least one.

(ii) Recall the alternating matchings on a ring dynamic graph from Section 6.5.1.1. Now take any set V of a number of nodes that is a multiple of 4 (this is just for simplicity and is not necessary) and partition it into two sets V_1, V_2 s.t. $|V_1| = |V_2| = n/2$. If each part is an alternating matchings graph for $|V_1|/2$ rounds then every u say in V_1 influences 2 new nodes in each round and similarly for V_2. Clearly we can keep V_1 disconnected from V_2 for $n/4$ rounds without violating oit $= 1$. □

The following is a comparison of the ct of a dynamic graph with its dynamic diameter D.

Proposition 6.5 (Michail, Chatzigiannakis, and Spirakis 2014) *ct* $\leq D \leq (n-1)$*ct.*

Proof. ct $\leq D$ follows from the fact that in time equal to the dynamic diameter, every node causally influences every other node and thus, in that time, there must have been an edge in every cut (if not, then the two partitions forming the cut could not have communicated with one another). $D \leq (n-1)$ct holds as follows. Take any node u and add it to a set S. In ct rounds, u influences some node from $V\backslash S$ which is then added to S. In $(n-1)$ct rounds S must have become equal to V, thus this amount of time is sufficient for every node to influence every other node. Finally, we point out that these bounds cannot be improved in general as for each of ct $= D$ and $D = (n-1)$ct there is a dynamic graph realizing it. ct $= D$ is given by the dynamic graph that has no edge for ct $- 1$ rounds and then becomes the complete graph while $D = (n-1)$ct is given by a line in which every edge appears at times ct, 2ct, 3ct, …. □

Note that the ct metric has been defined as an underapproximation of the dynamic diameter. Its main advantage is that, it is much easier to compute than the dynamic diameter since it is defined on the union of the footprints and not on the dynamic adjacency itself.

6.5.2 Fast Propagation of Information under Continuous Disconnectivity

In Section 6.5.1.1, we presented a simple example of an always-disconnected dynamic graph, namely, the alternating matchings dynamic graph, with optimal oit (i.e., unit oit). Note that the alternating matchings dynamic graph may be conceived as simple as it has small fer (equal to 2). We pose now an interesting question: is there an always-disconnected dynamic graph with unit oit and fer as big as $n-1$? Note that this is harder to achieve as it allows of no edge to ever reappear in less than $n-1$ steps. Here, by invoking a geometric edge-coloring method, we arrive at an always-disconnected graph with unit oit and maximal fer; in particular, no edge reappears in less than $n-1$ steps.

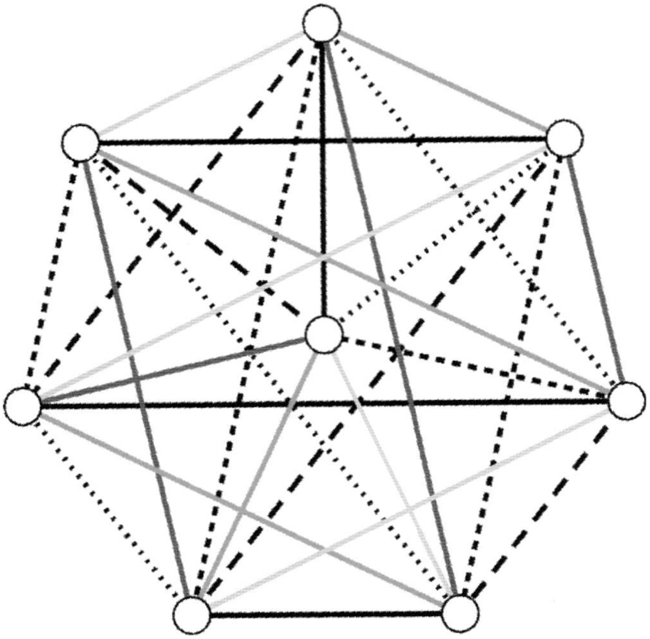

Figure 6.2 Soifer's dynamic graph for $n = 8$ and $t = 1, \ldots, 7$. In particular, in round 1, the graph consists of the black solid edges, then in round 2 the center becomes connected via a dotted edge to the next peripheral node clockwise and all edges perpendicular to it (the remaining dotted ones) become available, and so on, always moving clockwise.

To answer the above question, we define a very useful dynamic graph coming from the area of edge-coloring.

Definition 6.7 *We define the following dynamic graph S based on an edge-coloring method due to Soifer (2009):* $V(S) = \{u_1, u_2, \ldots, u_n\}$ *where* $n = 2l$, $l \geq 2$. *Place* u_n *on the center and* $u_1, \ldots, u_{n-1}$ *on the vertices of a* $(n-1)$*-sided polygon. For each time* $t \geq 1$ *make available only the edges* $\{u_n, u_{m_t(0)}\}$ *for* $m_t(j) := (t - 1 + j \bmod n - 1) + 1$ *and* $\{u_{m_t(-i)}, u_{m_t(i)}\}$ *for* $i = 1, \ldots, n/2 - 1$; *that is, make available one edge joining the center to a polygon-vertex and all edges perpendicular to it. (e.g., see Figure 6.2 for* $n = 8$ *and* $t = 1, \ldots, 7$*).*

In Soifer's dynamic graph, denote by $N_u(t) := i : \{u, u_i\} \in E(t)$, that is the index of the unique neighbor of u at time t. The following Lemma states that the next neighbor of a node is, in almost all cases (in addition to some trivial ones), the one that lies two positions clockwise from its current neighbor.

Lemma 6.4 (Michail, Chatzigiannakis, and Spirakis 2014) *For all times* $t \in \{1, 2, \ldots, n-2\}$ *and all* u_k, $k \in \{1, 2, \ldots, n-1\}$ *it holds that* $N_{u_k}(t+1) = n$ *if* $N_{u_k}(t) = (k - 3 \bmod n - 1) + 1$ *else* $N_{u_k}(t+1) = (k + 1 \bmod n - 1) + 1$ *if* $N_{u_k}(t) = n$ *and* $N_{u_k}(t+1) = (N_{u_k}(t) + 1 \bmod n - 1) + 1$ *otherwise.*

Proof. Since $k \notin \{n, t, t+1\}$ it easily follows that $k, N_k(t), N_k(t+1) \neq n$ thus both $N_k(t)$ and $N_k(t+1)$ are determined by $\{u_{m_t(-i)}, u_{m_t(i)}\}$ where $m_t(j) := (t - 1 + j \bmod n - 1) + 1$ and $k = m_t(-i)$. The latter implies $(t - 1 - i \bmod n - 1) + 1 = k \Rightarrow (t - 1 + i \bmod n - 1) + 1 + (-2i \bmod n - 1) = k \Rightarrow m_t(i) = k - (-2i \bmod n - 1)$; thus, $N_k(t) = k - (-2i \bmod n - 1)$. Now, let us see how the i that corresponds to some node changes as t increases. When t increases by 1, we have that $(t - 1 + i \bmod n - 1) + 1 = (t + i' \bmod n - 1) + 1 \Rightarrow i' = i - 1$, that is, as t increases i decreases. Consequently, for $t + 1$ we have $N_k(t + 1) = k - [-2(i - 1) \bmod n - 1] = (N_{u_k}(t) + 1 \bmod n - 1) + 1$. □

Theorem 6.9 (Michail, Chatzigiannakis, and Spirakis 2014) *For all $n = 2l$, $l \geq 2$, there is a dynamic graph of order n, with* oit *equal to 1,* fer *equal to $n - 1$, and in which every instance is a perfect matching.*

Proof. The dynamic graph is the one of Definition 6.7. It is straightforward to observe that every instance is a perfect matching. We prove now that the oit of this dynamic graph is 1. We focus on the set $\text{future}_{(u_n,0)}(t)$, that is the outgoing influence of the initial state of the node at the center. Note that symmetry guarantees that the same holds for all time-nodes (it can be verified that any node can be moved to the center without altering the graph). u_n at time 1 meets u_1 and thus $\text{future}_{(u_n,0)}(1) = \{u_1\}$. Then, at time 2, u_n meets u_2 and, by Lemma 6.4, u_1 meets u_3 via the edge than is perpendicular to $\{u_n, u_2\}$, thus $\text{future}_{(u_n,0)}(2) = \{u_1, u_2, u_3\}$. We show that for all times t it holds that $\text{future}_{(u_n,0)}(t) = \{u_1, \ldots, u_{2t-1}\}$. The base case is true since $\text{future}_{(u_n,0)}(1) = \{u_1\}$. It is not hard to see that, for $t \geq 2$, $N_{u_2}(t) = 2t - 2$, $N_{u_1}(t) = 2t - 1$, and for all $u_i \in \text{future}_{(u_n,0)}(t) \backslash \{u_1, u_2\}$, $1 \leq N_{u_i}(t) \leq 2t - 2$. Now, consider time $t + 1$. Lemma 6.4 guarantees now that for all $u_i \in \text{future}_{(u_n,0)}(t)$ we have that $N_{u_i}(t + 1) = N_{u_i}(t) + 2$. Thus, the only new influences at step $t + 1$ are by u_1 and u_2 implying that $\text{future}_{(u_n,0)}(t + 1) = \{u_1, \ldots, u_{2(t+1)-1}\}$. Consequently, the oit is 1.

The fer is $n - 1$ because the edges leaving the center appear one after the other in a clockwise fashion, thus taking $n - 1$ steps to any such edge to reappear, and, by construction, any other edge appears only when its unique perpendicular, that is incident to the center appears (thus, again every $n - 1$ steps). □

Note that Theorem 6.9 is optimal w.r.t. fer as it is impossible to achieve at the same time unit oit and fer strictly greater than $n - 1$. To see this, notice that if no edge is allowed to reappear in less than n steps then any node must have no neighbors once every n steps.

6.5.3
Termination and Computation

We now turn our attention to termination criteria that we exploit to solve the fundamental counting and all-to-all token dissemination problems. First, observe that if nodes know an upper bound H on the iit, then there is a straightforward optimal termination criterion taking time $D + H$, where D is the dynamic diameter. In every round, all nodes forward all ids that they have heard of so far. If a node does not hear of a new id for H rounds, then it must have already heard from all nodes. Keep in mind that nodes have no *a priori* knowledge of the size of the network.

6.5.3.1 Nodes Know an Upper Bound on the ct: An Optimal Termination Criterion

We here assume that all nodes know some upper bound T on the ct. We will give an optimal condition that allows a node to determine whether it has heard from all nodes in the graph. This condition results in an algorithm for counting and all-to-all token dissemination which is optimal, requiring $D + T$ rounds in any dynamic network with dynamic diameter D. The core idea is to have each node keep track of its past sets from time 0 and from time T and terminate as soon as these two sets become equal. This technique is inspired by Kuhn, Moses, and Oshman (2011), where a comparison between the past sets from time 0 and time 1 was used to obtain an optimal termination criterion in 1-interval connected networks.

Theorem 6.10 (Michail, Chatzigiannakis, and Spirakis 2014) *[Repeated Past] Node u knows at time t that* $\text{past}_{(u,t)}(0) = V$ *iff* $\text{past}_{(u,t)}(0) = \text{past}_{(u,t)}(T)$.

Proof. If $\text{past}_{(u,t)}(0) = \text{past}_{(u,t)}(T)$ then we have that $\text{past}_{(u,t)}(T) = V$. The reason is that $|\text{past}_{(u,t)}(0)| \geq \min\{|\text{past}_{(u,t)}(T)| + 1, n\}$. To see this, assume that $V \backslash \text{past}_{(u,t)}(T) \neq \emptyset$. At most by round T there is some edge joining some $w \in V\backslash\text{past}_{(u,t)}(T)$ to some $v \in \text{past}_{(u,t)}(T)$. Thus, $(w, 0) \rightsquigarrow (v, T) \rightsquigarrow (u, t) \Rightarrow w \in \text{past}_{(u,t)}(0)$. In words, all nodes in $\text{past}_{(u,t)}(T)$ belong to $\text{past}_{(u,t)}(0)$ and at least one node not in $\text{past}_{(u,t)}(T)$ (if one exists) must belong to $\text{past}_{(u,t)}(0)$ (see also Figure 6.3).

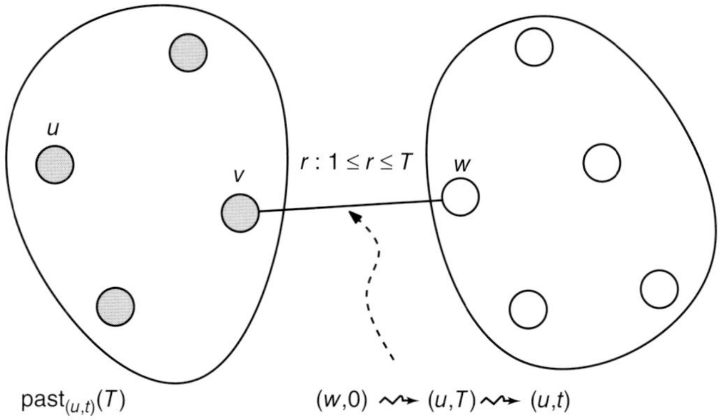

Figure 6.3 A partitioning of V into two sets. The left set is $\text{past}_{(u,t)}(T)$, that is, the set of nodes whose T-state has influenced u by time t. All nodes in $\text{past}_{(u,t)}(T)$ also belong to $\text{past}_{(u,t)}(0)$. Looking back in time at the interval $[1, T]$, there should be an edge from some v in the left set to some w in the right set. This implies that v has heard from w by time T and as u has heard from the T-state of v it has also heard from the initial state of w. This implies that $\text{past}_{(u,t)}(0)$ is a strict superset of $\text{past}_{(u,t)}(T)$ as long as the right set is not empty.

For the other direction, assume that there exists $v \in \text{past}_{(u,t)}(0)\backslash\text{past}_{(u,t)}(T)$. This does not imply that $\text{past}_{(u,t)}(0) \neq V$ but it does imply that even if $\text{past}_{(u,t)}(0) = V$ node u cannot know it has heard from everyone. Note that u heard from v at some time $T' < T$ but has not heard from v since then. It can be the case that arbitrarily many nodes were connected to no node until time $T - 1$ and from time T onwards were connected only to node v (v in some sense conceals these nodes from u). As u has not heard from the T-state of v it can be the case that it has not heard at all from arbitrarily many nodes, thus it cannot decide on the count. □

We now give a time-optimal algorithm for counting and all-to-all token dissemination that is based on Theorem 6.10.

Protocol A. All nodes constantly forward all 0-states and T-states of nodes that they have heard of so far (in this protocol, these are just the ids of the nodes accompanied with 0 and T timestamps, respectively) and a node halts as soon as $\text{past}_{(u,t)}(0) = \text{past}_{(u,t)}(T)$ and outputs $|\text{past}_{(u,t)}(0)|$ for counting or $\text{past}_{(u,t)}(0)$ for all-to-all dissemination.

For the time-complexity, notice that any state of a node needs D rounds to causally influence all nodes, where D is the dynamic diameter. Clearly, by time $D + T$, u must have heard of the 0-state and T-state of all nodes, and at that time $\text{past}_{(u,t)}(0) = \text{past}_{(u,t)}(T)$ is satisfied. It follows that all nodes terminate in at most $D + T$ rounds. Optimality follows from the fact that this protocol terminates as long as $\text{past}_{(u,t)}(0) = \text{past}_{(u,t)}(T)$ which by the "only if" part of the statement of Theorem 6.10 is a necessary condition for correctness (any protocol terminating before this may terminate without having heard from all nodes).

6.5.3.2 Known Upper Bound on the oit

Now, we assume that all nodes know some upper bound K on the oit. Then, one can show that if a node u has at some point heard of l nodes, then u hears of another node in $O(Kl^2)$ rounds (if an unknown one exists).

Theorem 6.11 (Michail, Chatzigiannakis, and Spirakis 2014) *In any given dynamic graph with* oit *upper bounded by K, take a node u and a time t and denote* $|\text{past}_{(u,t)}(0)|$ *by l. It holds that* $|\{v : (v, 0) \rightsquigarrow (u, t + Kl(l + 1)/2)\}| \geq \min\{l + 1, n\}$.

Proof. Consider a node u and a time t and define $A_u(t) := \text{past}_{(u,t)}(0)$ (we only prove it for the initial states of nodes but easily generalizes to any time), $I_u(t') := \{v \in A_u(t) : A_v(t')\backslash A_u(t) \neq \emptyset\}$, $t' \geq t$, that is $I_u(t')$ contains all nodes in $A_u(t)$ whose t'-states have been influenced by nodes not in $A_u(t)$ (these nodes know new info for u), $B_u(t') := A_u(t)\backslash I_u(t')$, that is all nodes in $A_u(t)$ that do not know new info, and $l := |A_u(t)|$. The only interesting case is for $V\backslash A_u(t) \neq \emptyset$. Since the oit is at most K we have that at most by round $t + Kl$, (u, t) influences some node in $V\backslash B_u(t)$ say via some $u_2 \in B_u(t)$. By that time, u_2 leaves B_u. Next, consider $(u, t + Kl + 1)$. In $K(l - 1)$ steps it must influence some node in $V\backslash B_u$ since now u_2 is not in B_u. Thus, at most by round $t + Kl + K(l - 1)$ another node, say for example, u_3, leaves B_u. In general, it holds that $B_u(t' + K|B_u(t')|) \leq \max\{|B_u(t')| - 1, 0\}$. It is not hard to see that at most by round $j = t + K(\sum_{1\leq i\leq l} i)$, B_u becomes empty, which by definition implies that u has been influenced by the initial state of a new node. In summary, u is influenced by another initial state in at most $K(\sum_{1\leq i\leq l} i) = kl(l+1)/2$ steps. □

The good thing about the upper bound of Theorem 6.11 is that it associates the time for a new incoming influence to arrive at a node only with an upper bound on the oit, which is known, and the number of existing incoming influences which is also known, and thus the bound is locally computable at any time. So, there is a straightforward translation of this bound to a termination criterion and, consequently, to an algorithm for counting and all-to-all dissemination based on it.

Protocol B. All nodes constantly broadcast all ids that they have heard of so far. Each node u keeps a set $A_u(r)$ containing the ids it knows at round r and a termination bound $H_u(r)$ initially equal to K. If, at round r, u hears of new nodes, it inserts them in $A_u(r)$ and sets $H_u(r) \leftarrow r + Kl(l+1)/2$, where $l = |A_u(r)|$. If it ever holds that $r > H_u(r)$, u halts and outputs $|A_u(r)|$ for counting or $A_u(r)$ for all-to-all dissemination.

In the worst-case, u needs $O(Kn)$ rounds to hear from all nodes and then another $Kn(n+1)/2 = O(Kn^2)$ rounds to realize that it has heard from all. So, the time complexity is $O(Kn^2)$.

Note that the upper bound of Theorem 6.11 is loose. The reason is that if a dynamic graph has oit upper bounded by K then in $O(Kn)$ rounds all nodes have causally influenced all other nodes and clearly the iit can be at most $O(Kn)$. We now show that there is indeed a dynamic graph that achieves this worst possible gap between the iit and the oit.

Theorem 6.12 (Michail, Chatzigiannakis, and Spirakis 2014) *There is a dynamic graph with oit k but iit $k(n-3)$.*

Proof. Consider the dynamic graph $G = (V, E)$ s.t. $V = \{u_1, u_2, \dots, u_n\}$ and u_i, for $i \in \{1, n-1\}$, is connected to u_{i+1} via edges labeled jk for $j \in \mathbb{N}_{\geq 1}$, u_i, for $i \in \{2, 3, \dots, n-2\}$, is connected to u_{i+1} via edges labeled jk for $j \in \mathbb{N}_{\geq 2}$. and u_2 is connected to u_i, for $i \in \{3, \dots, n-1\}$ via edges labeled k. In words, at step k, u_1 is only connected to u_2, u_2 is connected to all nodes except from u_n and u_n is connected to u_{n-1}. Then, every multiple of k there is a single linear graph starting from u_1 and ending at u_n. At step k, u_2 is influenced by the initial states of nodes $\{u_3, \dots, u_{n-1}\}$. Then at step $2k$ it forwards these influences to u_1. Since there are no further shortcuts, u_n's state needs $k(n-1)$ steps to arrive at u_1, thus there is an incoming-influence-gap of $k(n-2)$ steps at u_1. To see that oit is indeed k we argue as follows. Node u_1 cannot use the shortcuts, thus by using just the linear graph it influences a new node every k steps. u_2 influences all nodes in addition to u_n at time k and then at time $2k$ it also influences u_n. All other nodes do a shortcut to u_2 at time k and then for all multiples of k their influences propagate to both directions from two sources, themselves and u_2, influencing 1 to 4 new nodes every k steps. □

Next, we show that the $Kl(l+1)/2$ ($l := |\text{past}_{(u,t)}(0)|$) upper bound (of Theorem 6.11), on the time for another incoming influence to arrive, is optimal in the following sense: A node cannot obtain a better upper bound based solely on K and l. We establish this by showing that it is possible that a new incoming influence needs $\Theta(Kl^2)$ rounds to arrive, which excludes the possibility of a $o(Kl^2)$-bound to be correct as a protocol based on it may have nodes terminate without having heard of arbitrarily many other nodes. This, additionally, constitutes a tight example for the bound of Theorem 6.11.

Theorem 6.13 (Michail, Chatzigiannakis, and Spirakis 2014) *For all n, l, K s.t. $n = \Omega(Kl^2)$, there is a dynamic graph with* oit *upper bounded by K and round r such that, a node that has heard of l nodes by round r does not hear of another node for $\Theta(Kl^2)$ rounds.*

Proof. Consider the set $\text{past}_{(u,t)}(0)$ and denote its cardinality by l. Take any dynamic graph on $\text{past}_{(u,t)}(0)$, disconnected from the rest of the nodes, that satisfies oit $\leq K$ and that all nodes in $\text{past}_{(u,t)}(0)$ need $\Theta(Kl)$ rounds to causally influence all other nodes in $\text{past}_{(u,t)}(0)$; this could, for example, be the alternating matchings graph from Section 6.5.1.1 with one matching appearing in rounds that are odd multiples of K and the other in even. In $\Theta(Kl)$ rounds, say in round j, some *intermediary* node $v \in \text{past}_{(u,t)}(0)$ must get the outgoing influences of nodes in $\text{past}_{(u,t)}(0)$ outside $\text{past}_{(u,t)}(0)$ so that they continue to influence new nodes. Assume that in round $j-1$ the adversary directly connects all nodes in $\text{past}_{(u,t)}(0)\backslash\{v\}$ to v. In this way, at time j, v forwards outside $\text{past}_{(u,t)}(0)$ the $(j-2)$-states (and all previous ones) of all nodes in $\text{past}_{(u,t)}(0)$. Provided that $V\backslash\text{past}_{(u,t)}(0)$ is sufficiently big (see later), the adversary can now keep $S = \text{past}_{(u,t)}(0)\backslash\{v\}$ disconnected from the rest of the nodes for another $\Theta(Kl)$ rounds (in fact, one round less this time) without violating oit $\leq K$ as the new influences of the $(j-2)$-states of nodes in S may keep occurring outside S. The same process repeats by a new *intermediary* $v_2 \in S$ playing the role of v this time. Each time the process repeats, in $\Theta(|S|)$ rounds the intermediary gets all outgoing influences outside S and is then removed from S. It is straightforward to observe that a new incoming influence needs $\Theta(Kl^2)$ rounds to arrive at u in such a dynamic network. Moreover, note that $V\backslash\text{past}_{(u,t)}(0)$ should also satisfy oit $\leq K$ but this is easy to achieve by, for example, another alternating matchings dynamic graph on $V\backslash\text{past}_{(u,t)}(0)$ this time. Also, $n - l = |V\backslash\text{past}_{(u,t)}(0)|$ should satisfy $n - l = \Omega(Kl^2) \Rightarrow n = \Omega(Kl^2)$ so that the time needed for a $w \in V\backslash\text{past}_{(u,t)}(0)$ (in an alternating matchings dynamic graph on $V\backslash\text{past}_{(u,t)}(0)$) to influence all nodes in $V\backslash\text{past}_{(u,t)}(0)$ and start influencing nodes in $\text{past}_{(u,t)}(0)$ is asymptotically greater than the time needed for S to extinct. To appreciate this, observe that if $V\backslash\text{past}_{(u,t)}(0)$ was too small, then the outgoing influences of some $w \in V\backslash\text{past}_{(u,t)}(0)$ that occur every K rounds would reach u before the $\Theta(Kl^2)$ bound was achieved. Finally, we note that whenever the number of nodes in $V \backslash S$ becomes odd we keep the previous alternating matchings dynamic graph and the new node becomes connected every K rounds to an arbitrary node (the same in every round). When $|V \backslash S|$ becomes even, again, we return to a standard alternating matchings dynamic graph. □

We now show that even the criterion of Theorem 6.10, that is optimal if an upper bound on the ct is known, does not work in dynamic graphs with a known upper bound K on the oit. In particular, we show that for all times $t' < K(n/4)$, there is a dynamic graph with oit upper bounded by K, a node u, and a time $t \in \mathbb{N}$ s.t. $\text{past}_{(u,t)}(0) = \text{past}_{(u,t)}(t')$ while $\text{past}_{(u,t)}(0) \neq V$. In words, for any such t' it can be the case that while u has not been yet causally influenced by all initial states its past set from time 0 may become equal to its past set from time t', which violates the termination criterion of Theorem 6.10.

Theorem 6.14 (Michail, Chatzigiannakis, and Spirakis 2014) *For all n, K and all times $t' < K(n/4)$ there is a dynamic graph with* oit *upper bounded by K, a node u, and a time $t > t'$ s.t.* $\text{past}_{(u,t)}(0) = \text{past}_{(u,t)}(t')$ *while* $\text{past}_{(u,t)}(0) \neq V$.

Proof. For simplicity, assume that n is a multiple of 4. As in Proposition 6.4 (ii), by an alternating matchings dynamic graph, we can keep two parts V_1, V_2 of the network, of size $n/2$ each, disconnected up to time $K(n/4)$. Let $u \in V_1$. At any time t, s.t. $t' < t \leq K(n/4)$, the adversary directly connects $u \in V_1$ to all $w \in V_1$. Clearly, at that time, u learns the t'-states (and thus also the 0-states) of all nodes in V_1 and, due to the disconnectivity of V_1 and V_2 up to time $K(n/4)$, u hears (and has heard up to then) of no node from V_2. It follows that $\text{past}_{(u,t)}(0) = \text{past}_{(u,t)}(t')$ and $|\text{past}_{(u,t)}(0)| = n/2 \Rightarrow \text{past}_{(u,t)}(0) \neq V$ as required. □

6.5.3.3 Hearing the Future

In contrast to the previous negative results, we now present an optimal protocol for counting and all-to-all dissemination in dynamic networks with a known upper bound K on the oit, that is based on the following termination criterion. By definition of oit, if $\text{future}_{(u,0)}(t) = \text{future}_{(u,0)}(t+K)$ then $\text{future}_{(u,0)}(t) = V$. The reason is that if there exists uninfluenced nodes, then at least one such node must be influenced in at most K rounds, otherwise, no such node exists and $(u, 0)$ must have already influenced all nodes (see also Figure 6.4). So, a fundamental goal is to allow a node to know its future set. Note that this criterion has a very basic difference from all termination criteria that have so far been applied to worst-case dynamic networks: Instead of keeping track of its past set(s) and waiting for new incoming influences, a node now directly keeps track of its future set and is informed by other nodes of its progress. We assume, for simplicity, a unique leader ℓ in the initial configuration of the system (this is not a necessary assumption and we will soon show how it can be dropped).

Protocol *Hear_from_known* .We denote by r the current round. Each node u keeps a list $Infl_u$ in which it keeps track of all nodes that first heard of $(\ell, 0)$ (the initial state of the leader) by u (u was between those nodes that first delivered $(\ell, 0)$ to

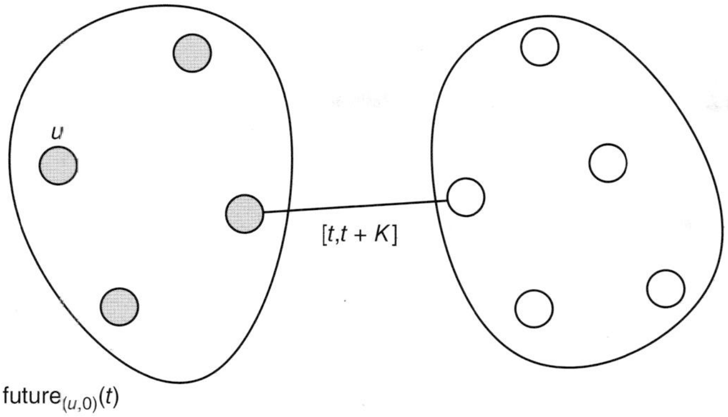

Figure 6.4 If there are still nodes that have not heard from u, then if K is an upper bound on the oit, in at most K rounds another node will hear from u (by definition of the oit).

nodes in $Infl_u$), a set A_u in which it keeps track of the $Infl_v$ sets that it is aware of initially set to $(u, Infl_u, 1)$, and a variable *timestamp* initially set to 1. Each node u broadcast in every round (u, A_u) and if it has heard of $(\ell, 0)$ also broadcasts $(\ell, 0)$. Upon reception of an id w that is not accompanied with $(\ell, 0)$, a node u that has already heard of $(\ell, 0)$ adds (w, r) to $Infl_u$ to recall that at round r it notified w of $(\ell, 0)$ (note that it is possible that other nodes also notify w of $(\ell, 0)$ at the same time without u being aware of them; all these nodes will write (w, r) in their lists). If it ever holds at a node u that $r > \max_{(v \neq u, r') \in Infl_u} \{r'\} + K$, then u adds (u, r) in $Infl_u$ (replacing any existing $(u, t) \in Infl_u$) to denote the fact that r is the maximum known time until which u has performed no further propagations of $(\ell, 0)$. If at some round r a node u modifies its $Infl_u$ set, it sets $timestamp \leftarrow r$. In every round, a node u updates A_u by storing in it the most recent $(v, Infl_v, timestamp)$ triple of each node v that it has heard of so far (its own $(u, Infl_u, timestamp)$ inclusive), where the "most recent" triple of a node v is the one with the greatest *timestamp* between those whose first component is v. Moreover, u clears multiple (w, r) records from the $Infl_v$ lists of A_u. In particular, it keeps (w, r) only in the $Infl_v$ list of the node v with the smallest id between those that share (w, r). Similarly, the leader collects all $(v, Infl_v, timestamp)$ triples in its own A_ℓ set. Let *tmax* denote the maximum timestamp appearing in A_l, that is the maximum time for which the leader knows that some node was influenced by $(\ell, 0)$ at that time. Moreover, denote by I the set of nodes that the leader knows to have been influenced by $(\ell, 0)$. Note that I can be extracted from A_ℓ by $I = \{v \in V : \exists u \in V, \exists timestamp, r \in \mathbb{N} \text{ s.t. } (u, Infl_u, timestamp) \in A_\ell \text{ and } (v, r) \in Infl_u\}$. If at some round r it holds at the leader that for all $u \in I$ there is a $(u, Infl_u, timestamp) \in A_\ell$ s.t. $timestamp \geq tmax + K$ and $\max_{(w \neq u, r') \in Infl_u} \{r'\} \leq tmax$ then the leader outputs $|I|$ or I depending on whether counting or all-to-all dissemination needs to be solved and halts (it can also easily notify the other nodes to do the same in $K \cdot |I|$ rounds by a simple flooding mechanism and then halt).

The above protocol can be easily made to work without the assumption of a unique leader. The idea is to have all the nodes begin as leaders and make all nodes prefer the leader with the smallest id that they have heard of so far. In particular, we can have each node keep an $Infl_{(u,v)}$ only for the smallest v that it has heard of so far. Clearly, in $O(D)$ rounds all nodes will have converged to the node with the smallest id in the network.

Theorem 6.15 (Michail, Chatzigiannakis, and Spirakis 2014) *Protocol Hear_from_known solves counting and all-to-all dissemination in $O(D+K)$ rounds by using messages of size $O(n(\log K + \log n))$, in any dynamic network with dynamic diameter D, and with* oit *upper bounded by some K known to the nodes.*

Proof. In time equal to the dynamic diameter D, all nodes must have heard of ℓ. Then, in another $D + K$ rounds all nodes must have reported to the leader all the direct outgoing influences that they performed up to time D (nodes that first heard of ℓ by that time) together with the fact that they managed to perform no new influences in the interval $[D, D + K]$. Thus, by time $2D + K = O(D + K)$, the leader knows all influences that were ever performed, so no node is missing from its I set, and also knows that all these nodes for K consecutive rounds performed no further influence, thus outputs $|I| = n$ (for counting) or $I = V$ (for all-to-all dissemination) and halts.

Can these termination conditions be satisfied while $|I| < n$, which would result in a wrong decision? Thus, for the sake of contradiction, assume that *tmax* is the time of the latest influence that the leader is aware of, that $|I| < n$, and that all termination conditions are satisfied. The argument is that if the termination conditions are satisfied then (i) $I = \mathtt{future}_{(\ell,0)}(tmax)$, that is the leader knows precisely those nodes that have been influenced by its initial state up to time *tmax*. Clearly, $I \subseteq \mathtt{future}_{(\ell,0)}(tmax)$ as every node in I has been influenced at most by time *tmax*. We now show that additionally $\mathtt{future}_{(\ell,0)}(tmax) \subseteq I$. If $\mathtt{future}_{(\ell,0)}(tmax)\backslash I \neq \emptyset$, then there must exist some $u \in I$ that has influenced a $v \in \mathtt{future}_{(\ell,0)}(tmax)\backslash I$ at most by time *tmax* (this follows by observing that $\ell \in I$ and that all influence paths originate from ℓ). But now, observe that when the termination conditions are satisfied, for each $u \in I$ the leader knows a $timestamp_u \geq tmax + K$, thus the leader knows all influences that u has performed up to time *tmax* and it should be aware of the fact that $v \in \mathtt{future}_{(\ell,0)}(tmax)$, that is, it should hold that $v \in I$, which contradicts the fact that $v \in \mathtt{future}_{(\ell,0)}(tmax)\backslash I$. (ii) The leader knows that in the interval $[tmax, tmax + K]$, no node in $I = \mathtt{future}_{(\ell,0)}(tmax)$ performed a new influence. These results in a contradiction as $|\mathtt{future}_{(\ell,0)}(tmax)| = |I| < n$ and a new influence should have occurred in the interval $[tmax, tmax + K]$ (by the fact that the oit is upper bounded by K).

Optimality follows from the fact that *a node u can know at time t that* $\mathtt{past}_{(u,t)}(0) = V$ *only if* $\mathtt{past}_{(u,t)}(K) = V$. This means that u must have also heard of the K-states of all nodes, which requires $\Theta(K + D)$ rounds in the worst case. If $\mathtt{past}_{(u,t)}(K) \neq V$, then it can be the case that there is some $v \in V\backslash\mathtt{past}_{(u,t)}(K)$ s.t. u has heard v's 0-state but not its K-state. Such a node could be a neighbor of u at round 1 that then moved far away. Again, similarly to Theorem 6.10, we can have arbitrarily many nodes to have no neighbor until time K (e.g., in the extreme case where oit is equal to K) and then from time K onwards are only connected to node v. As u has not heard from the K-state of v, it also cannot have heard of the 0-state of arbitrarily many nodes. □

An interesting improvement is to limit the size of the messages to $O(\log n)$ bits probably by paying some increase in time to termination. We almost achieve this by showing that an improvement of the size of the messages to $O(\log D + \log n)$ bits is possible (note that $O(\log D) = O(\log Kn)$) if we have the leader initiate *individual conversations* with the nodes that it already knows to have been influenced by its initial state. We have already successfully applied a similar technique in Section 6.4. The protocol, that we call *Talk_to_known*, solves counting and all-to-all dissemination in $O(Dn^2 + K)$ rounds by using messages of size $O(\log D + \log n)$, in any dynamic network with dynamic diameter D, and with oit upper bounded by some K known to the nodes.

We now describe the *Talk_to_known* protocol by assuming again for simplicity a unique leader (this, again, is not a necessary assumption).

Protocol *Talk_to_known*. As in *Hear_from_known*, nodes that have been influenced by the initial state of the leader (i.e., $(\ell, 0)$) constantly forward it and whenever a node v manages to deliver it, then it stores the id of the recipient node in its local $Infl_v$ set. Nodes send in each round the time of the latest influence (i.e., the latest new influence of a node by $(\ell, 0)$), call it *tmax*, that they know to have been performed so far. Whenever the leader hears of a greater *tmax* than the one stored in its local memory, it reinitializes the process of collecting its future set. By this, we mean that it waits K rounds and then starts again from the beginning, talking to the nodes that it has influenced itself, then to the nodes that were influenced by these nodes, and so on. The goal is for the leader to collect precisely the same information as in *Hear_from_known*. In particular, it sorts the nodes that it has influenced itself in ascending order of id and starts with the smallest one, call it v, by initiating a $talk(\ell, v, current\_round)$ message. All nodes forward the most recent *talk* message (w.r.t. their timestamp component) that they know so far. Upon receipt of a new $talk(\ell, v, timestamp)$ message (the fact that it is "new" is recognized by the timestamp), v starts sending $Infl_v$ to the leader in packets of size $O(\log n)$, for example a single entry each time, via $talk(v, \ell, current\_round, data\_packet)$ messages. When the leader receives a $talk(v, \ell, timestamp, data\_packet)$ message where $data\_packet = END\_CONV$ (for "END of CONVersation") it knows that it has successfully received the whole $Infl_v$ set and repeats the same process for the next node that it knows to have been already influenced by $(\ell, 0)$ (now also including those that it learned from v). The termination criterion is the same as in *Hear_from_known*.

Theorem 6.16 (Michail, Chatzigiannakis, and Spirakis 2014) *Protocol* Talk_to_known *solves counting and all-to-all dissemination in* $O(Dn^2 + K)$ *rounds by using messages of size* $O(\log D + \log n)$*, in any dynamic network with* *dynamic diameter D, and with* oit *upper bounded by some K known to the nodes.*

Proof. Correctness follows from the correctness of the termination criterion proved in Theorem 6.15. For the bit complexity, we notice that the timestamps and *tmax* are of size $O(\log D)$ (which may be $O(\log Kn)$ in the worst case). The data packet and the id components are all of size $O(\log n)$. For the time complexity, clearly, in $O(D)$ rounds the final outgoing influence of $(\ell, 0)$ will have occurred and thus the maximum *tmax* that will ever appear is obtained by some node. In another $O(D)$ rounds, the leader hears of that *tmax* and thus reinitializes the process of collecting its future set. In that process and in the worst case, the leader must talk to $n - 1$ nodes each believing that it performed $n - 1$ deliveries (this is because in the worst case it can hold that any new node is concurrently influenced by all nodes that were already influenced and in the end all nodes claim that they have influenced all other nodes) thus, in total, it has to wait for $O(n^2)$ data packets each taking $O(D)$ rounds to arrive. The K in the bound is from the fact that the leader waits K rounds after reinitializing in order to allow nodes to also report whether they performed any new assignments in the $[tmax, tmax + K]$ interval. □

6.6 Local Communication Windows

We assume here an underlying *communication network*, which is modeled by an undirected, simple, connected graph $C = (V, A)$, where V is a set of nodes and A is a set of undirected edges. We associate each $u \in V$ with a finite integer $c_u \geq 1$, called u's *neighborhood cover time* (or *cover time* for simplicity), and let $c \in \mathbb{N}^V$ be a vector of cover times indexed by nodes. We denote by $N(u)$ the *neighborhood* of node u in C, that is $N(u) = \{v \in V : \{u, v\} \in A\}$.

A dynamic graph $G = (V, E)$ has now $E : \mathbb{N}_{\geq 1} \to \mathcal{P}(A)$. We say that G *respects* a vector of cover times $c \in \mathbb{N}^V$ if for all $r \in \mathbb{N}$ and all $u \in V$ it holds that $\{v \in V : \{u, v\} \in \bigcup_{i=r}^{r+c_u-1} E(i)\} = N(u)$ (or $\geq$ in case we would allow a node to possibly communicate outside its underlying neighborhood); that is, each node u must cover all its possible neighbors in any c_u-window of the dynamic graph. Note that, again, we do not require the instantaneous graphs to be connected. Nonetheless, it is not hard to see that this definition guarantees that each node may eventually influence every other node in the network. We are interested in protocols that are correct for all possible pairs (G, c), where G is a dynamic graph that respects the vector of cover times c.

First, note that if nodes do not know their cover times nor some upper bound on them, then non-trivial halting computations are impossible. To see this, consider any protocol that terminates in k steps on some dynamic graph G. Now, augment G with some dynamic graph D that has its first communication with G at time $k + 1$ and notice that termination on G occurs at time k without any consideration of D.

We focus on the case in which each node u knows its precise cover time c_u. First of all, notice that for all cover times c there is a dynamic graph that respects c, namely, the static graph in which $E(r) = A$ for all $r \in \mathbb{N}$. However, not all cover times c *admit a worst-case dynamic graph*, that is one in which for all $u \in V$ there is an $r \in \mathbb{N}$ such that $|\{v \in V : \{u, v\} \in \bigcup_{i=r}^{r+c_u-2} E(i)\}| < |N(u)|$. It is not hard to see that a cover-time vector c admits a worst-case dynamic graph G iff $\forall u \in V, \exists v \in N(u)$ such that $c_v \geq c_u$.

An interesting question is whether nodes can verify if a given vector of cover-times admits a worst-case dynamic graph. In fact, we want nodes to accept if all cover-times are consistent and fix inconsistent cover-times otherwise. Let C_u be an upper bound on c_u. Each node u must check whether there is some $v \in N(u)$ such that $C_v \geq C_u$. u broadcasts C_u for C_u rounds. If $C_v < C_u$ for all $v \in N(u)$, then u sets C_u to $\max_{v \in N(u)}\{C_v\}$, otherwise it accepts.

We now deal with the problem of information dissemination and counting and present a protocol for the latter problem.

Let $u = u_0, u_1, \ldots, u_k = v$ be a simple path p joining nodes u, v. The worst case time for u to influence v by messages traversing p is $l(p) = \sum_{i=1}^{k-1} \min\{c_{u_i}, c_{u_{i+1}}\}$ (called *length* or *maximum delay*). Extend l as $l(u, v) = \min_{p \in P(u,v)} l(p)$, where $P(u, v)$ is the set of all simple paths joining u, v. In the dynamic networks under consideration, we have that the dynamic diameter is $D = \max_{u,v \in V} l(u, v)$. It is obvious that if all nodes knew some upper bound $H \geq D$ then each node could halt after H rounds knowing that it has influenced and been influenced by all other nodes. A natural question is whether nodes can achieve this without knowing D in advance. For example, is there a terminating algorithm for *counting* (i.e., for computing n) if nodes only know their exact cover times? In the sequel, we answer this question in the affirmative.

Let

- $\mathrm{psum}_{(u,t')}(t) := \sum_{v \in \mathrm{past}_{(u,t')}(t)} c_v$, and
- $\mathrm{fsum}_{(u,t)}(t') := \sum_{v \in \mathrm{future}_{(u,t)}(t')} c_v$.

Lemma 6.5 *For all times t, t' such that $t \leq t'$, all nodes u, v, and all $k \geq 1$, if $v \in \mathrm{past}_{(u,t')}(t)$ for all E then $v \in \mathrm{past}_{(u,t'+k)}(t + k)$.*

Proof. Take any $v \in \mathrm{past}_{(u,t')}(t)$. To show that $v \in \mathrm{past}_{(u,t'+k)}(t + k)$, we notice that for any dynamic edge function E' there exists E such that $E'(r + k) = E(r)$ for all $t \leq r \leq t'$. □

Lemma 6.6 $\mathrm{past}_{(u,t)}(0) \subseteq \mathrm{past}_{(u,\mathrm{psum}_{(u,t)}(0)-c_v)}(0)$.

Proof. We show that $v \in \mathrm{past}_{(u,t)}(0)$ implies $v \in \mathrm{past}_{(u,\mathrm{psum}_{(u,t)}(0)-c_v)}(0)$. The time-node $(v, 0)$ has influenced (u, t) via a simple path p that only visits nodes from $\mathrm{past}_{(u,t)}(0)$ since $(v, 0) \rightsquigarrow (w, j) \rightsquigarrow (u, t)$ for any intermediate node w implies $w \in \mathrm{past}_{(u,t)}(0)$; to see this note that $(w, 0) \rightsquigarrow (w, j)$ for all $j \geq 0$. Clearly, the longest such path p' is a path that is Hamiltonian in $C[\text{past}_{(u,t)}(0)]$[6] beginning from u and ending at v. Since $l(p) \leq l(p') \leq \text{psum}_{(u,t)}(0) - c_v$ and $(v,0) \rightsquigarrow (u, l(p))$ it also holds that $(v,0) \rightsquigarrow (u, \text{psum}_{(u,t)}(0) - c_v)$ or equivalently $v \in \text{past}_{(u,\text{psum}_{(u,t)}(0)-c_v)}(0)$. □

Lemma 6.7 *For all nodes $u \in V$ and times $t \geq 0$ we have:*

1) $|\text{past}_{(u,\text{psum}_{(u,t)}(0))}(0)| \geq \min\{|\text{past}_{(u,t)}(0)| + 1, n\}$ *and*
2) $|\text{future}_{(u,0)}(\text{fsum}_{(u,0)}(t))| \geq \min\{|\text{future}_{(u,0)}(t)| + 1, n\}$.

Proof. We only prove the first statement since the second is symmetric. The only interesting case is when $|\text{past}_{(u,t)}(0)| < n$ in which case there exists $w \in V \backslash \text{past}_{(u,t)}(0)$. By Lemma 6.6, $\text{past}_{(u,t)}(0) \subseteq \text{past}_{(u,\text{psum}_{(u,t)}(0)-c_v)}(0) \subseteq \text{past}_{(u,\text{psum}_{(u,t)}(0))}(0)$. So we just need to show that there is a $w \in \text{past}_{(u,\text{psum}_{(u,t)}(0))}(0) \backslash \text{past}_{(u,t)}(0)$. Connectivity ensures that there is some $\{w, v\} \in A$, for $w \in V \backslash \text{past}_{(u,t)}(0)$ and $v \in \text{past}_{(u,t)}(0)$. Clearly $(w,0) \rightsquigarrow (v, c_v)$. Since $(v,0) \rightsquigarrow (u, \text{psum}_{(u,t)}(0) - c_v)$, by Lemma 6.5 $(v, c_v) \rightsquigarrow (u, \text{psum}_{(u,t)}(0))$. Transitivity ensures that $(w,0) \rightsquigarrow (u, \text{psum}_{(u,t)}(0))$ and $w \in \text{past}_{(u,\text{psum}_{(u,t)}(0))}(0)$. □

Lemma 6.7 provides us with the following criterion for a node to detect when it has been causally influenced by all other nodes: $|\text{past}_{(u,\text{psum}_{(u,t)}(0))}(0)| = |\text{past}_{(u,t)}(0)| \Rightarrow |\text{past}_{(u,t)}(0)| = V$. That is, at any time t, any new influence of the state of u by some initial state must occur at most by time $\text{psum}_{(u,t)}(0)$. If this time elapses without any new influence, then u knows that it has been causally influenced by all other nodes. An easier to perform but equivalent test is the following: $t = \text{psum}_{(u,t)}(0) \Rightarrow |\text{past}_{(u,\text{psum}_{(u,t)}(0))}(0)| = |\text{past}_{(u,t)}(0)| \Rightarrow |\text{past}_{(u,t)}(0)| = V$. In the following proposition, we use the latter criterion to solve counting.

But, first, define an edge weight $w(e)$ for each edge $e \in A$ as $w(e) := minc_u, c_v$. We are then guaranteed that an edge e appears at least once in every time interval of length $w(e)$. This implies that within time $W := \sum_{e \in D(C)} w(e)$, where $D(C)$ is a diameter of C (that is within time equal to the weighted diameter of C), everyone hears from everyone else and then another $\sum_{u \in V} c_u - W$ rounds are needed for the nodes to know that they are done.

Proposition 6.6 *Counting can be solved in $O(\sum_{u \in V} c_u)$ rounds using messages of size $O(n \log n + \sum_{u \in V} c_u)$.*

Proof. Each node u maintains a set of unique ids A_u, where initially $A_u(0) = \{u\}$, and a vector $c_u[]$ of cover times indexed by unique ids in A_u, where initially $c_u = (c_u)$. In each round r, u sends (A_u, c_u) to all its current neighbors, stores in A_u all received unique ids and, for each new unique id $v \notin A_u(r-1)$, u stores t_v in c_u. Moreover, nodes keep track of the round number. At the end of each round r, if $r = \sum_{v \in A_u(r)} c_u[v]$ node u halts and outputs $|A_u|$; otherwise, u continues on to the next round □

6) We denote by $G[V']$ the subgraph of a graph G induced by nodes in V'.

6.7 Conclusions

In this chapter, we discussed several recently introduced models and problems regarding computational network analysis, which we treated from a theoretical point of view. In Section 6.4, we studied the fundamental naming and counting problems (and some variations) in networks that are anonymous, unknown, and possibly dynamic. Network dynamicity was modeled by the 1-interval connectivity model (Kuhn, Lynch, and Oshman, 2010), in which communication is synchronous and a (worst-case) adversary chooses the edges of every round subject to the condition that each instance is connected. We first focused on static networks with broadcast where we proved that, without a leader, counting is impossible to solve and that naming is impossible to solve even with a leader and even if nodes know n. These impossibilities carry over to dynamic networks as well. We also showed that a unique leader suffices in order to solve counting in linear time. Then, we focused on dynamic networks with broadcast. We conjectured that dynamicity renders nontrivial computation impossible. In view of this, we allowed the nodes know an upper bound on the maximum degree that will ever appear and showed that in this case the nodes can obtain an upper bound on n. Finally, we replaced broadcast with *one-to-each*, in which a node may send a different message to each of its neighbors. Interestingly, this natural variation was proved to be computationally equivalent to a full-knowledge model, in which unique names exist and the size of the network is known.

Then, in Section 6.5, we discussed the model of Michail, Chatzigiannakis, and Spirakis (2014), which was the first in the literature to consider worst-case dynamic networks that are free of any connectivity assumption about their instances. To enable a quantitative study, we introduced some novel generic metrics that capture the speed of information propagation in a dynamic network. We proved that fast dissemination and computation are possible even under continuous disconnectivity. In particular, we presented optimal termination conditions and protocols based on them for the fundamental counting and all-to-all token dissemination problems.

There are many open problems and promising research directions related to the above findings. We would like to achieve satisfactory lower and upper bounds for counting and information dissemination. Techniques from Haeupler *et al.* (2012) or related ones may be applicable to achieve quick token dissemination. It would be also important to refine the metrics proposed in this section so that they become more informative. For example, the oit metric, in its present form, just counts the time needed for another outgoing influence to occur. It would be useful to define a metric that counts the number of new nodes that become influenced per round, which would be more informative, w.r.t. the speed of information spreading. Note that in our work (and all previous work on the subject), information dissemination is only guaranteed under continuous broadcasting. How can the number of redundant transmissions be reduced in order to improve communication efficiency? Is there a way to exploit visibility to this end? Does predictability help (i.e., some knowledge of the future)?

References

Albers, S. and Henzinger, M. (2000) Exploring unknown environments. *SIAM Journal on Computing*, **29** (4), 1164–1188.

Angluin, D. (1980) Local and global properties in networks of processors (extended abstract), *Proceedings of the 12th Annual ACM Symposium on Theory of Computing (STOC)*, ACM, New York, pp. 82–93.

Angluin, D., Aspnes, J., Diamadi, Z., Fischer, M.J., and Peralta, R. (2006) Computation in networks of passively mobile finite-state sensors. *Distributed Computing*, **18** (4), 235–253.

Angluin, D., Aspnes, J., Eisenstat, D., and Ruppert, E. (2007) The computational power of population protocols. *Distributed Computing*, **20** (4), 279–304.

Aspnes, J., Fich, F.E., and Ruppert, E. (2006) Relationships between broadcast and shared memory in reliable anonymous distributed systems. *Distributed Computing*, **18** (3), 209–219.

Aspnes, J. and Ruppert, E. (2007) An introduction to population protocols. *Bulletin of the European Association for Theoretical Computer Science*, **93**, 98–117.

Attiya, H., Snir, M., and Warmuth, M.K. (1988) Computing on an anonymous ring. *Journal of the ACM*, **35** (4), 845–875.

Attiya, H. and Welch, J. (2004) *Distributed Computing: Fundamentals, Simulations, and Advanced Topics*, vol. 19, Wiley-Interscience.

Augustine, J., Pandurangan, G., Robinson, P., and Upfal, E. (2012) Towards robust and efficient computation in dynamic peer-to-peer networks. *Proceedings of the Twenty-Third Annual ACM-SIAM Symposium on Discrete Algorithms (SODA)*, SIAM, pp. 551–569.

Avin, C., Koucký, M., and Lotker, Z. (2008) How to explore a fast-changing world (cover time of a simple random walk on evolving graphs). *Proceedings of the 35th international colloquium on Automata, Languages and Programming (ICALP), Part I*, Springer-Verlag, Berlin, Heidelberg, pp. 121–132.

Awerbuch, B., Goldreich, O., Vainish, R., and Peleg, D. (1990) A trade-off between information and communication in broadcast protocols. *Journal of the ACM*, **37** (2), 238–256.

Baumann, H., Crescenzi, P., and Fraigniaud, P. (2009) Parsimonious flooding in dynamic graphs, *Proceedings of the 28th ACM Symposium on Principles of Distributed Computing (PODC)*, ACM, pp. 260–269.

Berman, K.A. (1996) Vulnerability of scheduled networks and a generalization of Menger's theorem. *Networks*, **28** (3), 125–134.

Boldi, P. and Vigna, S. (1999) Computing anonymously with arbitrary knowledge. *Proceedings of the 18th Annual ACM Symposium on Principles of Distributed Computing (PODC)*, ACM, pp. 181–188.

Bollobás, B. (1998) *Modern Graph Theory*, Springer, corrected edition.

Casteigts, A., Flocchini, P., Quattrociocchi, W., and Santoro, N. (2012) Time-varying graphs and dynamic networks. *International Journal of Parallel, Emergent and Distributed Systems*, **27** (5), 387–408.

Chalopin, J., Métivier, Y., and Morsellino, T. (2012) On snapshots and stable properties detection in anonymous fully distributed systems (extended abstract), *Structural Information and Communication Complexity*, LNCS, vol. 7355, Springer, pp. 207–218.

Chatzigiannakis, I., Michail, O., Nikolaou, S., Pavlogiannis, A., and Spirakis, P.G. (2011) Passively mobile communicating machines that use restricted space. *Theoretical Computer Science*, **412** (46), 6469–6483.

Clementi, A.E., Macci, C., Monti, A., Pasquale, F., and Silvestri, R. (2008) Flooding time in edge-Markovian dynamic graphs. *Proceedings of the 27th ACM Symposium on Principles of Distributed

Computing (PODC)*, ACM, New York, pp. 213–222.

Dolev, S. (2000) *Self-Stabilization*, MIT Press, Cambridge, MA.

Doty, D. (2014) Timing in chemical reaction networks. *Proceedings of the 25th Annual ACM-SIAM Symposium on Discrete Algorithms (SODA)*, pp. 772–784.

Deng, X. and Papadimitriou, C. (1990) Exploring an unknown graph. *31st Annual Symposium on Foundations of Computer Science (FOCS)*, IEEE, pp. 355–361.

Dutta, C., Pandurangan, G., Rajaraman, R., Sun, Z., and Viola, E. (2013) On the complexity of information spreading in dynamic networks. *Proceedings of the 24th Annual ACM-SIAM Symposium on Discrete Algorithms (SODA)*.

Fraigniaud, P., Pelc, A., Peleg, D., and Pérennes, S. (2000) Assigning labels in unknown anonymous networks (extended abstract). *Proceedings of the 19th Annual ACM Symposium on Principles of Distributed Computing, PODC '00*, ACM, pp. 101–111.

Haeupler, B. (2011) Analyzing network coding gossip made easy. *Proceedings of the 43rd Annual ACM Symposium on Theory of Computing (STOC)*, ACM, pp. 293–302.

Haeupler, B., Cohen, A., Avin, C., and Médard, M. (2012) Network coded gossip with correlated data. *CoRR*, abs/1202.1801.

Holme, P. and Saramäki, J. (2012) Temporal networks. *Physics Reports*, **519** (3), 97–125.

Kempe, D., Kleinberg, J., and Kumar, A. (2000) Connectivity and inference problems for temporal networks. *Proceedings of the 32nd Annual ACM Symposium on Theory of Computing (STOC)*, pp. 504–513.

Kuhn, F., Lynch, N., and Oshman, R. (2010) Distributed computation in dynamic networks. *Proceedings of the 42nd ACM Symposium on Theory of Computing, STOC '10*, ACM, pp. 513–522.

Kuhn, F., Moses, Y., and Oshman, R. (2011) Coordinated consensus in dynamic networks. *Proceedings of the 30th Annual ACM SIGACT-SIGOPS Symposium on Principles of Distributed Computing (PODC)*, pp. 1–10.

Kuhn, F. and Oshman, R. (2011) Dynamic networks: models and algorithms. *SIGACT News*, **42**, 82–96. Distributed Computing Column, Editor: Idit Keidar.

Lamport, L. (1978) Time, clocks, and the ordering of events in a distributed system. *Communications of the ACM*, **21** (7), 558–565.

Lynch, N.A. (1996) *Distributed Algorithms*, 1st edn, Morgan Kaufmann.

Mertzios, G.B., Michail, O., Chatzigiannakis, I., and Spirakis, P.G. (2013) Temporal network optimization subject to connectivity constraints. *40th International Colloquium on Automata, Languages and Programming (ICALP)*, Lecture Notes in Computer Science, vol. 7966, Springer, pp. 663–674.

Michail, O., Chatzigiannakis, I., and Spirakis, P.G. (2011a) Mediated population protocols. *Theoretical Computer Science*, **412** (22), 2434–2450.

Michail, O., Chatzigiannakis, I., and Spirakis, P.G. (2011b) in *New Models for Population Protocols*, Synthesis Lectures on Distributed Computing Theory (ed. N.A. Lynch), Morgan & Claypool.

Michail, O., Chatzigiannakis, I., and Spirakis, P.G. (2012) Brief announcement: naming and counting in anonymous unknown dynamic networks. *Proceedings of the 26th International Conference on Distributed Computing (DISC)*, Springer-Verlag, Berlin, Heidelberg, pp. 437–438.

Michail, O., Chatzigiannakis, I., and Spirakis, P.G. (2013) Naming and counting in anonymous unknown dynamic networks. *15th International Symposium on Stabilization, Safety, and Security of Distributed Systems (SSS)*, Springer, pp. 281–295.

Michail, O., Chatzigiannakis, I., and Spirakis, P.G. (2014) Causality, influence, and computation in possibly disconnected synchronous dynamic networks. *Journal of Parallel and Distributed Computing*, **74** (1), 2016–2026.

Michail, O. and Spirakis, P.G. (2014) Simple and Efficient Local Codes for Distributed Stable Network Construction.

Proceedings of the 33rd ACM Symposium on Principles of Distributed Computing (PODC), Paris, France, ACM, vol. 10, pp. 76–85, 978-1-4503-2944-6, http://doi.acm.org/10.1145/2611462.2611466 (accessed 13 February 2015).

O'Dell, R. and Wattenhofer, R. (2005) Information dissemination in highly dynamic graphs. *Proceedings of the 2005 Joint Workshop on Foundations of Mobile Computing (DIALM-POMC)*, pp. 104–110.

Orlin, J.B. (1981) The complexity of dynamic languages and dynamic optimization problems. *Proceedings of the 13th Annual ACM Symposium on Theory of Computing (STOC)*, ACM, pp. 218–227.

Scheideler, C. (2002) Models and techniques for communication in dynamic networks. *Proceedings of the 19th Annual Symposium on Theoretical Aspects of Computer Science (STACS)*, pp. 27–49.

Soifer, A. (2009) *The Mathematical Coloring Book: Mathematics of Coloring and the Colorful Life of its Creators*, 1st edn, Springer.

Yamashita, M. and Kameda, T. (1996) Computing on anonymous networks. I. Characterizing the solvable cases. *Parallel IEEE Transactions on and Distributed Systems*, **7** (1), 69–89.

7
Visualization and Interactive Analysis for Complex Networks by means of Lossless Network Compression

Matthias Reimann, Loïc Royer, Simone Daminelli, and Michael Schroeder

7.1
Introduction

Complex networks play a crucial role in many fields like biology, ecology, social sciences, communication, economics, or software engineering. Due to their complexity, classic graph, and network theory problems have been handled through the development and implementation of algorithms by computer scientists. Thus, in the last decades a plethora of computational methods have been applied to analyze properties or mine patterns in such networks.

Still, one of the open problems in computational network theory is network visualization. In this chapter, we will tackle the problem of visualize, and therefore analyze complex networks proposing a lossless transformation of graphs, through the power graph algorithm.

Large and dense networks are especially hard to layout and analyze since the amount of edges and the unavoidable amount of crossings between them bring existing methods to their limits. The network in Figure 7.1a with only 279 nodes and 4349 edges appears as black lump hiding almost all details. Even in less dense networks like the network shown in Figure 7.1b, where clusters and hubs can be detected easily, details in highly connected regions are hidden.

An experimental analysis by Purchase (1997) suggests that edge crossing minimization is the most important task for the graph visualization (Battista *et al.*, 1999). An approach that tries to remove edge crossings in graph drawings by visualizing non-planar graphs in a planar way is called confluent drawing (Dickerson *et al.*, 2005) that groups edges together. This method has been developed for graphs of low average degree like software interaction diagrams. For complex networks, the reduction of edge crossings is a more challenging endeavor.

Here, we propose a computational method named power graphs, which provide a solution by minimizing the number of edges in the network. Power graph analysis has been shown to be a powerful tool in the biological domain where, in various complex networks, the number of edges was reduced by up to 85 and 56% on average (Royer *et al.*, 2008, 2012; Daminelli *et al.*, 2012).

Computational Network Theory: Theoretical Foundations and Applications, First Edition.
Edited by Matthias Dehmer, Frank Emmert-Streib, and Stefan Pickl.

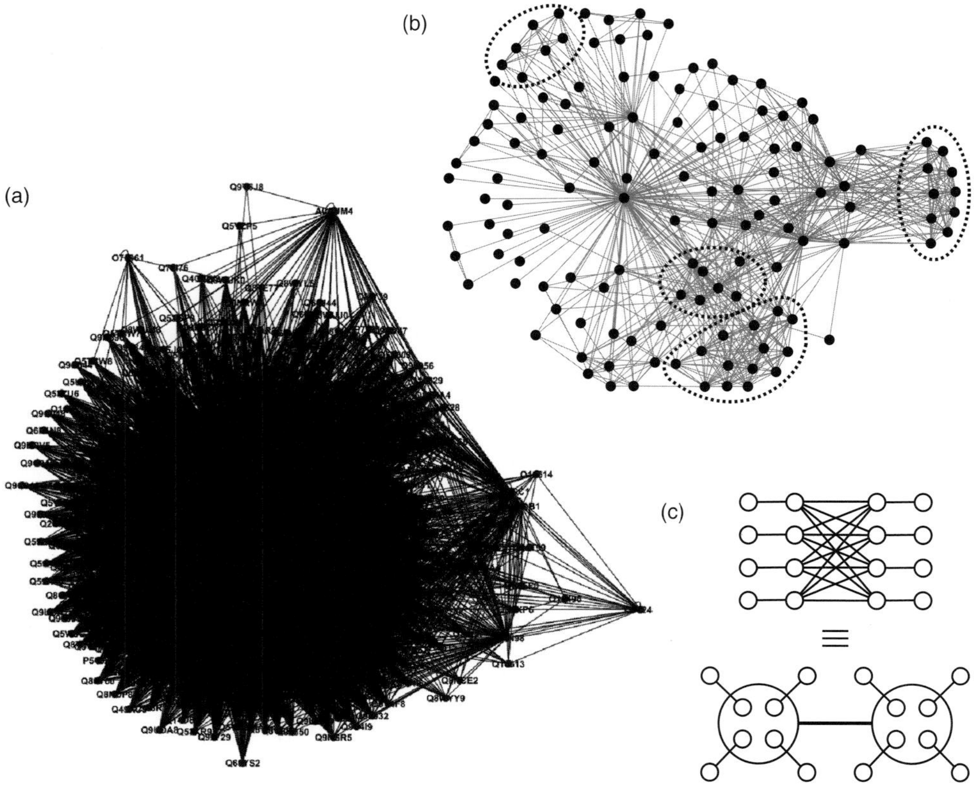

Figure 7.1 Difficulties of current graph drawing approaches. (a) Network with 279 nodes and 4849 edges appears as black lump. (b) Co-occurrence network of the New Testament. Details in the appearing clusters (encircled) are difficult to see. (c) Small graph with a biclique that would not be detected with modular decomposition, together with an equivalent power graph representation.

The last two decades, graph drawing methods that detect and use motifs within networks received increasing attention in the literature (Battista *et al.*, 1999). One approach in particular, modular decomposition, has been heavily employed to improve graph layouts (Gallai, 1967; Gagneur *et al.*, 2004; Papadopoulos and Voglis, 2006). Modular decomposition identifies modules as sets of nodes having exactly the same neighbors and builds a tree representation of modules. However, modular decomposition only finds modules with strictly identical neighborhoods. The small graph in Figure 7.1c contains a biclique motif which would not be found because no two nodes have an identical neighborhood.

Despite the broad effort invested in the improvement of graph drawing algorithms (Koren, 2004; Frishman and Tal, 2008; Holten and van Wijk, 2009; Shi *et al.*, 2009), the success is marginal (Chen, 2005). The emerging field of visual analytics investigates alternative ways to handle these massive networks by integrating human judgment by means of the visual interaction techniques in the analysis process (Keim *et al.*, 2008).

On the one hand, we show that power graphs are a more powerful approach for graph visualization. On the other hand, power graphs cannot directly completely overcome the major obstacle in graph drawing – edge clutter, but open up novel possibilities for interactive visual analytics.

7.1.1 Illustrative Example

Power graphs can be applied to a wide range of graphs. While we focused on protein-protein interaction networks in Royer *et al.* (2008), drug–target–disease networks in Daminelli *et al.* (2012), in the present work, we examine the wider applicability of power graph analysis.

Figure 7.2a shows a power graph representation of the co-occurrence network of the *New Testament* – the example from Figure 7.1b (Figure 7.2c shows the graph again, Figure 7.2d shows the Power graph semantics). Jesus is the hub and connected to 90% of all other characters and therefore, part of a star motif that covers almost the whole power graph. For this hub, it might be interesting to know what nodes it does not connect to. While this is difficult in the graph, in the power graph those nodes are clearly identifiable outside the power nodes connected to Jesus. Not appearing with Jesus are Old Testament figures such as Zerubbabel. The magnification in Figure 7.2b shows, that the 12 apostles can be found in a clique of 14, where the two outer ones are the father of Judas and the son of James. They are organized in an onion-ring fashion, with the most important apostles at the center: Peter, Andrew, Philip, John, and James. While we can check manually the relevance of the indicated structure for networks for which the underlying domain is well known, we need a quality measure for assessing power graphs of less understood networks.

7.2 Power Graph Algorithm

7.2.1 Formal Definition of Power Graphs

Given a graph $G = (V, E)$ where V is the set of nodes or vertices and $E \subseteq V \times V$ is the set of edges that are unordered pairs of distinct nodes, a *power graph* $G' = (V', E')$ is a graph defined on the power set of nodes $V' \subseteq \mathcal{P}(V)$ whose elements – *power nodes* – are connected to each other by *power edges*: $E' \subseteq V' \times V'$. The two power nodes of a power edge must be disjoint or identical: $\forall_{(u,v)\in E'}\ u \cap v = \emptyset \vee u = v$. A power edge is a set of edges. Hence, power graphs are defined on the power set of nodes $V' \subseteq \mathcal{P}(V)$ and the power set of edges $E' \subseteq \mathcal{P}(E)$. The set V of G is the union of all power nodes: $V = \bigcup_{v' \in V'} v'$. The set V' of power nodes is required to be minimal – each power node in V' must participate in at least one power edge in E' or be a singleton set.

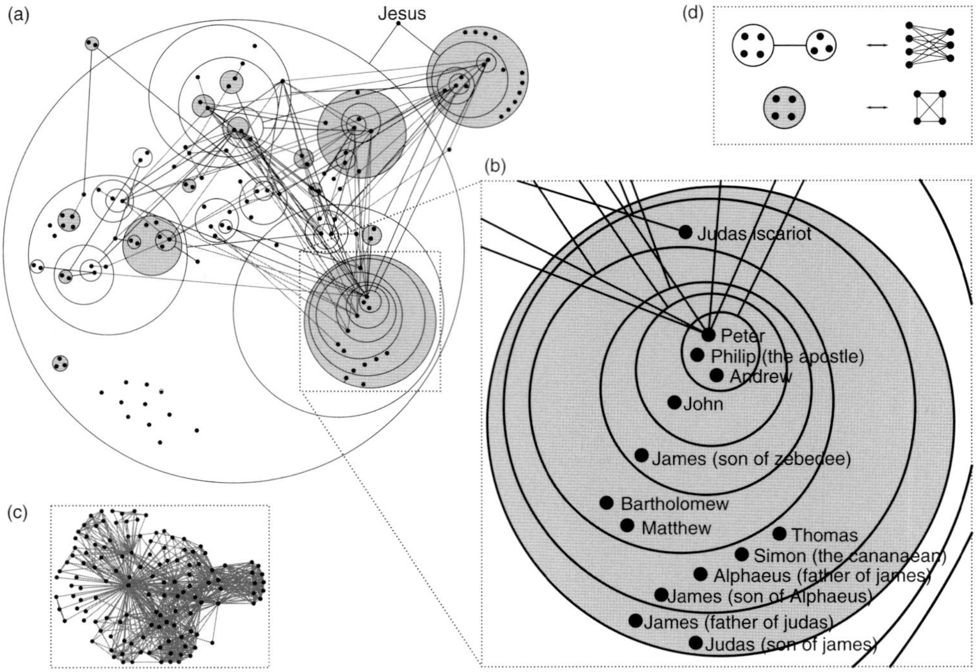

Figure 7.2 Co-occurrence network of the *New Testament* (characters that appear together in verses are connected). The network data comes from the Stanford GraphBase (Knuth, 1993). (a) Power graph, Jesus appears as central hub, nodes that are not connected to Jesus can be identified easily. (b) The 12 apostles can be found in an onion-like structured clique of 14. (c) Underlying network. (d) Legend. Cliques are colored grey for readability (instead of drawing the reflexive power edges as loops).

7.2.2
Semantics of Power Graphs

Figure 7.3a shows motifs that appear in power graphs. If two power nodes are connected by a power edge in G', this signifies that in G, all nodes of the first power node are connected to all nodes of the second power node, thus, the two sets form a complete connected bipartite subgraph. It does not imply that the nodes inside each power node are connected among each other. A special case of biclique is the star where one of the two power nodes is a singleton node. If a power node in G' is connected to itself by a reflexive power edge, this means that all nodes in the power node are connected to each other by edges in G, thus, the set is a complete connected subgraph.

7.2.3
Power Graph Conditions

Since we want to draw power graphs in the plane, two conditions are required:

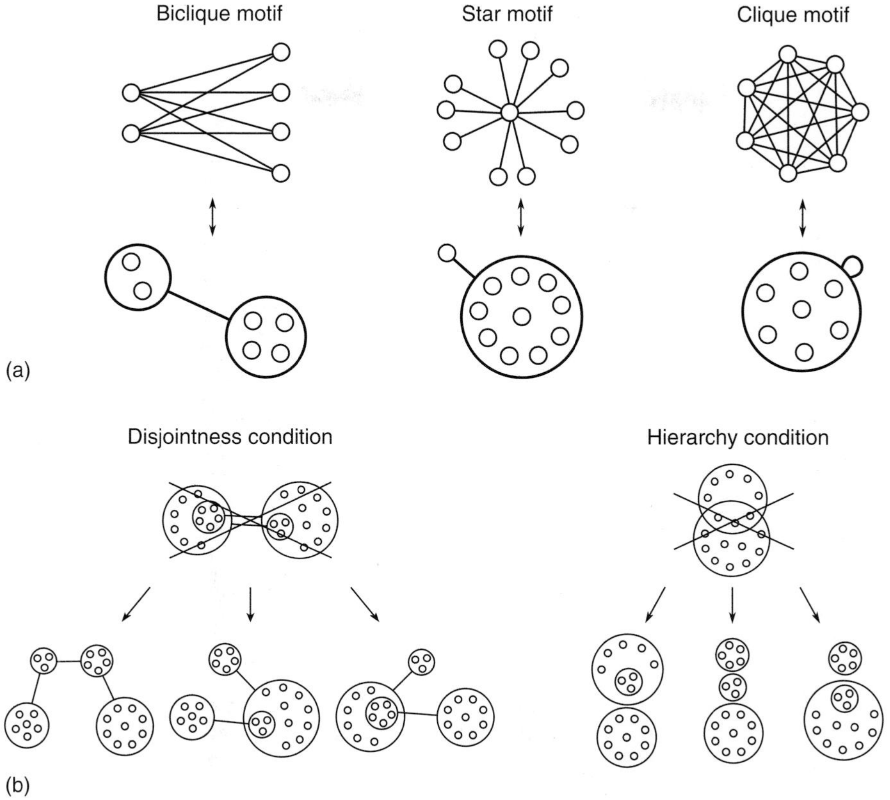

Figure 7.3 (a) Power graph semantics: biclique, star, and clique motifs. (b) Power graph conditions and their equivalent decompositions.

- *Disjointness Condition of Power Edges.* Each edge of the original graph G is represented by one and only one power edge in G'.
- *Hierarchy Condition of Power Nodes.* Any two power nodes in G' are either disjoint or one is included in the other.

Figure 7.3b shows the possible decompositions to fulfill these conditions. Relaxing the previous two conditions leads to abstract power graphs that are difficult to visualize.

7.2.4
Edge Reduction and Relative Edge Reduction

For a given graph G, different power graphs exist. Figure 7.4 shows for a small network of four nodes and five edges, several possible power graphs. For this particular network, the power graph in Figure 7.4k is preferable because it preserves the symmetries in the graph and has only two power edges.

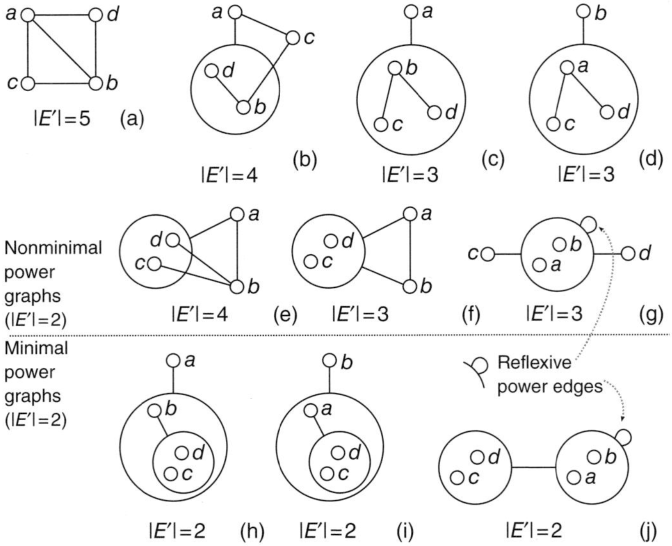

Figure 7.4 A graph can be transformed in different power graph representations, power graphs (b-k) are a selection for graph (a) (which is a power graph representation by itself). (h–k) are minimal, no equivalent power graph with fewer power edges exists.

For more complex networks, it is harder to decide which power graph representation to choose. We discussed in the introduction that the biggest improvement of a graph's visualization can be gained by reducing edge crossings in the drawing. The amount of edge crossings can be reduced by minimizing the number of edges, hence, we decided to search for power graphs with the least power edges – the *minimal power graphs*. From a combinatorial point of view, the problem of finding a minimal power graph for a graph G amounts to finding a minimum partition of the set of edges of G into disjoint cliques and bicliques. The complexity of similar problems – finding the minimal partition of a graph into cliques and finding the minimal biclique partition – are known to be NP-hard (Duh and Fürer, 1997) and NP-complete (Kratzke, Reznick, and West, 1988), respectively. Unfortunately, the complexity of the power graph algorithm cannot be estimated by these results. For the purpose of analyzing complex networks, we assume that finding near-minimal power graphs is sufficient.

To assess the quality of a power graph, we use two measures. First, the *edge reduction* $\mathcal{R} = (|E| - |E'|)/|E|$, the amount by which the number of edges in G' is smaller than in G, relative to the number of edges in G. The edge reduction is nothing else than the compression rate achieved by the power graph transformation. It assesses the improvement of a power graph representation over the plain graph, without considering the meaning of the indicated structures. Second, to show that the structures (cliques and bicliques) revealed by power graphs distinguish networks significantly from random and are meaningful, we introduce the notion of *relative edge reduction*.

The *relative edge reduction* $\mathcal{R}_{rel}$ measures an original network's edge reduction in relation to an average random network of same topology. To compute the relative edge reduction, we generate 1000 random networks following the null model (see later) and compute the average edge reduction. The relative edge reduction measures by how much the original network's edge reduction differs from the average random edge reduction: $\mathcal{R}_{rel} = \mathcal{R} - \overline{\mathcal{R}_{random}}$, where $\overline{\mathcal{R}_{random}}$ is the mean of the edge reductions for the random networks. For example, a relative edge reduction of 0.1 means that the edge reduction is 0.1–10% points – higher than the average edge reduction of equivalent random networks. The relative edge reduction is a more relevant measure than the edge reduction because a certain level of compressibility is always expected, even from random networks.

Network null model – degree preserving random rewiring: For a given network, we generate a large population (1000) of randomly rewired networks. These random networks have the same number of nodes and edges, as well as the same degree distribution as the original network. These networks are generated by randomly re-wiring the original network (Maslov and Sneppen, 2002). Two randomly chosen edges A–B and C–D are replaced by two new edges A–C and B–D. This preserves the number of edges per node. This operation is repeated a number of times which is a multiple of the number of edges in the network, thus, ensuring that almost all edges are rewired at least once. Moreover, each random network is generated from a previously rewired network and, thus, correlation with the original network is unlikely.

7.2.5
Power Graph Extraction

Our improved algorithm for computing near-minimal power graph representations from graphs supports weighted graphs and a minimum similarity threshold. The algorithm consists of a first phase that collects candidate power nodes and a second phase that uses these to search for power edges. Table 7.1 summarizes the algorithm in pseudocode.

In the first phase (Table 7.1, ll. 5–14), the candidate power nodes are identified with hierarchical clustering (Eisen *et al.*, 1998) based on neighborhood similarity. A candidate power node is a set of nodes that have neighbors in common. The similarity of two neighborhoods is a generalized Jaccard Index (Rasmussen, 1992) on weighted sets:

$$s(N(u), N(v)) = \frac{\sum_{x \in N(u) \cap N(v), x \neq u, x \neq v} \min(w_{xu}, w_{xv}) + \alpha}{\sum_{x \in N(u) \cup N(v), x \neq u, x \neq v} \max(w_{xu}, w_{xv}) + \alpha}$$

where $N(u)$ is the neighborhood of cluster u, w_{xu} the weight of x in the weighted neighborhood of u, and α the clique contribution to the similarity:

$$\alpha = \begin{cases} \frac{1}{2}(w_{uv} + w_{vu} + w_{uu} + w_{vv}) & \text{if } \min(w_{uv}, w_{vu}, w_{uu}, w_{vv}) > 0 \\ 0 & \text{otherwise} \end{cases}.$$

Table 7.1 Pseudocode for the power graph algorithm.

1:	**Input**: A graph $G = \{V, E\}$, Minimum similarity s_{min},
2:	Weight w_{uv} for each edge (u, v) in E
3:	**Output**: A power graph $G' = \{V', E'\}$
4:	**Algorithm**:
5:	├ Initialize C and C' to empty sets, and M to an empty numeric matrix
6:	├ Add for each node v in V the singleton cluster $\{v\}$ to C and to C'
7:	├ Calculate for each pair (U, W) of clusters in C its neighborhood similarity $s(U, W)$ using the edge weights and put it in the matrix M
8:	├ **While** $\lvert C' \rvert > 1$ and $s_{max}(M) \geq s_{min}$:
9:	│ ├ Find one pair of clusters (U, W) with the maximal similarity s_{max} in M
10:	│ ├ Remove the two clusters U and W from C'
11:	│ ├ Add the union of the two clusters $U_{new} = U \cup W$ to C and C'
12:	│ └ Update neighborhood similarity matrix M: First, remove columns and rows of U and W, second, calculate and add column and row for U_{new}
13:	├ For each cluster U in C add neighborhood $N(U)$ to C if $s(N(U)) > s_{min}$
14:	├ Again, for each cluster U in C neighborhood $N(U)$ to C ifs $(N(U)) > s_{min}$
15:	├ Initialize V' and E' to empty sets, and L to an empty list
16:	├ Add for each node v in V a singleton set $\{v\}$ to V'
17:	├ **For** all unordered pairs (U, W) of node sets U and W in C:
18:	│ ├ **If** $U \cap W = \varnothing$ and $(U \cup W, U \times W)$ isasub-graph in G:
19:	│ │ ├ Add the power edge (U, W) to the list L
20:	│ │ └ Compute for (U, W) its *size* $\sum_{u \in U, v \in V} w_{uv}$
21:	│ └ **If** $U = W$ and the U-induced graph in G isa clique:
22:	│ ├ Add the power edge (U, U) to the list L
23:	│ └ Compute for (U, U) its *size*: $\frac{1}{2} \sum_{u \in U, v \in U, u \neq v} w_{uv}$
24:	├ **While** list L is not empty:
25:	│ ├ Sort the list L in descending order of power edge sizes
26:	│ ├ Remove the first candidate power edge (U, W) from list L
27:	│ ├ **If** the size of power edge (U, W) is two and $U = W$ then do nothing
28:	│ ├ **Else if** there is a S in V' such that: $U \cap S \neq \varnothing$ but $U \not\subset S$ and $S \not\subset U$:
29:	│ │ └ Add to L the candidate power edges $(U \setminus S, W)$ and $(U \cap S, W)$
30:	│ ├ **Else if** there is a S in V' such that: $W \cap S \neq \varnothing$ but $W \not\subset S$ and $S \not\subset W$:
31:	│ │ └ Add to L the candidate power edges $(U, W \setminus S)$ and $(U, W \cap S)$
32:	│ ├ **Else if** there is a (S, T) in E such that: $(U \times W) \cap (S \times T) \neq \varnothing$:
33:	│ │ └ **If** (S, T) covers not all edges of (U, W): $((U \times W) \not\subset (S \times T))$:
34:	│ │ ├ **If** $U \subset S$: Add to L the candidate power edge $(U, W \setminus T)$
35:	│ │ ├ **Else if** $U \subset T$: Add to L the candidate power edge $(U, W \setminus S)$
36:	│ │ ├ **Else if** $W \subset S$: Add to L the candidate power edge $(U \setminus T, W)$
37:	│ │ └ **Else if** $W \subset T$: Add to L the candidate power edge $(U \setminus S, W)$
38:	│ ├ **Else if** (U, W) isa clique $(U = W)$:
39:	│ │ └ Add power node U to V' **and** power edge (U, U) to E'
40:	│ └ **Else**: Add power nodes U and W to V' **and** power edge (U, W) to E'
41:	└ **For** each edge (u, v) in E not yet covered by any power edgein E'
42:	└ Add the singleton power edge $(\{u\}, \{v\})$to E'

The input is a graph $G = \{V, E\}$ and the output a power graph $G' = \{V', E'\}$. The first step (ll. 5–14) is the search for candidate power nodes. Hierarchical clustering on the set of nodes V is done using neighborhood similarity on node clusters. After line 14, the collection C contains these clusters, as well as for each cluster U its neighbors set and second-order neighbors set (added at ll. 13 and 14). The second step (ll. 15–42) is the greedy search for power edges. All cliques and bicliques induced by node sets in C are enumerated and their size is calculated (ll. 17–23). Power edges are then incrementally decomposed and eventually added to the power graph until all edges from G are covered by one and only one power edge.

The weight of a neighbor n in the weighted neighborhood of cluster c is the average over all nodes in cluster c: $w_{nc} = \sum_{x \in c} w_{nx}$. For the identification of stars and other highly asymmetric bicliques, we add for each node v two sets to the candidate power nodes: its neighborhood set $N(v)$ and the set of common neighbors of the nodes in $N(v)$, $\bigcap_{v' \in N(v)} N(v')$), that contains at least v. Each of those clusters u is only added if its accumulated neighborhood similarity is above the given threshold: $s(N(u)) > s_{min}$ (Table 7.1, ll. 13 and 14).

In the second phase (Table 7.1, ll. 15–42), power edges are searched. The minimal power graph problem is to be seen as an optimization problem to find the power graph achieving the highest edge reduction. The greedy power edge search follows the heuristic of making the local optimum decision at each step with the aim of finding the global optimum, or at least to come close to it. Among the candidate power nodes found in phase one, each pair that forms a complete connected (bipartite) subgraph in G is a candidate power edge. The candidates abstracting the most edges are added successively to the power graph. If necessary, candidates are decomposed (Figure 7.3B and Table 7.1, ll. 28–40).

7.3 Validation – Edge Reduction Differs from Random

We collected nine networks from the network science literature derived from complex systems of interacting entities (Table 7.2). These networks were chosen for their accuracy and completeness: The Internet network, software module dependencies in Java and Cytoscape, North American airport network, ownership relationships of American corporations, a food web in South Florida, co-appearance relationships between characters in the Bible, North American power grid network, and the neural network of *Caenorhabditis elegans* (roundworm).

A similar analysis has been proposed to evaluate the quality of protein-protein interaction networks, obtained by various experimental methods having distinct quality (Royer *et al.*, 2012).

The networks have between 130 and 23 000 nodes, between 700 and 50 000 edges, an average degree between 1.8 and 11.9, and an edge reduction between 45 and 82% is achieved.

The relative edge reduction of 8 of these 9 networks is in the range 15–50%, that is, the edge reduction achieved with power graphs drops significantly for randomized version of the same networks. This result shows that the structures explicitly shown by those power graphs distinguish the networks from random and thus are meaningful. The South Florida Ecosystem network is strict bipartite and, thus, has a clustering coefficient of zero. However, it has a relative edge reduction of 48%, which means that the relative edge reduction is not solely measuring clique content and clustering in networks.

Table 7.2 Edge reduction $\mathcal{R}$ and relative edge reduction $\mathcal{R}_{rel}$ of diverse networks of complex systems (d denotes average degree). Edge reduction $\mathcal{R}$ is generally in the range of 45–82%, relative edge reduction $\mathcal{R}_{rel}$ in the range 15–50%.

Network	Year	Source	Nodes	Edges	d	$\mathcal{R}$	$\mathcal{R}_{rel}$
South Florida Ecosystem	2000	Heymans, Ulanowicz, and Bondavalli (2002)	381	2,137	11.2	0.77	0.48
Cytoscape class dependencies	2009		615	3,463	11.2	0.80	0.47
Bible co-appearance network	1993	Knuth (1994)	130	743	11.4	0.82	0.33
US Airports	2007	Colizza *et al.* (2006)	500	2,980	11.9	0.65	0.21
Corporate Ownership	2002	Norlen *et al.* (2002)	7,253	6,711	1.8	0.79	0.20
Java library class dependencies	2006		1,538	7,817	10.1	0.54	0.17
Internet (autonomous systems)	2006	Leskovec, Kleinberg, and Faloutsos (2005)	22,963	48,436	4.2	0.68	0.17
C. elegans neural network	1986	White *et al.* (1986)	297	2,148	14.4	0.45	0.15
Power Grid (USA)	1998	Watts and Strogatz (1998)	4,941	6,594	2.6	0.45	0.04

The relative edge reduction of the network of the power grid of the United States is much lower, showing that the indicated structures in the power graph are less meaningful. The power grid network is an exception among the examples as its entities and their interactions are strongly constrained in two dimensions. In the other networks, the interacting entities are embedded in higher dimensional spaces and have more freedom to interact.

7.4 Graph Comparison with Power Graphs

Complex networks can be compared using power graphs in terms of Precision, Recall and F-Measure. The corespective power graphs of two distinct graphs G_1 and G_2 are confronted, searching for each power node (group of nodes) in the G_1 the best overlapping power node in the G_2. Three measures derive from this procedure (summarized in Figure 7.5): PG-Precision, PG-Recall, and PG-F-Measure.

These measures compare edges only indirectly, by matching the induced structure, resulting advantageous when analyzing networks with distinct number of edges or distinct degree distribution, since the power graphs are only little dependent on the edge density. Such approaches have been applied to show that power graphs are robust to noise (Royer *et al.*, 2008).

PG-Precision and PG-Recall, when comparing PG_1 to PG_2, are identical and inverted when changing the direction of the comparison (PG_2 to PG_1). The PG-F-Measure instead is invariant of the direction, therefore we consider it as a representative of the similarity of two networks.

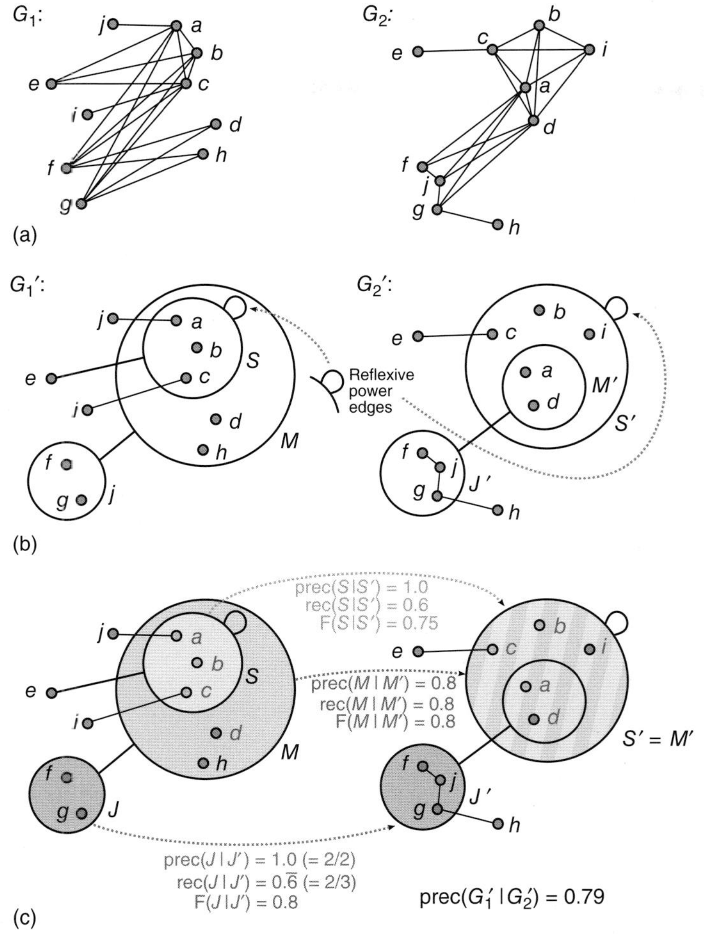

Figure 7.5 Power graph similarity (PG-similarity). (a) Two distinct graphs G_1 and G_2 on the same set of nodes. (b) Power graphs G_1' and G_2' for G_1 and G_2, respectively, after applying the power graph algorithm. (c) Power node matching as basis of the similarity measure. Each power node in G_1' is matched with the power node in G_2' with the highest F-measure, and vice versa. Precision and recall of those matchings are summarized to precision, recall and F-measure between G_1' and G_2'. (See inset for color representation of this figure.)

7.5 Excursus: Layout of Power Graphs

Reducing the number of edges by 45–82%, power graphs provide an optimal basis for improving the visualization of networks. But power graphs present new challenges in terms of layout.

While minimizing the number of edges, at the same time power graphs introduce power nodes. Power nodes form a hierarchy in which two power nodes u and

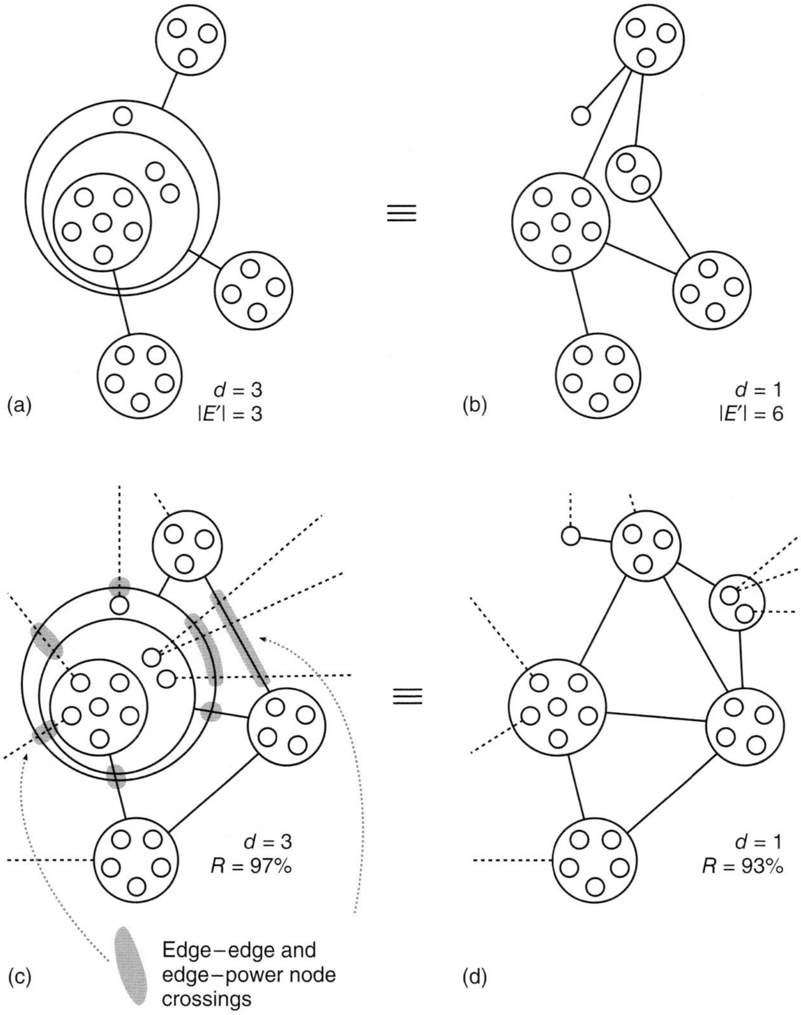

Figure 7.6 The influence of nesting level d (depth) in power graphs on the layout, the edge–edge and edge–power node crossing count, and the edge reduction $\mathcal{R}$.

v can be either disjoint or proper subsets of each other ($u \cap v = \emptyset \vee u \subset v \vee v \subset u$). This hierarchy has a *depth d* (the longest path from the set V' to any other power node in the hierarchy). The depth – in power graphs also called *nesting level* – of a power graph and its effect on a power graphs' layout is illustrated in Figure 7.6. In the simple example, the power graph with higher nesting (Figure 7.6a) is a good choice since it shows the structure of the graph, which consists of three bicliques, clearly by three power edges, whereas in the equivalent representation in Figure 7.6b two of the power edges are decomposed into two and three power edges, respectively. In more complex graphs nesting complicates the layout and

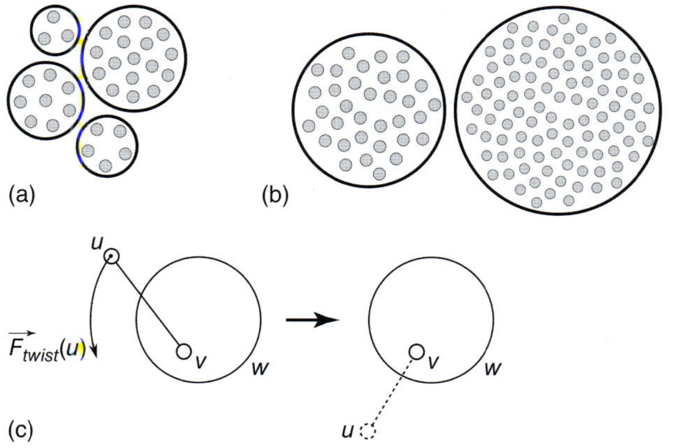

Figure 7.7 (a/b) Deterministic patterns for power nodes without outgoing edges (a – circular patterns, b – phyllotactic patterns (Shipman and Newell, 2004)). (c) Additional to attractive and repulsive forces, *twisting* forces are applied.

induces edge-edge and edge-power-node crossings (Figure 7.6c). Reducing the nesting level of a power graph helps but reduces at the same time its edge reduction (Figure 7.6d).

Standard layout algorithms (Battista *et al.*, 1999; Fruchterman and Reingold, 1991; Wiese, Eiglsperger, and Kaufmann, 2002) must be extended to handle the layout of a hierarchy of power nodes. Our algorithm is force-directed and uses the structure of power graphs to reduce the number of necessary force calculations. Instead of calculating repulsive forces for all pairs of power nodes, only pairs of power nodes which have the same parent (power nodes u and v have the same parent in $G' = (V', E')$ if $u \cap v = \emptyset \vee \forall_{t \in V'} t \supseteq u \Leftrightarrow t \supseteq v$) must be considered. The algorithm starts to calculate all pair-wise forces on the set of power nodes that are not part of any other power node and continues recursively on the power node hierarchy. Deterministic patterns are used for the layout of single nodes inside of power nodes without outgoing edges (Figure 7.7a/b). Additional to attractive and repulsive forces, *twisting* forces are applied for minimizing the length of edges that cross the border of a power node (Figure 7.7c).

7.6
Interactive Visual Analytics

Figures 7.8 and 7.9 illustrate with an example a key advantage of power graphs for visual analytics of complex networks. In their network analysis of the South Florida Everglades ecosystems (Heymans, Ulanowicz, and Bondavalli, 2002), the authors compiled a highly detailed and complex network of carbon exchanges between species.

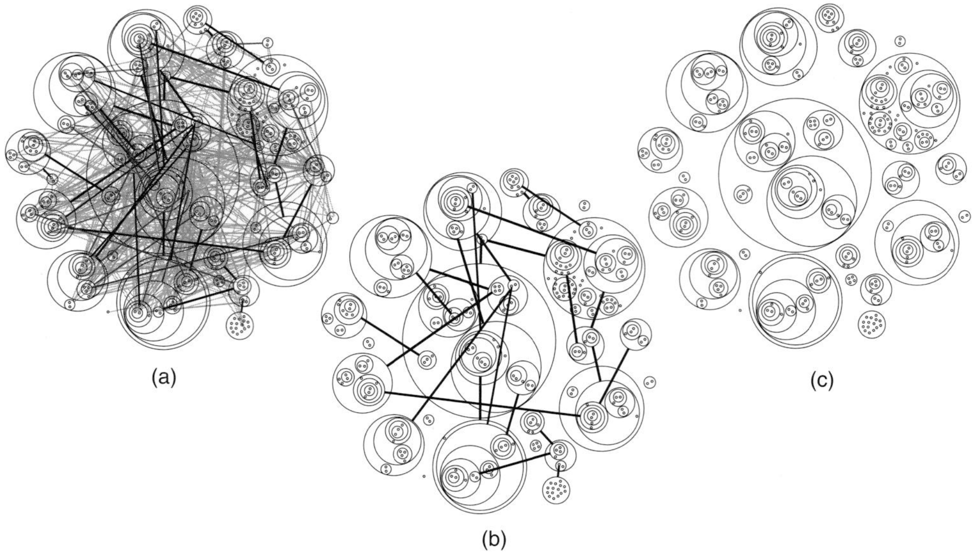

Figure 7.8 Power edge filtering. (a) Unfiltered power graph. (b) Filtered by size, only power edges abstracting at least 13 edges are kept. (c) All power edges are removed, only power nodes remain which provide information on the architecture of the network. In average, each power node is derived from the information of 11 edges.

7.6.1 Power Edge Filtering

After reducing the number of edges by lossless transformation into a power graph, the edge clutter can still hinder visual analytics (Figure 7.8a). To solve this problem, power graph analysis allows to further reduce complexity by filtering the power edges according to characteristics such as their type (biclique, clique, or star) or the number of edges and nodes they represent. Network visual analytic tasks that are facilitated by power edge filtering are:

Coarse-grained understanding of the overall architecture of the network by highlighting the main connectivity backbones in the network represented by large bicliques, cliques, and stars.

De-noising by filtering out *small* power edges that abstract few edges. Noise in networks can be modeled as edges uncorrelated to any other structures in the network. Ignoring these power edges reduces noise in the representation.

Motif mining by selecting only power edges representing bicliques, cliques, stars above a threshold size.

Figure 7.8b shows a filtering of power edges that represent at least 13 edges – thus highlighting the connectivity backbone of the network. Figure 7.8c shows that

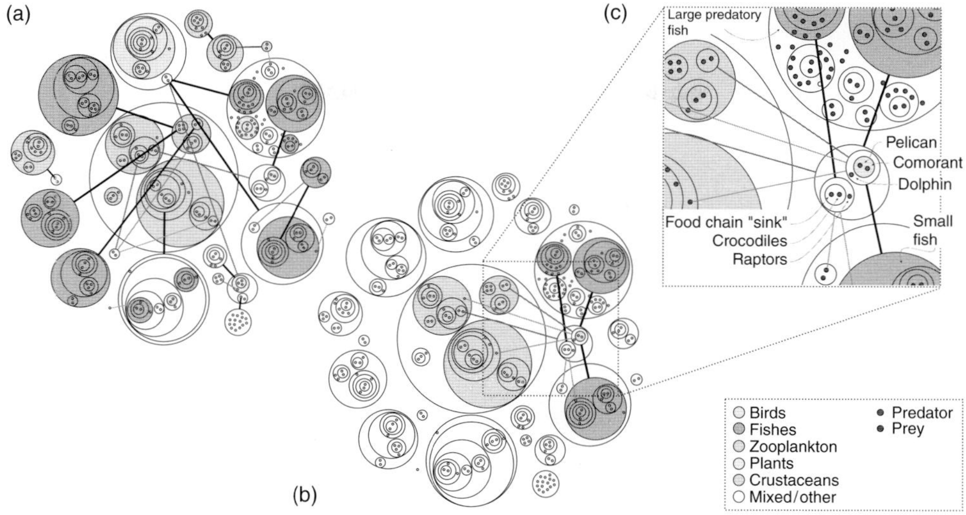

Figure 7.9 Interactive visual analysis of the Florida Food Chain Network. (a) The largest power nodes correspond to relevant groups of animals in the food chain. (b) Selecting power edges around a region of interest – for example here a group of predators – helps to locally explore the network. (c) These predators share many fish species in their diet and are thus in competition in the food chain. Yet it can be seen that crocodiles and raptors prefer larger predatory fish and pelicans, cormorants and dolphins prefer smaller fish. Note: The *food chain sink* corresponds to the output of carbon from the ecological system studied, it is thus not a species but represent exchanges with the outside. (See inset for color representation of this figure.)

the removal of all power edges still provides a useful picture since power nodes convey information about the organization of the network. Compared with a traditional graph representation, effective power edge filtering is facilitated by the explicit hierarchical structure provided by power nodes and the different characteristics of power edges such as size (number of edges in power edge or number of nodes in corresponding power nodes) and type (clique, biclique, star).

7.6.1.1 **Zooming and Network Expansion**

Power edge filtering reduces edge clutter when looking at the network in its entirety. Figure 7.9a shows that the power nodes correspond to meaningful groups. These groups correspond to animals that have a similar diet or that share predators.

When analyzing a network with a reductionist mindset, the next step is to identify one or several regions of interest and explore their organization and structure in detail. Figure 7.9b shows a part of the corresponding power graph, centered around five predators: Raptors, Crocodiles, Pelicans, Cormorants, and Dolphins (red nodes). These five predators all eat medium-sized fish such as Parrotfish and Needlefish. In the power graph representation, these five predators are further divided into two groups. Dolphins share with Pelicans and Cormorants (two birds) a particular taste for much smaller fish such as Goby and Silversides. Whereas Raptors and Crocodiles prefer much larger predatory fishes such as Baracudas, Catfish, and Snapper.

7.7 Conclusion

In this chapter, we presented the first detailed description of the power graph algorithm, which we also adapted for weighted graphs. The algorithm compresses redundant information in undirected, weighted graphs based on cliques and bicliques without detail loss, allowing its use for a broad spectrum of applications.

We showed by the notion of relative edge reduction that the re-occurring motifs explicitly shown in power graphs contain information that distinguishes the networks significantly from random. Moreover, power graphs allow the comparison of distinct graphs by means of F-measure, Precision, and Recall.

We gave a sketch of our layout algorithm and showed how power graph representations improve the visualization of complex networks by reducing the number of edges in the network and the number of unavoidable crossings in the layout.

We showed on a complex example that power graphs provide new possibilities for efficient meaningful edge filtering and local network expansion, thus a powerful tool for interactive visual analysis.

References

Battista, G.D., Eades, P., Tamassia, R., and Tollis, I.G. (1999) *Graph Drawing. Algorithms for the Visualization of Graphs*, Prentice-Hall, NJ.

Chen, C. (2005) Top 10 unsolved information visualization problems. *IEEE Computer Graphics and Applications*, **25** (4), 12–16.

Colizza, V., Barrat, A., Barthélemy, M., and Vespignani, A. (2006) The role of the airline transportation network in the prediction and predictability of global epidemics. *Proceedings of the National Academy of Sciences of the United States of America*, **103** (7), 2015–2020, doi: 10.1073/pnas.0510525103.

Daminelli, S., Haupt, V.J., Reimann, M., and Schroeder, M. (2012) Drug repositioning through incomplete bi-cliques in an integrated drug-target-disease network. *Integrative Biology (Cambridge)*, **4** (7), 778–788, doi: 10.1039/c2ib00154c.

Dickerson, M., Eppstein, D., Goodrich, M.T., and Meng, J.Y. (2005) Confluent drawings: visualizing non-planar diagrams in a planar way. *Journal of Graph Algorithms and Applications*, **9** (1), 31–40.

Duh, R.C. and Fürer, M. (1997) Approximation of K-set cover by semi-local optimization, Proceedings of the 29th annual ACM symposium on Theory of computing, ACM Press, pp. 256–264.

Eisen, M.B., Spellman, P.T., Brown, P.O., and Botstein, D. (1998) Cluster analysis and display of genome-wide expression patterns. *Proceedings of the National Academy of Sciences of the United States of America*, **95** (25), 14/,863–14/,868.

Frishman, Y. and Tal, A. (2008) Online dynamic graph drawing. *IEEE Transactions on Visualization and Computer Graphics*, **14**, 727–740, doi: ieeecomputersociety.org/10.1109/TVCG.2008.11.

Fruchterman, T.M.J. and Reingold, E.M. (1991) Graph drawing by force-directed placement. *Software - Practice and Experience*, **21** (11), 1129–1164.

Gagneur, J., Krause, R., Bouwmeester, T., and Casari, G. (2004) Modular decomposition of protein-protein interaction networks. *Genome Biology*, **5** (8), R57, doi: 10.1186/gb-2004-5-8-r57.

Gallai, T. (1967) Transitiv orientierbare graphen. *Acta Mathematica Academiae Scientiarum Hungaricae*, **18**, 25–66, doi: 10.1007/BF02020961.

Heymans, J.J., Ulanowicz, R.E., and Bondavalli, C. (2002) Network analysis of the south Florida everglades graminoid marshes and comparison with nearby cypress ecosystems. *Ecological Modelling*, **149** (1-2), 5–23, doi: 10.1016/S0304-3800(01)00511-7.

Holten, D. and van Wijk, J.J. (2009) Force-directed edge bundling for graph visualization. *Computer Graphics Forum*, **28** (3), 983–990, doi: 10.1111/j.1467-8659.2009.01450.x.

Keim, D., Mansmann, F., Schneidewind, J., Thomas, J., and Ziegler, H. (2008) Visual analytics: scope and challenges, in *Visual Data Mining, Lecture Notes in Computer Science*, Vol. 4404 (eds S. Simoff, M. Böhlen, and A. Mazeika), Springer Berlin / Heidelberg, pp. 76–90.

Knuth, D.E. (1993) *The Stanford GraphBase: A Platform for Combinatorial Computing*, ACM, New York.

Knuth, D.E. (1994) *The Stanford GraphBase. A Platform for Combinatorial Computing*, ACM Press and Addison-Wesley Publishing Company.

Koren, Y. (2004) Graph drawing by subspace optimization, in *Joint Eurographics - IEEE TCVG Symposium on Visualization* (eds O. Deussen, C. Hansen, D. Keim, and D. Saupe, Eurographics Association Aire-la-Ville, Switzerland.

Kratzke, T., Reznick, B., and West, D. (1988) Eigensharp graphs: decomposition into complete bipartite subgraphs. *Transactions of the American Mathematical Society*, **308** (2), 637–653.

Leskovec, J., Kleinberg, J., and Faloutsos, C. (2005) *Graphs over time: densification laws, shrinking diameters and possible explanations*, pp. 177–187, doi: doi.acm.org/10.1145/1081870.1081893.

Maslov, S. and Sneppen, K. (2002) Specificity and stability in topology of protein networks. *Science*, **296** (5569), 910–913, doi: 10.1126/science.1065103.

Norlen, K., Lucas, G., Gebbie, M., and Chuang, J. (2002) Visualization and analysis of the telecommunications and media ownership network. Proceedings of International Telecommunications Society 14th Biennial Conference.

Papadopoulos, C. and Voglis, C. (2006) Drawing graphs using modular decomposition, in *Graph Drawing 2005, LNCS*, Drawing Graphs using Modular Decomposition, vol. 3843 (eds P. Healy and N.S. Nikolov), Springer Berlin Heidelberg, pp. 343–354.

Purchase, H. (1997) Which aesthetic has the greatest effect on human understanding? in *Graph Drawing*, Drawing Graphs using Modular Decomposition, LNCS, Vol. 1353 (ed. G.D. Battista), Springer Berlin Heidelberg, pp. 248–261.

Rasmussen, E. (1992) Clustering algorithms, in *Information Retrieval - Data Structures and Algorithms*, Chapter 16 (eds W.B. Frakes and R. Baeza-Yates), Prentice-Hall, pp. 419–442.

Royer, L., Reimann, M., Andreopoulos, B., and Schroeder, M. (2008) Unraveling protein networks with power graph analysis. *PLoS Computational Biology*, **4** (7), e1000 108, doi: 10.1371/journal.pcbi.1000108.

Royer, L., Reimann, M., Stewart, A.F., and Schroeder, M. (2012) Network compression as a quality measure for protein interaction networks. *PLoS ONE*, **7** (6), e35 729, doi: 10.1371/journal.pone.0035729.

Shi, L., Cao, N., Liu, S., Qian, W., Tan, L., Wang, G., Sun, J., and Lin, C.Y. (2009) Himap: adaptive visualization of large-scale online social networks. Visualization Symposium, 2009. PacificVis '09. IEEE Pacific, pp. 41–48.

Shipman, P.D. and Newell, A.C. (2004) Phyllotactic patterns on plants. *Physical Review Letters*, **92** (16), 168 102.

Watts, D.J. and Strogatz, S.H. (1998) Collective dynamics of 'small-world' networks. *Nature*, **393** (6684), 440–442, doi: 10.1038/30918.

White, J., Southgate, E., Thomson, J., and Brenner, S. (1986) The structure of the nervous system of the nematode Caenorhabditis elegans. *Philosophical Transactions of the Royal Society of London, Series B: Biological Sciences*, **314** (1165), 1.

Wiese, R., Eiglsperger, M., and Kaufmann, M. (2002) yFiles: visualization and automatic layout of graphs, *Graph Drawing, LNCS*, vol. 2265, Springer Berlin Heidelberg, pp. 588–590.

Index

Computational Network Theory: Theoretical Foundations and Applications, First Edition.
Edited by Matthias Dehmer, Frank Emmert-Streib, and Stefan Pickl.